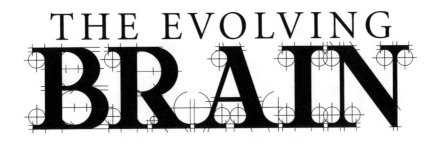

THE EVOLVING BRAIN

THE EVOLVING

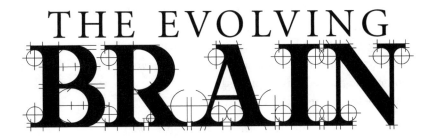

BRAIN

THE KNOWN AND THE UNKNOWN

R. GRANT STEEN, PhD

Prometheus Books

59 John Glenn Drive
Amherst, New York 14228–2197

Published 2007 by Prometheus Books

Inquiries should be addressed to
Prometheus Books
59 John Glenn Drive
Amherst, New York 14228–2197
VOICE: 716–691–0133, ext. 207
FAX: 716–564–2711
WWW.PROMETHEUSBOOKS.COM

11 10 09 08 07   5 4 3 2 1

Library of Congress Cataloging-in-Publication Data

Steen, R. Grant.
    The evolving brain : the known and the unknown / R. Grant Steen.
        p.   cm.
    Includes bibliographical references (p.   ) and index.
    ISBN-13: 978–1–59102–480–4 (hardcover : alk. paper)
    ISBN-10: 1–59102–480–3 (hardcover : alk. paper)
    1. Brain. I. Title.

QP376.S7635 2006
612.8'2—dc22

                                                                2006022767

Printed in the United States of America on acid-free paper

Dedicated to Professor D. O. Hebb of McGill University:

Who answered more questions than I knew to ask; whose boyish enthusiasm at 68 remains an inspiration long after his passing; who was dismissed by his peers but vindicated by his ideas; who was greatly admired by a young man sitting toward the back of the class.

"All that is comes from the mind; it is based on the mind, it is fashioned by the mind."

The Pali Canon, ~500 BC
Suttipitaka, Dhammapada, chap. 1, verse 1
(Sacred scriptures of Thuravada Buddhists)

CONTENTS

8 CONTENTS

PREFACE

The brain poses an exceedingly difficult problem for modern science. Science is wedded to the notion that every system, no matter how complex, can eventually be fully understood in terms of its component parts. This idea, called reductionism, is an intellectual strategy that enables the scientist to study complex systems in an experimental fashion. Reductionism has been hugely successful in advancing science over the past 400 years. There is real cause for hope that the reductionist approach will enable us to conquer human disease, to solve the problem of poverty, to extend human capacity, to build efficient engines that use alternative fuel sources, and so on. Virtually every scientist is committed to the idea that reductionism is the pathway to a brighter future. Yet whether such an approach can enable us to understand the human brain fully is another question entirely. While scientists hope that by studying individual nerve cells it will eventually become possible to understand the function of the brain, it is not certain that this is feasible.

Can the brain ever truly understand itself? Is there a distinction between understanding the brain and knowing the mind? Are those traits that we perceive to be uniquely human, traits such as creativity or altruism or spirituality, dependent upon attributes unique to the human brain? Is it consciousness that makes us fully human, or do we share consciousness with many other creatures? How is it possible that the human brain, probably the most complex and sophisticated object in the universe, could have arisen in a world so often characterized by chaos? Has the human brain changed over time and will it continue to change? Did the brain evolve or was it necessarily created in its final form? Are the mechanisms of mutation and natural selection adequate to explain the human brain, or do they fall woefully short? How can we explain what we are?

Clearly, we cannot claim to know the world if we do not even know our own minds. Yet it has been very challenging to understand even relatively simple things: how can a nerve cell stimulate a muscle to contract or a heart to beat? An enormous gap remains between comprehending how an individual nerve cell works and how the brain as a whole works: how can a mass of neurons enable us to learn and remember? And this gap in our knowledge is trivial compared to our current inability to understand how the brain gives rise to behavior and culture: how can a mass of brain tissue produce art or science or theology? Our lack of insight into how the brain works is nowhere more apparent than in our fundamental incomprehension of the creative act itself.

Creativity may be the single greatest challenge to modern reductionist science, yet science has already yielded fresh insights into innovation and imagination. It is our view that evolution can now explain the human brain in broad strokes, and will eventually explain human behavior in detail. Our understanding of mutation and natural selection provides useful insights into the evolution of the human brain. We contend that, because of the selective advantage of having a highly complex brain, the emergence of sophisticated behavior is inevitable; creativity is simply the fruit of emergent complexity, an unintended consequence of the evolved mind. Nevertheless, the fact that Handel could conceive *The Messiah* seems no less remarkable than if ants in rural Pennsylvania resolved to outdo Frank Lloyd Wright, to build another, better Fallingwater.

ACKNOWLEDGMENTS

Every scientist knows that you cannot work in isolation, that every achievement rests on a foundation provided by others. Probably every writer knows this as well, though the culture is different and the contributions of others are often not so willingly acknowledged. As a writer-scientist, it can seem like every idea and every word belonged first to someone else, so I want to acknowledge with profound gratitude the contributions of many others:

My parents, Ralph and Noreen Steen, who expected so much;

My mentors, Arthur Gilmore, Henry Leeds, Donald Hebb, Leonard Muscatine, and James Langston, who started me on this path;

My friends, Wil O'Loughlin, Fred Gundersen, Merritt Lindgren, Linda Porter, Hilary Cain, and Margaret McFall-Ngai, who didn't laugh when I started to write;

My children, Alena and Mariel Steen, who helped me to keep my balance;

My editors, Linda Regan and Brian McMahon, who helped me to focus my thoughts and fight my way through the jungle of commas that I naturally tend to use.

CHAPTER 1

THE ANTHILL
OF THE BRAIN

A brain is like an anthill. Tiny neurons are the ants, performing their circumscribed tasks in a manner determined by the role they were born to serve. They look no farther than their nearest neighbors, they understand nothing of the larger structure that results from their persistent action. Although there are an enormous variety of tasks to perform, each neuron, like each ant, can do a limited number of tasks, each in a limited way. And yet complexity emerges, in an anthill as in a brain, in ways that cannot be predicted.

The subtle and adaptive response of the brain to its environment is possible, in part, because of the high degree of specialization brain cells demonstrate. In a similar way, the complex response of the anthill to its environment is partially a result of the astonishing degree of specialization in different ant castes. Most people are familiar with the idea that a typical ant colony has workers, soldiers, and a queen, with an apportionment of tasks among the castes that is appropriate to the bodily form of the members of that caste. But even this simple separation can astonish: the Asian marauder ant shows a division of body types so extreme that a soldier can be as much as 500-fold larger than a worker ant. A worker ant could easily ride on a soldier of the same species, like an infant clinging to an elephant.

THE POWER OF SPECIALIZATION

The human brain is a marvel of specialization: certain neurons in the visual cortex will only respond to straight lines oriented at a specific angle in the visual field. The specialization of ants is also far more specific than simply a division of castes into workers, soldiers, and a queen. For example, leafcutter ants perform a complex task that has been apportioned among worker ants in a particular way. Certain ants make foraging trips to growing plants, where they clip off leaves or leaf fragments. These leaf bits are brought back to the colony and macerated into pulp, and the leaf pulp is used as fertilizer for a fungus. This fungus, which grows into a gray blob the size of a kitchen sponge, is tended as carefully as an asparagus farmer tends his crop, providing food for the entire colony. The whole process, from clipping and macerating of leaves to the careful tending of the fungus, is done in assembly-line fashion, with each step of the process done by a specific type of worker ant. Each ant is slightly different in form and function from the other ants elsewhere on the assembly line.

Specialization in an anthill or in a brain allows the component tasks of life to be performed with greater efficiency. The foraging leafcutter ants that venture from the nest to cut leaves are the largest ants involved in the whole process; they are nearly three times larger than the smallest ants at the other end of the production line. Forager ants are large enough to bring huge leaf pieces back to the nest, carrying them above their heads like green parasols. When the forager ant reaches the nest chamber, leaf pieces are passed off to worker ants of slightly smaller size, who clip the large leaf piece into tiny fragments. As soon as this process of creating leaf confetti has been completed, smaller ants begin to mulch the confetti into moist pellets of leaf mush. Workers even smaller in size then take the moist pellets and insert them into the growing mass of mold, like a farmer fertilizing a crop. Macerated leaf pellets are placed near active sites of fungal growth so that fungal strands can be loosened and draped over the fresh leaf mulch. The very smallest and most abundant worker ants then tend the fungal mass, plucking out stray spores and strands of unwanted fungal types so that the desired fungus can grow as quickly as possible. Dwarf worker ants actually move through tiny corridors that penetrate deep within the growing fungal mass. Occasionally, the dwarf ants will tear away tufts of fungus, which resemble tiny stalked cabbages, and bring these morsels out to their nest-mates. It is these fungal tufts that provide food for the rest of the colony.[1]

To be clear, we are not implying that the human brain is a mass of fungus or that our brains are filled with cells that crawl about, carrying knowledge like ants carry leaf mulch. It is perhaps too easy to abuse the analogy of the anthill in thinking about the complexity of the human brain. Yet the anthill and the brain have some very deep parallels, and it is possible that careful consideration of the anthill analogy will lead us to new ideas about the human brain. At a prosaic level, the anthill is nourished by worker ants that forage and bring food back to the colony; this seems analogous to how oxygen and glucose are brought to the brain by the circulatory system. But there are far deeper parallels.

PARALLELS BETWEEN THE ANTHILL AND THE BRAIN

Both the anthill and the brain behave as a unit, are composed of sub-units of varying types, undergo a cycle of growth and senescence, and respond to their environment in adaptive ways. Both anthill and brain show subtle yet significant idiosyncrasies of structure and function that make each individual unlike any other of the same type. Both anthill and brain are rather ancient, at least in their incipient form, so there has been ample time for a relentless process of change to produce what is now a highly sophisticated structure.

Both anthill and brain lack a clear chain of command: there is no "boss" in a brain, just as there is no "king" in an anthill. The ant queen merely produces young, rather than controlling the moment-to-moment action of her subjects. All decisions are made at a local level, and decisive action can be taken without any centrally placed or "integrative" functions. A clear example of this is how an insect intruder into an anthill can be attacked and destroyed without a general alarm being sounded. The soldier ants already in the area coalesce to dispose of the intruder. In much the same way, an immune response can be mounted in the brain, relying upon locally available immune cells to attack an intruder.

Local responses can be amplified, in both anthill and brain, and the speed and scale of a response can be increased in both. Soldier ants in a colony can be broadly stimulated to respond to a major threat to the colony, just as a systemic immune response can be mobilized to fight a potent threat to the brain. The only way to insure that a local response remains adaptive is for there to be efficient communication among the responding units, as well as some form of feedback that can turn off a response when that response is no longer needed. Feedback loops, especially those negative feedback loops that help diminish a response before it escalates out of control, are central to maintaining a balance within the whole—in the anthill and in the brain.

It is an astonishing property of the collective, whether that collective is a brain or an anthill, that a local response can generate action appropriate at the systemic level. Moreover, decisions, made at the level of the individual, can nonetheless benefit the whole. For example, you can withdraw your hand from scalding water before a message of pain reaches your brain because a local reflex successfully mediates the systemic response well before pain is perceived. The same property, of a local response benefiting the whole, is also seen in an anthill. If the colony comes under attack from a slave-making species of ant, larval ants in the nursery can be saved by the workers tending them, even if there is no general alarm that recruits the colony to respond. The actions of individual ants may be nonadaptive, as some ants will pick up larvae and run about in a seemingly random fashion, but the response of ants in the aggregate is adaptive, so that the larvae are quickly saved. An adaptive response at the systemic level is therefore produced by autonomy at the individual level, modulated by appropriate feedback controls.

Another key property of the collective is that the reliability of individual units, be they ants or neurons, is lower than the reliability of the whole. The enhanced reliability of the collective is possible only because of functional redundancy: if a single ant or a single neuron fails to perform a required function, there are other units still capable of generating an appropriate response. Redundancy of function not only makes the anthill and the brain more flexible and more reliable, it may also free the collective to experiment somewhat. Over time, new functions and new ways of interacting can arise, and this may explain why both anthill and brain have been so successful through time.

THE CONCEPT OF EMERGENT COMPLEXITY

Perhaps the most exciting parallel between the anthill and the brain is that both entities can demonstrate emergent complexity—a synergy in the system that has grown over time. Emergent complexity is shown by those properties of the whole that could not have been predicted from an intimate knowledge of the parts. No matter how deep our understanding of the brain at the level of neurons, it is simply not possible to predict that a collection of neurons could write *For Whom the Bell Tolls*. It may be that emergent complexity is a real-world outcome of chaos theory, in that unpredictability is inherent to any highly complex system, even if that system runs by rigid rules. But, whatever the cause, it is clear that the whole can be far more than the sum of its parts.

We are all familiar with examples of emergent complexity in human endeavor, but examples of emergent complexity in an anthill are much less familiar. Nevertheless, emergent complexity seems to be the only way to explain how an ant colony is able to perform certain remarkable feats. Ants, for example, prefer a warm environment, avoiding temperature extremes; most ant species are unable to remain active if the air temperature dips below about 20° centigrade (which is roughly equivalent to 70° Fahrenheit). Certain ant species that live in either cool or very hot climates can sometimes find a congenial microclimate by digging their nests deeply into the soil. But other ant species are able to maintain a congenial nest temperature by a much more sophisticated means than by merely exploiting the constant temperature found several meters below the soil surface. The European wood ant builds a mound nest that projects above the ground surface, yet the nest maintains a stable internal temperature as much as 10° C (22° F) warmer than the air above it or the soil below it. This constant temperature is maintained by stuffing the nest full of decaying vegetation. As the vegetation decomposes, it releases heat, just as the middle of a haymow is warm in a frigid Siberian winter. Furthermore, the nest is built to retain heat, so that the heat of decomposition and the heat created by the ants as they move about is retained within the mound. What this means, in essence, is that ants adapt their behavior to maintain a constant nest temperature despite changing environmental temperatures.

Emergent complexity is postulated to be a property not of ants, but of the colony, and not of neurons, but of the neuronal collective. Individual neurons are unable to generate the complex behavior possible for a brain, just as individual ants cannot build an anthill. The complex behavior so characteristic of humans cannot be produced by a simple brain but, rather, requires the sophistication of a mind that is able to perceive the subtleties of the world. In short, emergent complexity is a property of the mind, not of the brain. A brain is simply a collection of neurons, each of which behaves in a certain way. A mind is what arises when a huge number of neurons are free to perform their specific functions. The mind is what interposes between the brain and behavior; mind may be evidence of brain, but brain is the source of mind. A mind has thoughts and feelings, opinions and beliefs, strengths and weaknesses: a mind has a personality.

How Can We Explain the Parallels between Ants and Neurons?

We have noted the similarities between ants and neurons without asking why such parallels exist. There can be one reason and one reason alone why ants in an anthill function so much like neurons in a brain.

The energetic cost of having a brain is extremely high; the energy required for brain function in a resting adult is roughly 20% of the total energy intake for the entire body. This high demand for energy arises because, while it may be inefficient to maintain the brain in a state of constant readiness to respond when no threat is approaching, it is catastrophic to be unprepared for such peril. Even if a neuron is not conveying information, it must stand ready to do so, and the energetic cost of readiness is extraordinarily high. This places a very strong selective pressure on the brain, since resources are limited in virtually every environment. If a brain were to become too inefficient, if the benefits of having such a complex structure did not outweigh the costs, then the cockroach brain might well be the most advanced brain in the world. Similarly, ants that waste precious resources will not survive for long, and every anthill must act to conserve space, materials, time, and energy. An anthill that profligately wastes resources would lose out over time to another more efficient ant species.

In short, ants and neurons, as different as they seem, are under similar selective pressures, and they have evolved similar responses. The more we learn about the brain, the more we come to appreciate that it was shaped by the same subtle evolutionary pressures that act on an anthill.[2] This is one of the strongest possible demonstrations of the power of evolution to explain widely disparate and seemingly unrelated observations.

EVOLUTION: THE BASICS

Some people are uncomfortable with the idea that the human brain arose by an evolutionary process. The standard objection is that a random process like evolution cannot have given rise to an engineering marvel like the brain. Yet this objection is based on a misunderstanding of the true nature of evolution. Evolution is far from a random process. Evolution is a result of two complementary processes: mutation and natural selection. Mutation is a random process of change at the level of the genes. However, these genetic mutations are winnowed out in the most nonrandom process imaginable—natural selection. This process is the merciless culling of unsuccessful mutations through predation or premature death.

To discuss evolution with the rigor and the dispassionate fairness that it deserves, we must have a clear understanding of exactly what it is. Evolution is simply a gradual process of change. To a biologist, evolutionary change takes place by a mechanism that can be explained in fairly simple terms. The following four precepts are sufficient to generate gradual change in an animal stock that any biologist would recognize as evolution:

1) **Variation exists.** This observation is so simple that it seems incontrovertible. Any large gathering of people will include people who are slim and fat, old and young, tall and short, weak and strong, healthy and ill. It could perhaps be argued that most variations are meaningless in an evolutionary sense, especially in a human gathering, and this may well be true. Yet variation exists in all organisms of all species in all places.

2) **Some variants are more successful than others.** Imagine a herd of antelope in which some are slim and some are fat, some are old and some are young, some are weak and some are strong. Clearly, if a lion were stalking that herd, then the fat or the old or the weak would be more likely to die. Predation is not random; lions risk injury every time they hunt, so they always seek the weakest prey. Even then, lions are not always successful, since they are sometimes unable to kill any prey. Nevertheless, over time, there is a stronger selection pressure working against the weak than against the strong.

3) **Variation is heritable.** Everything we know about our own families convinces us that certain traits are likely to run in families: tall parents tend to have tall children just as near-sighted

parents tend to have near-sighted children. Everything we know about genetics concurs that certain traits are passed down to offspring, often with a high degree of fidelity. This simple truth can have terrible consequences, as certain families are ravaged by hereditary illness.

4) **Successful variants tend to become more abundant over time.** Because certain individuals are more likely to survive long enough to reproduce, and because these individuals are able to pass specific traits on to their offspring, these traits will tend to be well represented in ensuing generations. In contrast, other individuals may have traits that are more likely to lead to premature death, so these traits are less likely to be passed down. Over time, the successful traits will tend to increase in the population whereas the unsuccessful traits will gradually decrease. While this process is inexorable, it is not as effective as one might imagine. For example, heart disease can kill people at an early age, but usually not so early that those with a weakened heart are unable to have children. Thus, there is effectively no selective pressure against heart disease in humans, unless the survival of the children is somehow impaired.

These four precepts are entirely sufficient to explain evolution. If the opponents of evolution are unable to disprove at least one of these precepts, then they are in the untenable position of having to deny reason in order to preserve their faith.

THE ENGINE OF EVOLUTION

The "inventiveness" of evolution emerges from a combination of processes that would seem to be polar opposites: a random process of change and a nonrandom process of testing change. Mutation generates few changes that are ultimately successful, just as a blind watchmaker would rarely be able to alter a watch so that it would keep better time. Just as we would not accept the alterations of a blind watchmaker without verifying that the watch still works, random changes to an antelope are subjected to the stringent selection pressure of a hungry lion. If mutation makes a change to the genome of an organism, that organism must then survive long enough to pass that change down to later generations, or there can be no evolution. The process of natural selection thus removes unsuccessful mutations from a population in the most unforgiving way imaginable. The lame, the halt, the weak, and the maladapted—all are slaughtered with a fierce egalitarianism that is inexorable. Evolution is not a random process at all.

Any consideration of how the mind arose from the brain must acknowledge the engine of evolution. In fairness, we don't know all the details of how this happened, although by looking at the range of brains in simple animals we can get a fairly good idea of the broad strokes. The earliest "brain" was simply a bit of reactive protoplasm, a cell or a part of a cell that responded to change in the environment such that the organism with this "brain" was more likely to survive. Perhaps this simple "brain" helped the organism to eat another or to avoid being eaten, but it is clear that the ability to respond to the environment in even the most primitive way would have had exceptional survival value. Later, when organisms became larger, there was a concentration of reactive cells, which formed the first true brain.

Over evolutionary time there would have been a selective advantage to any organism that had more reactive tissue so long as that organism was better able to survive. A conglomeration of protoneurons might provide redundancy so that some protoneurons could be freed from the dictates of necessity. In this way, established neurons could take on novel functions. Although such experimentation is random, each "experiment" is evaluated by the inexorable process of natural selection. Change would happen very gradually and would require an enormous amount of time, but change would be unavoidable, especially if the environment was also changing.

If change produced many neurons of flexible function, then we postulate that the emergence of mind, given sufficient time, may have been inevitable. This is because evolution tends to tinker, to co-opt parts evolved for one purpose and to use them for another purpose. Certain preconditions would be required for the evolution of a complex brain: an enormous number of neurons must be available for selection to act upon and there must be an enormous amount of time for selection to occur. Such preconditions provide means and opportunity for the evolution of complexity, but what could be the motivating force for evolving a highly complex brain?

In other words, why is the human brain so complex when very simple organisms with very simple brains have been so successful? The answer must be that some sort of selective pressure has favored the elaboration of human behavior. Were this not so, were there no selective advantage to such sophistication, there would be no reason for humans to have a brain any more elaborate than that of a cockroach. The nature of this selective pressure may never be known for certain, but we will explore many stimulating and provocative possibilities.

SETTING THE AGENDA

In our quest to understand how the human brain evolved, we will describe those principles needed to understand the basic workings of the brain. We will then build upon these principles in an effort to comprehend human behavior and how it may have evolved. A key issue is identifying the forces that drove the emergence and elaboration of human behavior. That we do not yet know or understand all of these forces does not argue decisively against the effort to grasp them. A thorough understanding of the human brain will likely engage scientists for generations to come, and a deep understanding of the human mind may be the single greatest achievement of the human brain.

CHAPTER 2

CONDUCTION
AND CONNECTION

To understand the brain, to have even the faintest hope of understanding human behavior, it is necessary first to understand the
function of the neurons that comprise the brain. Yet many of the simplest and most useful experiments that would enable us to understand
neuronal function cannot ethically be done in a human being. This
consideration has required scientists to study neurons in very creative
ways. Yet, too often, our understanding of a problem is limited by the
ways in which we are able to study it. Like the blind men trying to
describe an elephant, what we can describe is critically limited by
what we are able to perceive. For those scientists interested in the
physiology of the nervous system, this has been a difficult problem,
since individual neurons are too tiny to study easily.

To approach a difficult problem, scientists are trained to use model
systems, to do simple experiments, to make testable predictions, and
to amass careful measurements. This accounts for why so many scientists have studied the squid giant axon; not from a particular interest
in the way that a squid swims but, rather, because the giant axon that
controls squid swimming is one of the few nerves that is large enough
to study easily. Over the years, a great many scientists have used the

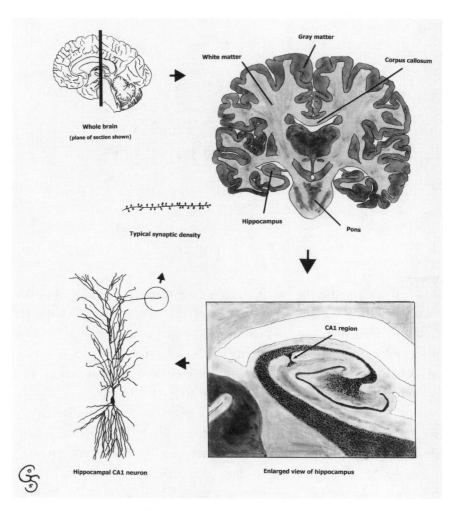

Figure 1. Details of a hippocampal neuron. A view of the entire brain is shown (top left), with the plane of section shown as a black bar. This section through the brain is then seen in cross-sectional view, with the hippocampus (which is symmetrical on both sides) labeled on the left side (top right). The left hippocampus is then enlarged (bottom right), and the CA1 region is shown with one of many hippocampal neurons; this neuron is drawn somewhat larger than scale to show its location clearly (for clarity, all other neurons are omitted). A hippocampal CA1 neuron is then redrawn (bottom left) to show the complexity of the branching pattern typical for this type of neuron. Finally, a typical synaptic density on a single dendrite is shown (center left); this same structure is enlarged in figure 2.

squid giant axon as a tool to clear away the underbrush of intractability from the forests of our ignorance. But it is easy to over-generalize from a simple model; in fact, once a consensus has been reached that an experimental model is worth studying, there is a temptation for the inexperienced scientist to assume that the model is a faithful rendition of the problem. In time, the phenomenon under study can begin to seem as simple as the model would make it: this is a crucial mistake.

The problem is that the human brain may well be the most complex object in the universe, and our current models are far too simple to be a reasonable surrogate for it. A squid giant axon is one neuron that makes one synapse, whereas the human brain is composed of 100 billion (10^{11}) neurons, each of which makes 10 thousand (10^4) synapses. When scientists study the squid giant axon, they are evaluating the function of only one synapse, but they are using the quadrillion (10^{15}) synapses of their own brain to do it. Some of this complexity is suggested by a diagram that shows a human brain at ever-finer levels of resolution (fig. 1); first a slice from the whole brain, then an enlargement of part of that slice, then a single neuron from that part, and finally the density of synapses on that one neuron. Bear in mind that, since we have illustrated only one neuron, the brain is at least 100,000,000,000 times more complex than the diagram.

SIMPLE BEGINNINGS

In the 1830s, physiologists believed that it would simply not be possible to measure the speed of conduction of an impulse down a nerve.[1] By then, it was already known that a nerve impulse, which is often called an "action potential," has properties that are similar to those of an electrical impulse moving down a wire. In fact, this had been shown conclusively in about 1792 when Alessandro Volta demonstrated that a primitive battery could stimulate a nerve and cause the associated muscle to contract. But it was thought that a nerve impulse might move with a speed approaching that of a current down a wire, which, at that time, would have been far too fast to measure with their crude instruments.

Yet, in a triumph of technique over limitation and ambition over fatalism, the speed of an action potential was accurately measured just ten years later. Hermann von Helmholtz used an elegantly simple apparatus, which can easily be replicated in a freshman Biology lab today, to make a measurement that others had believed impossible. A frog leg muscle was removed intact, together with the nerve that innervated it, and the muscle was attached to a simple apparatus that could record the time course of a muscle contraction. Basically, the muscle was connected to a lever so that the lever arm made a tracing as the muscle moved. The nerve was then stimulated with a small electrical current administered through an electrode attached to the nerve. The muscle would then contract and the lever arm move, making a mark on a piece of paper attached to a rotating drum.

This experiment, by itself, cannot reveal the time course of nerve conduction, since the lever arm records so many separate events, each of which takes a finite period of time. These separate events include initiation of the action potential, conduction of the action potential to the end of the nerve, transmission of that impulse from the end of the nerve to the muscle itself, and the slow building of force within the muscle that is necessary to move the lever arm. Yet here is where experimental ingenuity comes in: von Helmholtz repeated the experiment, moving the electrode closer to the muscle. The only difference between the first and second experiment was the distance that the action potential had to travel down the nerve before it could induce the muscle to contract. When this experiment was done, simple calculations showed that the action potential moved at about 30 meters (98 feet) per second. While this may seem rapid indeed, it is still about seven orders of magnitude (or about 10 million times) slower than the speed of electrical conduction through a wire. Because of this finding, von Helmholtz correctly concluded that conduction of an action potential down a nerve is far more complex than conduction of an electrical current down a wire.

To understand how an action potential is possible, it is essential to understand some basics about how the charged particles called ions behave. An ion is an electrically charged atom in solution; everyone knows from experience that a few grains of salt put into a glass of warm water will eventually dissolve. When table salt is put in water, the crystalline molecule dissolves into its component atoms. Table salt, which is sodium chloride, thus dissociates into sodium ions and chloride ions. The sodium ion, which can be written in chemical notation as Na^+, is positively charged, whereas the chloride ion, written as Cl^-, is negatively charged. As the Na^+ and Cl^- ions go into solution, the individual ions begin to interact with the water molecules around them. Ions gradually move away from the undissolved salt grains in a random process called diffusion. Diffusion is simply the tendency of substances to move haphazardly from a region of greater concentration to a region of lesser concentration. If you were to open a bottle of perfume and put it in the corner of a room, eventually the whole room would be faintly scented with perfume because the odorant molecules would have diffused away from the bottle to the far corners of the room.

Ions are present in and around nerve cells, and it is the movement of ions that causes an action potential to move down the length of a neuron. Were the movement of ions in and around nerve cells a completely random process, there would be no action potential; in fact, an action potential is the result of an ordered and asymmetric movement of ions. The asymmetry of ionic movement in an action potential is a result of specific properties of the cell membrane.

Ions cannot diffuse freely across the cell membrane of the neuron. Instead, this membrane is impermeable to ions most of the time; charged ions move across a neuronal membrane as much as 100 million times slower than they could move across an equivalent distance inside the cell. Ions can only move across a neuronal membrane in two ways: by active transport, which requires a tiny "pump" in the membrane, or by passive transport, which requires a "channel" or pore through the membrane.

Active transport can move ions across a membrane against a concentration gradient. This means that ions can become more concentrated, moving in a direction opposite to what would happen by diffusion alone. Because it opposes the natural tendency of ions to move down their gradient, active transport is a process that requires energy. In fact, the movement of ions against a gradient requires a great deal of energy, and this explains why the energetic cost of having a brain is so very high. The tiny membrane pumps that perform active transport are made of protein, and each is capable of moving a specific type of ion from one side of the membrane to the other.

Passive transport can move ions across a membrane only down their concentration gradient. Passive transport is not necessarily an energy-requiring process, except that energy may be needed to open channels in the membrane, which, when at rest, are generally closed. Were this not true, it would be impossible to generate any asymmetry of ions, as ions would simply move down their gradient to be equally concentrated inside and outside the cell, and such symmetry is antithetical to an action potential. Channels can be induced to open and allow ions to pass in several ways. There are voltage-gated channels, which open only when they experience a change in voltage, and there are chemically gated channels, which open in response to a specific chemical or protein. But whatever stimulates a channel to open, the result tends to be the same: channel opening usually results in the movement of ions down their concentration gradient. However, channel opening is typically a brief event; after a short period of time, the channel will return to its normal closed configuration.

The movement of ions across a neuronal membrane can happen only in the context of pumps and channels, which enable certain ions to pass while stopping others. Since not all ions are allowed to pass through a membrane, the neuronal membrane is said to be "selectively permeable" to ions. Selectivity of the neuronal membrane is conferred solely by the particular types of pumps and channels present in the membrane.

CONDUCTION: THE WAY A NEURON WORKS

How is an action potential generated by the movement of ions? We now know that an action potential is an electrochemical phenomenon: the movement of ions across the cell membrane of the nerve generates an electrical current. In other words, an action potential happens because a chemical change induces an electrical change. At rest, the inside of the neuronal membrane is relatively enriched in potassium ions while the outside of the membrane is relatively enriched in sodium ions. This asymmetry is generated by a key protein in the neuronal membrane, variously known as the sodium (Na^+) pump or the sodium/potassium (Na^+/K^+) exchange protein or the Na^+/K^+ ATPase. This protein pump moves three Na^+ ions out of the neuron, in exchange for two K^+ ions moving into the cell. There is an energetic cost for this movement because K^+ can become quite concentrated inside the neuron. Since there is a net movement of positive ions out of the cell, a chemical gradient is generated across the cell membrane, leaving the inside of the cell less positive (more negative) than the outside of the cell. It is this chemical nonequilibrium that accounts for the "resting potential" of the neuron. While the actual difference in concentration of Na^+ and K^+ ions between the inside and the outside of the cell is quite tiny, there still can be a potent local effect.

The local effect, which produces the resting potential of the neuron, arises because the cell membrane is so extraordinarily thin. Positive and negative ions interact with each other across the width of the membrane, with positive ions on the outside of the membrane (where they are in excess) attracting negative ions on the inside of the membrane. A stable situation is set up, with oppositely charged ions kept separate from each other by the insulating ability of the membrane. This insulating ability is remarkably effective, since ions cannot flow down their concentration gradients. Thus, potential energy is stored in much the same way that a battery stores a charge.

An action potential happens when the charge separation across the membrane temporarily breaks down. Protein channels in the membrane, which are normally closed, are stimulated to open. When these channels open, Na$^+$ leaks into the neuron and K$^+$ moves out, with these ionic movements opposing each other and happening quite rapidly. This movement of ions results in a local discharge of the potential energy that was stored in the asymmetric distribution of Na$^+$ and K$^+$ ions.

Let us summarize what happens at a specific point in a neuron during passage of an action potential. Initially, chemical energy is used by the Na$^+$/K$^+$ exchange protein to separate Na$^+$ and K$^+$ ions across the membrane and to generate potential energy in the form of an asymmetric distribution of charge. Potential energy is converted to electrochemical energy during the action potential as the system briefly returns to a state in which there is a more symmetric distribution of ions across the cell membrane. After the action potential has passed, the Na$^+$/K$^+$ exchange protein begins to transport Na$^+$ out of the neuron again. As this happens, the asymmetric distribution of ions that is typical of a resting neuron is reestablished, thereby restoring the resting potential.

This describes what happens at an imaginary point on an axon. However, there would be no action potential if such a local depolarization failed to induce a wave of depolarization capable of moving down the length of the neuron. But how does a local change in the permeability of a neuron generate an action potential? As we noted before, proteins in the neuronal membrane can respond to the local electrical potential. As one part of a membrane depolarizes, that part has an effect on the ionic gradient in the adjacent membrane. Voltage-gated channels in the adjacent membrane begin to open in response to the nearby voltage change across the membrane. In this way, a depolarization at one point in the membrane can induce a depolarization a bit farther down the membrane, and the newly depolarized region of the membrane will, in turn, have an effect even farther down the length of the neuron. In this sense, an axon behaves much like an electrical cable.

A wave of depolarization can be generated, which potentially moves down the entire length of the neuron. This depolarization is what carries information from one part of the cell to another, much as a wave carries energy from one part of the ocean to another. Waves of depolarization are characteristic of all nerve cells: an olfactory neuron stimulated by the scent of a rose; an auditory neuron stimulated by the plucking of a guitar string; a retinal neuron stimulated by the sight of a loved one—all these neurons ultimately respond in the same way, by carrying a wave of depolarization from one point to another. It is rather striking to realize that every thought is ultimately a wave of depolarization moving somewhere in the brain.

A final issue to note is that each neuron has a characteristic form, and this form may have an influence on the way that an action potential travels down the neuron. In fact, each type of neuron can have slightly different "cable properties" that affect how that neuron is able to carry an action potential. Imagine a small, perfectly spherical neuron. Were such a neuron to become depolarized at a single point, the wave of depolarization would spread equally in every direction, so that every part of the neuron would become depolarized to the same extent and at about the same time. Now imagine an infinitely long cable-thin neuron. Were this neuron to become depolarized at a single point, it is possible that one part of the neuron could be conducting an action potential while another part of the neuron is still at the resting potential. An infinitely long neuron might be able to carry many action potentials at the same time, with each action potential at a different point along the length of the cell. Real neurons are neither spheres nor cables, but the point remains: cell shape can have a defining effect on the cable properties of a neuron. For all we know, form could define function, and neuronal shape could determine the fate of an action potential.

Von Helmholtz's lovely experiment to measure conduction velocity in a neuron represents a triumph of reductionism over nihilism; an accurate measurement was made in a situation where conservative people thought it too difficult to try. This experiment is probably the first time that a real physiological measurement was made in nervous tissue, yet this experiment also oversimplifies a complex problem. It turns out that axons of different diameters conduct action potentials at different speeds and that axons of different types conduct action potentials in different ways. But both of these caveats are trivial compared to the real problem created by this wonderfully simple experiment. Over time, researchers gradually accepted the notion that isolated neurons are useful as a model system with which to study nerve conduction. In a larger sense, scientists began to think that isolated neurons could serve as an analogy for the way a simple brain might work. Certainly, no rational person made the mistake of thinking that a squid giant axon functions in the same way as neurons in the human brain. Nevertheless, use of simple model systems pushed us down a pathway that eventually limited our vision in profound ways.

CONNECTION: THE WAY A NEURON INDUCES OTHER NEURONS TO WORK

An action potential passing down a neuron would ultimately have no impact if that neuron were unable to induce nearby neurons to respond in some way. In a sense, an action potential that fails to induce an adjacent neuron to discharge would be functionally equivalent to the proverbial tree falling in a deserted forest. But an action potential cannot simply flit from one neuron to the next, as neurons are usually not in electrical connection with one another. The cable properties that enable an action potential to propagate down a neuron do not allow the action potential to "jump" to the next neuron. Instead, there is another type of connection between neurons in sequence, a connection that uses rather different mechanisms to induce membrane depolarization.

This connection between adjacent neurons is called a "synapse." The synapse forms a structural connection, albeit a "loose" one; although two neurons together form the structure, they are still physically separated from each other, with a tiny synaptic cleft between them. A synapse is a functional rather than a physical connection, even though the synapse typically has a characteristic physical appearance.

What we know of synapses has largely come from study of a model system that is technically not a synapse at all. The favored model for experimentation was, for many years, the junction between nerve and muscle in the frog leg. This model system offers a great many experimental advantages, most notably that muscle contraction reveals when a functional connection has been made between nerve and muscle. The frog leg muscle offers an easy, inexpensive way to study a very complex process, a process that would be impossible to study in humans and very difficult to study in other, more familiar experimental animals. Standard wisdom is that the frog neuromuscular junction behaves in much the same way as a synapse, with the exception that different stimulatory molecules or "neurotransmitters" are used and different quantitative relationships may prevail. In truth, we don't know for certain that the neuromuscular junction behaves like a typical synapse, but we do know that it fails to reflect the full degree of complexity in a synapse.

In any case, it is possible to describe the synapse in light of what we have learned from the neuromuscular junction. When an action potential reaches the synaptic cleft, the presynaptic neuron, often called the axon, is stimulated to release certain specific chemicals into the synaptic cleft. These chemicals can take any of a large number of forms, but all the chemicals function as neurotransmitters. After the action potential in the axon stimulates release of neurotransmitters into the synaptic cleft, these neurotransmitters diffuse across the cleft and interact with proteins in the cell membrane of the postsynaptic neuron, often called the dendrite (fig. 2). Proteins in the cell membrane of the dendrite form chemically gated channels, which open when a specific neurotransmitter binds to the channel. The opening of ionic channels in the dendrite allows ions to flow down their gradient, generating a fresh action potential. This fresh action potential can then propagate down the dendrite toward the next neuron in the series.

Subtle and sophisticated experiments have shown that neurotransmitter molecules are stored in tiny packets at the end of the axon. As an action potential reaches the synapse, these packets of neurotransmitter spill into the synaptic cleft. If the axon releases a lot of neurotransmitter into the synaptic cleft, then there will be a high likelihood

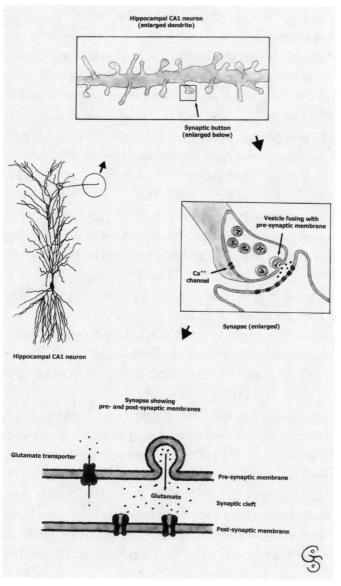

**Hippocampal CA1 neuron
(enlarged dendrite)**

**Synaptic button
(enlarged below)**

Hippocampal CA1 neuron

**Vesicle fusing with
pre-synaptic membrane**

**Ca++
channel**

Synapse (enlarged)

**Synapse showing
pre- and post-synaptic membranes**

Glutamate transporter

Pre-synaptic membrane

Glutamate

Synaptic cleft

Post-synaptic membrane

Figure 2. Details of a hippocampal synapse. A hippocampal CA1 neuron is shown (left of center), similar to the neuron shown in figure 1. Detail is shown of one dendrite, with the density of synaptic buttons suggested (top). A single synaptic button is then enlarged (right of center) to show the relationship between the pre- and post-synaptic membranes and to suggest how glutamate is released into the synaptic cleft. A highly magnified view of a single synapse is shown (bottom), with a single vesicle fusing to the presynaptic membrane to release glutamate into the synaptic cleft. A glutamate transporter protein is also shown to the left, which acts to reabsorb glutamate that has been released into the synaptic cleft.

that the dendrite will generate a fresh action potential. In contrast, if the axon releases little of the neurotransmitter, then there may be a low likelihood that the dendrite will discharge. Even when a dendrite does begin to respond to a neurotransmitter, this does not guarantee that an action potential will be generated; the dendritic potential must be strong enough to generate an action potential that can then propagate down the rest of the neuron. In certain circumstances, a postsynaptic potential that is not strong enough to generate an action potential can actually inhibit the generation of future action potentials. This means that a small stimulation may make later, larger stimulations less likely to generate an action potential. Similarly, if several synapses are close to one another on the same neuron, they could interact with each other in very sophisticated ways.

NEUROTRANSMITTERS

In order to prove that a certain substance is a neurotransmitter, it is necessary to show that the presynaptic axon can make the substance, that the synapse is filled with the substance when an action potential reaches the synapse, and that the postsynaptic dendrite responds to the substance.[2] It is hard to prove that a specific substance is a neurotransmitter because such evidence can be very difficult to obtain. Nevertheless, many neurotransmitters are known, with some of them being familiar to anyone who has read about brain function: dopamine and serotonin, for example, are nearly household names. Other neurotransmitters are somewhat less familiar (e.g., acetylcholine, epinephrine, norepinephrine, GABA, oxytocin, and vasopressin). Much less familiar are the neuropeptides (e.g., cholecystokinin, CCK, dynorphin, eledoisin, enkephalin, melanocortin, neurokinin, neurotensin, NPY, somatostatin, substance P, TRH, and VIP), which are small proteins that seem to coexist in the same neuron with neurotransmitters. Neuropeptides can apparently modulate the response of the dendrite to a neurotransmitter in ways that are, as yet, poorly understood.

The simple description given above of how a synapse functions hardly begins to acknowledge the full complexity of the interaction. The postsynaptic neuron is limited as to which chemicals it can respond to according to the particular chemically gated channels present in the synaptic cleft. But it would make no sense for an axon to release chemicals at the synaptic cleft if the neighboring dendrite was unable to respond to them. Thus, there is a form of cooperation between pre- and postsynaptic neurons in terms of which neurotransmitters are released and what kind of response is generated. This neuronal cooperation can apparently be stymied when certain diseases occur—an inadequate amount of neurotransmitter may be released by an axon or an inadequate response may be generated by a dendrite. Because a single neuron can have as many as 10,000 synapses, the overall response of a neuron is a function of the integrated response at many synapses. This becomes infinitely more complicated when one considers that certain synapses may be inhibitory in function and that stimulation of an inhibitory synapse may make other, adjacent synapses *less* likely to generate an action potential. Furthermore, the response to a neurotransmitter at a synapse may be modulated by a neuropeptide. Finally, there can be change over time in the strength of the relationship between neurons; one neuron can become more closely attuned to another neuron, in a simple analogy of learning.

Let us note, for the record, that the nervous system is enormously complicated and that there can also be electrical synapses between neurons. In an electrical synapse, action potentials can jump from one neuron to another without involving neurotransmitters. It is not known how common or how significant this type of synaptic transmission is in humans; it may be rare or it may be more common than we now suspect. We note further that virtually every simplification in describing the brain is an oversimplification, that every rule of thumb breaks down sooner or later, and that any axiom that claims to be always right is, by virtue of that claim, always wrong.

How the Simple Models Go Wrong

It should be clear that everything the brain does, everything the mind is, hinges upon cell-to-cell communication. If we are ever to understand the brain, or even vaguely to perceive what mind is, we must clearly comprehend the most basic function of brain: cellular interaction. To what extent do the simple models that are used to study neurons reflect our evolving knowledge of the human brain? How much of what we think we know is actually untrue, and how much is dangerously oversimplified? Of course, this question is unanswerable, since we don't know what we don't know . . . yet. But we can be sure that current experimental models do *not* reflect the level of self-organization and emergent complexity that is characteristic of the human brain. In fact, the current models scarcely reflect the level of emergent complexity characteristic of the insect brain. Is this the fault of the models or the fault of those scientists who use the models?

The squid giant axon has focused attention on the propagation of an action potential down a structurally simple, large-diameter, cable-like axon. The vertebrate axon is much smaller in diameter; all other things being equal, the conduction velocity should be slower. But all other things are not equal: the vertebrate neuron has an insulating layer called a myelin sheath, which surrounds the axon like rubber insulation wraps an electrical wire. Because there is no such sheath around a squid giant axon, it has taken time to appreciate the importance of the myelin sheath in the vertebrate neuron. The myelin sheath enables an action potential literally to jump along an axon, greatly enhancing the speed of propagation, while reducing the energetic cost of neuronal discharge.

The simple shape of the squid giant axon also drew attention away from the exceedingly complicated anatomy of vertebrate neurons. The giant axon resembles a length of thread, whereas vertebrate neurons often more closely resemble a ball of cotton. Thus, the cable properties of the vertebrate neuron are far harder to predict and remain very poorly understood. There is evidence that there is a theoretical limit as to how fine the brain's wiring can become. If the wiring is too fine, the system will be dominated by random variations called "noise," yet many neurons approach that limit without apparent harm.[3] Finally, studying a healthy squid axon gives little insight into how the function of an axon might be disturbed by disease. One can imagine that relatively tiny disturbances in the function of a channel protein or a simple change in the abundance of a channel protein could have a major effect on axonal function. Such changes could be associated with, or even causative of, the neurologic and psychiatric illnesses that take such a dreadful human toll.

The frog sciatic nerve model has focused attention on the propagation of an action potential across a structurally simple neuromuscular junction whose function is somewhat different from a true synapse. The frog neuromuscular junction is much larger than a synapse and, moreover, each frog sciatic nerve makes only a few junctions with a muscle. In contrast, a human cortical neuron makes, on average, about 10,000 synapses. Synapses in a human neuronal network can be very close to one another so that one synapse can potentially influence how strongly an adjacent synapse will respond to a stimulus. Furthermore, the morphology of the human neuron is far more complicated than that of the frog sciatic nerve. The particular arrangement of synapses with respect to one another could be important in determining how a given neuron will respond to a given stimulus. Finally, any process that has an impact upon the exquisite "tuning" of pre- and postsynaptic membranes to one another will have an impact upon the strength of connection of the two neurons. If learning occurs, this may mean that a synaptic connection has been facilitated and that one particular neuron is changing how it responds to another particular neuron. If disease has an effect at the synapse, a neuronal connection could be weakened, even if that connection is essential to continued health.

Another difficulty with the squid giant axon and the frog sciatic nerve is that these models have not forced us to consider the energetic cost of a neuronal response. When a scientist sets up an experiment, a fresh frog sciatic nerve is harvested and used until eventually the muscle fails to contract in the predicted way. This is often because the muscle or the nerve has been stimulated so many times that the energetic reserves of the tissue have been completely depleted. At this point, the scientist will simply discard the old preparation and harvest fresh tissue. But the fact that the energy reserves of the tissue *can* be depleted should be telling us something; just as it is possible to exercise a muscle to the point of fatigue, it is also possible to deplete energy reserves needed for the conduction of an action potential down a neuron. The energy required for brain function is so great that, in a resting adult, roughly 20% of all of the energy consumed is used to maintain the brain. In infants, the energy requirement of the brain is 60% because of the added cost of growing brain tissue.[4] The fact that "nerve fatigue" may never happen is a testament to how well the system is adapted to whatever energetic constraints do exist. Such energy demands likely place a very powerful constraint on the system, and it could be a grave error to forget this.

Another problem with the current models is that they implicitly assume that relations between neurons are stable and unchanging. However, we know that this assumption is false. In fact, memory may depend upon learning-induced changes in the wiring between neurons.[5] High-resolution optical imaging methods have shown that synapses can grow or regress within days, and that the rate of turnover is moderated by sensory experience. Rapid learning is apparently mediated by changes in the ease of transmission at established synapses, whereas learning that endures is mediated by growth of new synapses.[6] It is even possible that rapid and enduring learning occur at different sites in the brain; this could explain why lesions to certain parts of the brain block new learning, whereas lesions to other parts of the brain may block the recall of older memories. Even though change is clearly essential in our brains, dynamic changes at the synapse are not at all reflected in the current animal models.

A final difficulty with the current models is that they have led to the idea that neuronal function always occurs in a deterministic "on-off" fashion. We have formed the impression that when a particular event happens, another particular event will inevitably follow, just as one link in an unbroken chain follows another. This is a natural consequence of working with the giant axon; if the axon is exposed to a sufficiently strong depolarizing current, an action potential will propagate down the axon every time. In the brain, this may not be so. Recent evidence suggests that there is a rather large probability that a neuron will fail to generate an action potential in the next neuron down the line. Some recent data suggests that neurons respond to stimulation at the synaptic cleft as little as 10 to 50% of the time.[7]

Thus, it is probably wrong to think of a series of neurons as a rigidly deterministic system. It may be that neurons connect to one another in only a probabilistic way; the longer the series of neurons, the lower the probability that every neuron will respond in the predicted way. In other words, it may be that current models of neuronal function underestimate the role of chance and uncertainty. This is a difficult concept to fathom. We are used to thinking of the world as a rather predictable place; when we drop something, it will always fall down, not up. However, at the microscopic scale, life is not always so predictable. In fact, there is a growing appreciation that all of the laws of physics are probabilistic; an electron is not in one specific location in an atom but, rather, it occupies a probability shell around the nucleus. As the physics Nobelist Murray Gell-Mann wrote:

> [The laws of physics] are probabilistic . . . and they do not . . . determine by themselves the history of the universe. Instead, they co-determine it along with an inconceivably long sequence of accidents or chance events, the outcomes of which cannot be predicted in advance except for their probabilities. . . . Everything about the universe is attributable in principle to some combination of a contribution from the fundamental laws and a contribution from historical accidents. Since the laws are believed to be simple, we should attribute most of the complexity in the history of the universe to the results of accidents.[8]

NEW APPROACHES TO STUDYING THE BRAIN

Clearly, there are problems with the experimental models that have been used to study neural function. This is not to say that the use of simple systems should be curtailed: there will always be a place for simple systems because the brain is just too dauntingly complex to study as a whole. But scientists will need to be more careful about how the whole is broken apart for study.

Science is essentially the process of testing ideas against actuality by making measurements precise enough to put the lie to false theory. It is certainly true that simple systems can still be used to rigorously test new ideas, although it will likely be more productive in the future to utilize new and better model systems. One possibility in the future will be to put more emphasis on using whole organ preparations rather than simple nerve preparations. A goal of this work could be to see how neurons in a native assemblage can process information. Whole organ studies would be very exciting if they enable scientists to study emergent complexity, the synergy between neurons that is very poorly reflected in experimental models so far.

As an example of the power of using whole organ preparations, a recently published study described a series of experiments using the eye of a locust.[9] In the locust visual system, two specific neurons are sensitive to a looming visual target, and these neurons stimulate collision-avoidance behavior, probably as an adaptation to avoid predators. Yet, when the response of the locust eye to a looming visual stimulus was mapped, it was found that information was processed differently than predicted. Sensitivity of the neurons to a looming object could be predicted only by considering multiple factors, including neuronal shape, inhibition of key neurons by other neurons, and local characteristics of the neuronal pathways. This type of unpredictable relation between neurons could be evidence of emergent complexity in the locust eye.

Another useful approach to studying the brain will likely be provided by computer modeling of neuronal function. A computer model of a neuron would behave according to explicit rules set up by the programmer, even though these rules could be very complex. Such an approach can even be used to study probabilistic behavior by a neuron, since a computer simulation can be run many, many times, with a different probability value assigned during each iteration. Computer modeling was used to calculate the metabolic cost of neural information in the insect eye.[10] A mathematical description of the process of neural transmission in the blowfly eye was derived and a metabolic cost for each step in the process was calculated. Careful analysis of the model suggests that the blowfly eye requires up to 10% of all the energy expended by the entire body just to maintain the eye in a condition ready to respond to the environment. Roughly half of this energy expenditure is used simply to maintain gradients of sodium and potassium across the neural membrane. The energetic cost of the insect eye is thus higher than human muscle, gram for gram, and is comparable in energetic cost to the human brain. Computer modeling was also used to investigate the energetic efficiency of two types of neurons in the insect eye. A low-capacity neuron (55 information bits per second at the synapse) was compared to a high-capacity neuron (1,600 bits per second at the synapse) to determine which neuron was more efficient. Computer simulation showed that the low transmission rate was energetically cheaper, per unit of information. The reduced cost of an action potential is especially important if action potentials can be generated by accident. We know that such random "noise" is a problem in an electrical circuit; this is what makes a stereo hum if a circuit is not grounded properly. In any neuronal system with significant noise, a low-capacity system should be favored because noise would be less "costly" in an energetic sense. Evolution would select for multiple neuronal pathways, each with low data capacity, rather than a single pathway with high data capacity, since noise would be more costly in the latter system. This suggests that the multiple redundancy of the human brain is evolutionarily favored over a system with fewer, stronger connections.

What is becoming clear is that the brain can perform an enormous number of computations with truly remarkable efficiency. A clear demonstration of this astonishing efficiency is seen in the human sensory system, which is under very stringent selection pressure; an organism that is not exquisitely sensitive to its environment will not long survive, yet an organism that squanders energy in an inefficient manner cannot long survive either. These competing interests ensure that the human eye is a miracle of efficiency. The more we learn about the structure and function of the brain, the more we come to appreciate that all neurons and all neuronal circuits are shaped by subtle evolutionary forces that act to conserve space, materials, time, and energy.[11]

CHAPTER 3

THE SENSATION OF CHANGE

Consider for a moment, what is the basic purpose of a brain? Surely, it is to protect an individual from harm and to allow that individual to flourish in a given environment. There is also a selective pressure in favor of individuals who, sensing danger, are able to protect their offspring from harm. This is because offspring that can grow and reproduce successfully will guarantee that genes will be carried forward through the generations.[1]

But how does a brain accomplish its tasks of protection and perpetuation of the species? A brain must be able to extract meaningful patterns from a noisy, dangerous, and unpredictable blur of stimuli in the environment; it must control and coordinate the movement of muscles and bones to respond to the environment in a self-preservatory way; it must form memories and strategies to deal with the environment in a way that can prevent present and future harm; it must construct detailed and explicit mental maps of the environment to interpret the sense and meaning of events.[2] Above all, a brain must control behavior. Yet a brain cannot do any of these things unless it first becomes aware of the environment in a detailed and timely manner. Keen sense organs are therefore vital in protecting the organism from predation so as to minimize the impact of natural selection on the bloodline.

THE PUZZLE OF SENSORY ADAPTATION

If you are wearing a wedding ring, take a moment to think about the sensations that may be associated with that ring. There may be a feeling of rapture that comes from the knowledge that you are loved; there may be a feeling of claustrophobia that comes from the sense that you are trapped; there may be a feeling of peace that comes from the perception that you are part of a family; there may be a feeling of desperation that comes from the impression that your marriage is not going as you had planned. What is probably lacking is an actual sense of the metal band around your finger. Most of the time, the unchanging sensation of the ring around your finger simply fades into the sensory background and you are far more aware of things in your environment that are changing. For the same reason, you are probably unconscious of the clothing that is touching your body. Unless your clothing is uncomfortable, it is simply not worth much attention. Certainly, you can make yourself aware of these sensations, but you are not ordinarily so attentive.

A single-celled organism known as a *Stentor* shows a behavior that may reflect the same sort of sensory adaptation. This tiny, funnel-shaped creature lives in the muck at the bottom of ponds, filtering the pond water for microscopic food particles. If a *Stentor* is collected and brought into the laboratory, it will attach to the bottom of a glass bowl and begin to filter-feed again. However, if a scientist pokes at the *Stentor*, the cell will withdraw and lean away from the touch and will eventually release its hold on the bottom and swim away. At first, virtually every touch will elicit an evasive response, with a high degree of certainty. But if the experimentalist keeps poking at the *Stentor*, the organism will eventually cease to respond with such certainty. After 30 or 40 trials, a touch may evoke an evasive response less than half the time, as if the *Stentor* had simply gotten used to constant irritation. If a tiny electrode is inserted into the *Stentor* to investigate the nature of this adaptation, several striking findings emerge. The touch that evokes a withdrawal response first stimulates a "receptor potential," which is simply the flow of current induced as the cell senses the touch. This receptor potential then evokes an action potential, and it is the action potential that is associated with contraction of the cell. If the *Stentor* is touched repeatedly, the receptor potential becomes progressively smaller, even if the strength of the stimulus is unchanged. Finally, the receptor potential is so small that it fails to induce an action potential at all, so the *Stentor* does not withdraw and swim away. However, if the mechanical stimulation is discontinued for even a few minutes, the strength of the receptor potential regenerates and the *Stentor* will again swim away when touched.

Loss of sensitivity to an unchanging stimulus, a phenomenon known as "habituation," is adaptive. Certainly, it would not be useful to remain acutely aware of the wedding ring on your finger. If the ring were to become a stronger sensation, this might increase marital fidelity, but it could also distract you from something dangerous looming in your visual field. But there is another reason to become less aware of an invariant stimulus over time: such a stimulus, though it conveys nothing new about the environment, may engender an energetic cost of response anyway. The evasive response of the *Stentor* is energetically quite costly; it is counterproductive for the *Stentor* to waste energy if it is not actually at risk of being eaten.

Habituation is basically adaptation by a sensory organ. Eventually, an invariant stimulus can hardly be detected. What does habituation teach us about sensory physiology? Perhaps the most basic lesson is that what we are attuned to, what it is imperative to sense, is change, not stability. This could be to conserve energy or to be maximally attuned to stimuli that convey new information about the environment. Clearly, it would be advantageous to be sensitive to the new stimuli provided by the presence of predators or prey in the environment. But it is also possible that habituation is an unintended consequence of the way that our sense organs work.

THE BASIC FEATURES OF A SENSORY RECEPTOR

Detecting change in the environment is a function of specialized cells or organs known as sensory receptors—so called because these cells are receptive to stimulation by the environment. Any environmental change can, in principal, be detected: a sensory cell in the human eye can detect a single photon of light; the human ear can detect the movement of air particles that are vibrating as little as the diameter of a hydrogen atom; a moth can detect mating hormones released in the air miles away; a bat can echolocate and eat a flying mosquito in total darkness; a migrating bird can detect the Earth's magnetic field; a honeybee can "see" intricate patterns of polarized light reflected from bland-looking white flowers; and a rattlesnake can sense the radiant heat of a mouse from 40 centimeters (16 inches) away using infrared heat detectors that are sensitive to air temperature changes of as little as 0.002° C. It is probably fair to say that certain organisms are sensitive to environmental changes of which we, in our human arrogance, remain completely unaware.

Sensory receptors act as transducers—that is, they convert energy in one form into energy in another form. In a retinal cell detecting a photon of light, light energy is transduced into the electrochemical energy of an action potential. All sensory receptors transduce environmental energy into an action potential, whether that environmental energy is in the form of a photon of light or a vibrating air molecule. It is a remarkable property of receptor cells that so many forms of energy in the environment converge into a single form of energy in the nervous system: any form of energy that can be detected by humans is transduced into an action potential.

Sensory receptors are typically highly selective in what types of energy they can transduce. Hair cells in the ear are sensitive to incredibly tiny vibrations of air, but completely insensitive to the vibrations of light photons. A receptor cell in a human retina can respond to a single photon, but will not respond even to very energetic signals if those signals are not in the form of light photons. In fact, human retinal cells are tuned so exquisitely that they will not even respond to photons of the wrong wavelength; the human eye is thus insensitive to infrared and ultraviolet light.

Sensory receptors also act as powerful signal amplifiers. In the human eye, visual receptor cells known as rods respond, not to color, but to the presence of light. If a person is put in complete darkness long enough for his eyes to adjust, that person will be able to see a faint flash of light if as few as 10 rod cells are each stimulated by a single photon.[3] Assuming that 10 photons can generate an action potential within the eye, we can calculate the amplifying ability of the human eye. An action potential releases roughly 10^{-11} units (Watt-seconds per second) of electrical energy. On the other hand, 10 photons of light yield, at most, 10^{-16} units of radiant energy. The power amplification of the eye, in this case, is at least 100,000-fold (or 10^5 units), as radiant energy is transduced into electrical energy. To put this in perspective, military-quality night-vision binoculars amplify light to roughly the same degree, but they require an external power source—like a battery—to do so.

Sensory receptors typically have an enormous dynamic range over which they are effective. Biological receptors are often sensitive to very faint signals but will not be overwhelmed by very powerful signals, whereas electronic receptors often have a very narrow range over which they function well. For example, night-vision binoculars are remarkably sensitive on nights when there is ambient moonlight or starlight but rather ineffective on cloudy nights without ambient light, and completely ineffective under daylight conditions, even if the day is overcast. Yet the human eye can detect visual targets under all of these conditions. Similarly, a camera without a flash attachment is able to take photographs under a fairly narrow range of lighting conditions, whereas your eye can see well under far darker and far brighter conditions.

Sensory receptors typically have two different responses to an environmental stimulus. As we saw in the case of *Stentor*, a touch that evokes a withdrawal response will first stimulate a receptor potential, which, in turn, stimulates an action potential. The receptor potential is the flow of current induced, in the case of *Stentor*, by a mechanical stimulation. A receptor potential is graded so that a faint touch evokes a small potential, whereas a jab evokes a large potential. In contrast, an action potential, which is associated with contraction of the cell, is an all-or-nothing response. It is possible to separate a receptor potential from an action potential under laboratory conditions. This can be done using a stretch receptor in the abdominal muscle of a crayfish; if the muscle is stretched, a graded receptor potential will be generated, which induces an action potential if the muscle is stretched sufficiently. However, if the stretch receptor is poisoned with tetrodotoxin, which specifically blocks the sodium channel, the action potential can be prevented. Yet even when the action potential is blocked, there is still a graded receptor potential, since the receptor potential results from an independent but still poorly understood mechanism.

Sensory receptors typically make a graded response to an environmental stimulus. Although a nerve cell responds to a stimulus in an all-or-nothing fashion by generating an action potential, a sensory cell typically has a broader repertoire of responses. A human retinal cell responds more strongly to sunlight than to moonlight by inducing a stronger receptor potential. Ultimately, this will mean that there are more frequent action potentials going to the visual center in the brain. Sensory receptors can function in a fashion that is really a hybrid between a graded or "analog" response and an on-off or "digital" response. This is possible because the receptor potential, which is graded in response to the stimulus, is analog in nature, whereas the action potential, which is all-or-nothing, is usually thought of as being a digital response. Since both responses occur in the same cell or group of cells, sensory receptors offer an amalgam of advantages. The frequency of action potentials is a function of the magnitude of the receptor potential, and the magnitude of the receptor potential correlates with the stimulus strength. Thus, indirectly, the frequency of action potentials is a crude approximation of the stimulus strength.

Sensory receptors encode the environment as a series of action potentials. One feature of the environment can differ from another *only* in terms of which specific receptors are induced to respond and how vigorous that response is. A bright light is encoded as a series of action potentials, but so is a loud noise; the two stimuli differ *not* in any inherent characteristic of the action potentials but, rather, in the part of the brain to which those action potentials are routed.

Thus, the "connectivity" of a sensory receptor is crucial. This may account for one of the most bizarre and interesting abnormalities of the human sensorium. Synesthesia is the rare but involuntary ability to experience one stimulus via two senses. The most common form of synesthesia is for a sound to be experienced as both a note and a color. Recently, the case of a 27-year-old professional musician was described: whenever this woman heard a specific tone interval in a piece of music, she automatically and consistently experienced a particular taste in her mouth.[4] A minor third interval was experienced as salty while a major third interval was experienced as sweet. However, she did not hear tone intervals when exposed to specific tastes. This woman also experienced the more common (but still quite rare) synesthesia of hearing a particular tone linked to a particular color. Probing her case very carefully, scientists became convinced that she was neither lying nor irrational. The most plausible explanation for this peculiar abnormality is that there was a wiring abnormality in her brain such that auditory neurons somehow routed action potentials to both the auditory and the gustatory centers of her brain. This gives new meaning to the idea that music is food for the soul.

SENSING PRESSURE

Perhaps the easiest sense to understand is the sense of touch. Receptors sensitive to touch are known as mechanoreceptors, and these tend to be among the simplest types of sensors. The human skin has mechanoreceptors that are sensitive to four main things: touch, cold, warmth, and pain (including pricking, burning, itching, and perhaps tickling). Each modality is associated with at least one kind of receptor; it can be imagined that some stimuli activate more than one type of receptor and are thus perceived as complex stimuli (e.g., the sensation of wetness). Cold sensors can detect a temperature decrease of as little as $0.2°$ Centigrade, if the temperature change occurs quickly, whereas warm sensors require a temperature increase of at least $0.3°$ Centigrade.

The function of human touch receptors is relatively poorly understood, since it is so hard to put needle electrodes into human nerves without getting complaints. But animal models have been used extensively to study mechanoreception; a popular approach is to study such receptors in the abdominal muscle of a lobster. In a lobster stretch receptor, physical stretching increases the permeability of the receptor membrane to sodium. A simple explanation could be that stretching actually makes the Na^+ channels larger so that ions passively leak through more easily; if stretching is strong enough, an action potential would then be generated. The relatively small amount of energy needed to stretch the membrane would then be amplified by the large amount of potential energy stored in the gradient of ions across the membrane.

A more sophisticated type of mechanoreceptor, common in vertebrates, is called a hair cell. A hair cell typically takes the form of a single cell with one or several large hairs projecting from it. Pressure on the hair leads to mechanical deformation of the hair cell, and it is this deformation that leads to an increase in permeability of Na^+ channels in the hair cell. This simple system has been adapted in various ways to make a broad range of sense organs. Fish have a lateral line down the side of the body made up of hair cells set into the skin, and these hair cells are sensitive to pressure waves in water. Toads have hair cells covered with a gelatinous plug that prevents the hairs from drying out when the toad is out of water and which may amplify water pressure waves when submerged. Vertebrates (including humans) have hair cells in the inner ear, which are sensitive to fluid motion in the semicircular canals and are involved in sensing acceleration and maintaining balance. But the most sophisticated adaptation of hair cells to pressure is undoubtedly the sense of hearing in vertebrates.

Hearing results from the ability of hair cells in the cochlea of the ear to be deformed by sound waves. The cochlea is a tiny organ, less than a centimeter across and about the shape of a snail shell, which contains fluid. Airborne vibrations that impinge upon the eardrum are transmitted through the tiny bones of the middle ear to the wall of the cochlea, thereby setting up vibrations in the fluid that fills the cochlea. A pure tone starts a pressure wave in the cochlear fluid that travels up the spiral of the cochlea, with high tones traveling farther up the spiral than low tones. The pressure wave causes movement of hair cells in the cochlea, each of which can generate an action potential. Each hair cell along the length of the cochlea is displaced most effectively by a particular sound frequency. It may be that hair cells that respond to a particular frequency inhibit the response of adjacent hair cells, thereby "tuning" the response of the cochlea to a particular tone. In any case, a tone is encoded in the physical location of the hair cell that responds to the tone, while the volume of the sound is encoded in the frequency of action potentials generated by that hair cell. Sound is thus transduced to an electrochemical signal, which is then carried to the auditory portion of the brain. It remains a deeply mysterious process how a complex set of separate sounds can merge into the combined sound of a symphony, yet each instrument can still be perceived as retaining a separate musical identity.

SENSING CHEMICALS

The sense of smell is fundamentally different from the sense of pressure in that the stimulus perceived is actually a chemical or odorant molecule in air. Any receptor sensitive to chemicals is called a chemoreceptor, whether that receptor is located in a rat nasal cavity or a human tongue or an insect antenna. In 2004, the sense of smell moved to center stage when the Nobel Prize was awarded for seminal work that had been done more than a decade earlier.[5] The sense of smell is crucial to survival; not only does it alert the predator to the presence of prey, it also alerts the prey that its life may be in jeopardy; not only does it help animals to find food, it also alerts them when that food may no longer be safe to eat; not only can it help the sexes to find one another over great distances, it also makes reproduction more likely when they do find one another. The olfactory system of mammals is able to detect thousands of different odorant molecules, most of which are organic compounds of low molecular weight; these are the sorts of chemicals that carry the stench of life.

Even humans have a well-developed sense of smell. The major reason smell is relatively less important to humans than to other animals is that our erect posture has elevated our noses far above the rich stew of smells that wafts close to the ground. A human nose is so precise that it can distinguish molecules that differ by a single carbon atom, making it more sensitive than the best man-made device. We can easily discern between 6- and 7-carbon aliphatic aldehydes that smell, respectively, like grass and soap.[6] The scent of caramel is stimulated by a certain molecule known as maple furanone, even at a concentration of only one part in 10^{15}.

The sense of smell relies upon olfactory receptors that are embedded in the membranes at the back of the nasal cavity. There are over 1,000 different genes that code for olfactory receptors in mice; each olfactory receptor neuron expresses one and only one receptor molecule. But a given odorant can bind to multiple receptors, and a single receptor may be able to bind to multiple odorants; this may mean that a mouse can distinguish tens of thousands of odors. Olfactory receptor genes form the largest family of genes in the mammalian genome, as at least 3% of all mouse genes are dedicated to olfactory receptors. Even in humans, in whom the sense of smell is greatly atrophied compared to mice, there are still about 350 genes dedicated to olfactory receptors. It is particularly noteworthy that 350 genes code for olfactory receptors in people, whereas only 3 genes code for the visual pigments that enable us to have color vision.

Although the sense of taste is not as well understood as the sense of smell, it is conceptually rather similar. In practice, it can be difficult to separate the sense of smell from the sense of taste, since virtually any food with a flavor also has a scent. Anyone who has ever eaten a favorite food while having a bad cold is aware of how much what we sense as flavor is actually due to odor. Five tastes are recognized by human beings: sweet, bitter, sour, salty, and umami. Umami, the Japanese word for delicious, describes a complex, savory, smoky flavor (like soy sauce or ketchup) whose prototypical form is monosodium glutamate (MSG). All of the various tastes that we perceive in food are thus some combination of these five tastes, plus whatever odorants are associated with that food.

Chemoreceptors, in general, respond to chemicals in much the same way that the dendritic membrane of a neuron responds to neurotransmitter released into the synaptic cleft. When a receptor binds a specific chemical, whether that chemical is an odorant or a tastant, it induces a change in the sodium current across the receptor membrane. Sodium ions flow into the receptor cell, triggering a cascade of events that ultimately results in an action potential that goes from the receptor to the related region of the brain.

SENSING LIGHT

Vision is contingent upon receptor cells in the retina of the eye, known as photoreceptors, which are so sensitive that they can detect a single photon of light, and so robust that they can adapt to the enormously high photon flux of the noonday sun. The structure of the eye is merely a device that shines an image upon the retina with fidelity; the real business of the eye occurs in the retina.[7] Photoreceptor cells in the retina are of only two types: rods are receptive to light and dark and shades of gray while cones are receptive to color (fig. 3). Cones are found in the central portion of the retina, known as the fovea, which is an area of only about a square millimeter. The fovea is the most sensitive region of the eye in terms of visual acuity and the ability to distinguish angle of motion, but the fovea is much less sensitive to light and dark. Rods are more sensitive to the presence or absence of light, and are found outside the fovea, which explains why it is easier to see a faint star in the night sky if you don't look at it directly. This also explains why, in very dim light, all colors appear as shades of gray.

Both rods and cones function in more or less the same way, as each type of photoreceptor has a store of pigment molecules that absorb light energy. When light enters a photoreceptor cell, it causes these visual pigments to change their conformation. As the molecules change shape, they also change in volume, and it may be this increase in volume that transiently opens ion channels in the photoreceptor membrane. This induces a change in the flow of current across the cell membrane. However, there is a key difference in the way that a photoreceptor cell responds to light, compared to the way that a mechanoreceptor responds to touch. Touch results in an increased flow of current across the cell membrane of a mechanoreceptor, and it is this increase in current that eventually induces an action potential. In a photoreceptor, light actually decreases the flow of current across the membrane. In complete darkness, there is a substantial leakage of ions across the photoreceptor membrane; this is often called the "dark current." Some of the dark current is carried by ions other than Na^+, with calcium (Ca^{++}) being involved in both rods and cones.[8] A photon of light can block the dark current, meaning that the membrane becomes less "leaky" to ions. Then, as the light ceases, the membrane returns to its normal "leaky" state. Given that we are exposed to light most of the time, it would seem that this should result in an absence of action potentials, a form of light-induced blindness.

In reality, the dark current is a receptor potential that makes the retina more sensitive under low-light conditions.[9] Recall that a receptor potential is different from an action potential. In a photoreceptor, the receptor potential is simply the change in current flow induced as the cell senses light. A receptor potential can be generated by any of the 100 million rods or 5 million cones in a human retina. Even dim light tends to stimulate hundreds of receptor cells. As these rods or cones generate a receptor potential, this stimulates a receptor potential in bipolar cells, which are the next cell up in the chain of command. The bipolar cells integrate signals from many separate rods

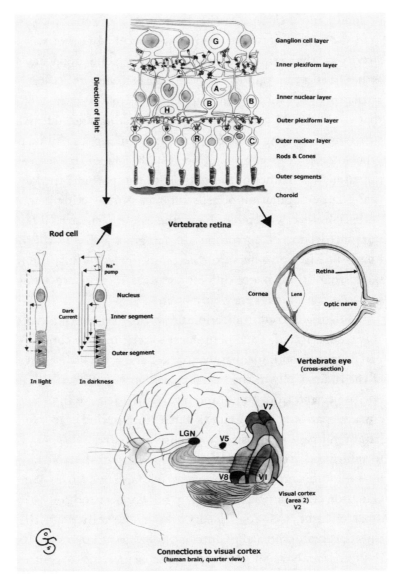

Figure 3. The human visual cortex. Rod cells in light and in darkness are shown (left of center), with the relationship of rod cells to various other cells in the human retina (top). The anatomy of the human eye is shown in cross-section (right of center), with the retina at the back wall of the eye and the optic nerve emerging from the middle of the retina. The connection from the retina through the optic nerve to the lateral geniculate nucleus (LGN), and thence to the human visual cortex, is also shown (bottom).

or cones, and a change in the dark current of rods or cones induces a graded receptor potential in the bipolar cell. In other words, the sensitivity of the retina to light does not depend upon any one cell—rather, it is a function of many cells interacting together. Even in the retina there is a form of emergent complexity, a synergy of the system that could not be produced by one cell acting alone. Bipolar cells, in turn, signal to ganglion cells, which receive input from many different bipolar cells; it is the ganglion cell that finally generates an action potential. The names of the cells are not important; the key concept is that a large population of cells interact in the retina before an action potential is ever generated. Nevertheless, it is the action potential, the integrated sum of excitations of ganglion cells in the retina, that moves up the optic nerve to the brain (fig. 4). Thus, the retina itself blends data from many different sources to generate a novel output, just as if the retina were a tiny brain.

Cell-to-cell communication networks in the retina can be remarkably sophisticated. Each photoreceptor cell may be connected to many ganglion cells and, ultimately, to thousands of cells in the visual cortex of the brain. Cells in the visual cortex do not receive a simple one-to-one projection from the retina but, rather, receive information that is already partially integrated. Such integration can, for example, heighten image contrast. If rod cells stimulated by a spot of light were to inhibit adjacent rod cells that were not similarly stimulated, such lateral inhibition could make the spot stand out with greater contrast. Ganglion cells may be able to map networks of rod cells stimulated by a bar of light. Cell-to-cell networks in the retina and the brain can thus integrate signals in interesting ways. Experiments with anesthetized cats have shown that certain neurons in the visual cortex of the brain are literally "tuned" to certain highly specific features of the environment, probably as a result of learning. For example, certain neurons in the cat cortex may respond maximally to bars of light that are oriented at 40 degree angles to the right of the vertical axis in the visual field.[10] Thus, neurons in the visual cortex are tuned so as to respond to simple—and rather abstract—features of a visual image.

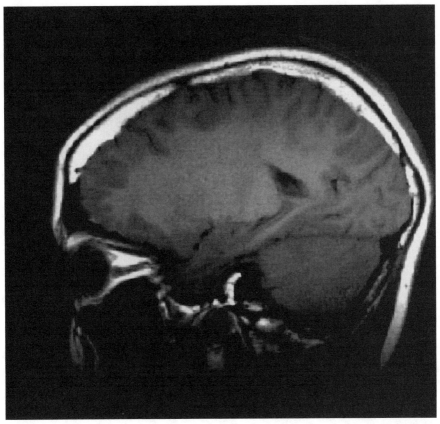

Figure 4. The optic nerve connects retina to brain. To the left is the globe of the eye, seen as a dark blob in the brightness of the periorbital fat, with the dark stem of the optic nerve passing from the back of the globe to the front of the brain. The integrated sum of excitations from ganglion cells in the retina pass up the optic nerve to the brain. Also seen in this view is the hippocampus, which is a dark gray mass of tissue directly behind the eye. The hippocampus is rather small in volume, considering the vital function it plays in memory consolidation.

THE DISTINCTION BETWEEN SENSING LIGHT AND SEEING

We now have a fairly good understanding of how the eye and, in turn, the brain senses light. In groundbreaking studies done nearly 50 years ago, with cats as an experimental model, it was proven that a retinal ganglion cell responds to light in the visual field by generating a graded receptor potential.[11] In awake and aware cats, this receptor potential causes an action potential to move up the optic nerve to the visual cortex of the brain. Each part of the visual field can be mapped to a corresponding part of the visual cortex, and every time there is a flash of light, there is a corresponding response in the brain. The functional connection between retinal cells and the visual cortex is established very early in development, and the exquisite specificity of these physical connections enables a cat to interpret visual stimuli. Such findings led to a key generalization: the richness of responses generated by the brain is a result of the complexity of physical connections or neural circuits and not a result of the complexity of signals, since all signals between eye and brain are simply action potentials.

Yet there is a critical distinction between sensing light and seeing. We have described how cells in the retina sense light but we have not illuminated the means by which light and shadow, color and grayscale, parallax and perspective, combine to make an interpretable image. To cells in your retina, your mother or your child is simply a mass of grayscale and color, which may be moving or may be stationary, but is not an emotionally freighted entity. How do we get from sensing light to actually seeing and interpreting an image?

Francis Crick, one of the pioneering biologists who discovered the structure of DNA, pointed out a fascinating conundrum of the visual system. The eye functions essentially like a television camera, with moving images transduced into electrical signals that move up the optic nerve and eventually reach the brain. The visual image is then mapped onto the visual cortex in what amounts to a point-for-point representation of reality. It is as if the visual world is projected onto a television screen in our brain. But what remains unknown is who is watching this television. How can we use data about facial shape, eye color, and hair length to recognize a familiar face? Even harder to explain: How can we, by merely looking at a face, interpret what someone else is thinking? How is it possible for us to recognize a person we have not seen in years when that person may look quite different than she did when last we saw her? For that matter, how can we recognize a person we have not seen in years when our own eyes have aged and may not see the image in the same way?

As we age, our eyes change. A visual image may be distorted by astigmatism or biased by a change in the color sensitivity of our retina. Even over the course of a single day, the color of an object may change, depending upon whether the object is illuminated by dawn light, noonday sun, or the light of late afternoon. Yet a color can be sensed as unchanging if the perceived color is in accord with the remembered color. In other words, we see fundamentally the same object, even though the image has changed. How do we maintain constancy in our perception despite a change in the signal? In the case of color constancy, it may be that the eye calculates the "real" color of an object based on the reflected color and other features of the visual environment.[12] Yet this implies that vision is a computational ability of the brain as much as it is a sensing ability of the retina.

A convincing demonstration of the sophistication of human vision is provided by the finding that people can recognize a face that has been inverted so as to appear upside down in the visual field.[13] Although the accuracy of recognition may suffer, people are quite good at recognizing a familiar inverted face even though the process may be rather slow. Perhaps it should not be surprising that the human brain is so good at recognizing the human face; recognition of a fellow tribe member or of a loved one may be one of the most basic survival skills for a social being.

AREAS OF IGNORANCE

The physiology of the eye has been described in bewildering detail, and a quick survey of the literature might leave the reader with a sense that there is little left to learn. But this would be the wrong impression. Scientists tend to focus on measurable things—such as ion flow across membranes in photoreceptor cells—simply because this forms a tractable problem. Yet we still have a very poor understanding of how perception emerges from sensation. How are we able to determine that a particular moving blob of light is a loved one and not a predator? The higher levels of integration that are needed to determine the identity, importance, and intent of a feature in the visual environment are, for the moment at least, beyond our understanding.

But there are other, simpler, and far more workable problems that also remain in the realm of the unknown. Some of these intriguing problems relate to senses that seem to exist in humans but that must rely upon sense organs that have not yet been identified while other problems relate to senses that exist in simpler vertebrates but have not been proven to exist in humans.

It is well known that animals can communicate with one another by scent, using volatile hormones to induce behavioral changes in other members of the same species.[14] The classic example is the sexual attractant pheromone released by the female *Bombyx* moth, which is so powerful that a single molecule can induce a male moth to begin searching for the female. But there is also clear evidence of pheromones in vertebrates. If pregnant mice are housed with males other than the original mate, these female mice are significantly more likely to miscarry within a few days. Following the failed pregnancy, the female will invariably mate with the new male. This behavior suggests that male mice can use pheromones to induce miscarriage and bring a female mouse to estrous quicker than would occur otherwise. Pheromones fall into two broad classes: "signal pheromones" that induce short-term behavioral changes and that could act as opposite-sex attractants (or same-sex repellants) and "primer pheromones" that cause long-term behavioral changes.

It is controversial whether humans are sensitive to pheromones, although it is well known that women who live in close proximity to one another will synchronize their menstrual periods, possibly through primer pheromones. If women are exposed to a male hormone, there can be a measurable effect on blood flow to the skin. Underarm secretions of sexually active, fertile women can induce more frequent sexual overtures to postmenopausal women if these secretions are used as perfume.[15]

Is it possible that the release of sexual pheromones is why some people are perceived to be "sexy"? Does an insensitivity to infant-recognition pheromones perhaps account for why some mothers fail to bond with and care for their infants? How does our modern obsession with cleanliness play into social interactions? Are the nasal membranes at the back of the nose the seat of sensitivity to pheromones or do humans have a vomeronasal organ, which is the organ sensitive to pheromones in animals? Recently, a tiny structure at the base of the nasal septum inside the human nose was described.[16] This structure, in a part of the nose that could easily be damaged by nasal surgery, was hypothesized to be the human vomeronasal organ.

What mediates the sensitivity of a person to falling levels of blood sugar? Most people know that if they are very hungry they become crabby and that if they eat even a small amount of sugar or chocolate they can become more placid, even if they are still hungry. Dietary intake of sugar is influenced by two brain mechanisms, one associated with regulation of energy balance and one associated with reward.[17] Neither mechanism is well understood, although both influence behavior in subtle ways. For example, chronic food restriction (at 80% of normal caloric intake) or acute food restriction (fasting for 48 hours) will induce rats to select a diet rich in fat in preference to a diet rich in carbohydrates or proteins. This could be because fat is a denser source of calories and fat consumption can therefore replenish depleted reserves more quickly. But what sense organ detects the effect of starvation, and how does this induce the selection of dietary fat? Can anorexia be partly explained as a disorder in sensing the effects of starvation?

Chronic ingestion of sugar by a laboratory animal has an effect on the body much like the effect of mild opioid dependence. Drugs of abuse, including cocaine, heroin, and amphetamine, cause an increase in the concentration of the neurotransmitter dopamine in certain brain regions. Chronic sugar intake has a similar, albeit milder, effect. It may be literally true that some people are addicted to chocolate, that food activates the same reward systems that are involved in drug addiction. In humans, there is good evidence of an increased preference for sweets among reformed drug addicts and alcoholics. This may be why national conventions of Alcoholics Anonymous will typically produce a citywide spike in consumption of chocolate and ice cream. Yet the mechanisms that sense blood glucose and induce changes in dietary intake are unknown.

Changes in behavior that occur on a 24-hour cycle are called circadian rhythms.[18] These rhythms attune us to the length of the day so that we become sleepy at night and feel most energetic in daylight. Many of these circadian rhythms are preserved even if a person is kept in complete darkness or brilliant light where there are no cues to day length. A small part of the mouse brain, known as the suprachiasmatic nucleus (SCN), has been implicated in establishing the circadian rhythm; if the SCN is damaged, the rhythm is destroyed. Mutation of a gene known as *Clock* can reset the circadian rhythm; in normal mice, the cycle of activity in an exercise wheel is about 23.7 hours long, even if the animal is kept in total darkness. If one copy of *Clock* is mutated, the rhythm of activity increases to about 24.8 hours. But if both copies of *Clock* are mutated, the rhythm becomes chaotic. Cells from the SCN can maintain a pattern of rhythmic firing, even if they are brought into cell culture, but the firing is more random than normal unless there are environmental light cues. These cells, which live in the center of the brain, are able to entrain to environmental light in an intact animal because they are connected to photoreceptors in the retina. Yet the biology of the human SCN is not well understood, even though humans show circadian rhythms. Certain neurologic illnesses (e.g, Alzheimer disease, Huntington disease, multiple sclerosis, stroke, and fatal familial insomnia) are often associated with disruption of the sleep-wake cycle whereas other illnesses (e.g, Alzheimer and epilepsy) tend to worsen at a particular time of the day. It has been proposed that all of these illnesses may involve damage to the human circadian sleep mechanism.

In addition to these well-defined questions, there are a number of other questions that have not yet been defined as appropriate for scientific research. For example, do animals sense things that we cannot detect, such as very low frequency sound waves, which enable them to predict natural disasters when we cannot? After the disastrous Southeast Asian tsunami of 2004, there were reports that wild goats and elephants had moved to higher ground where they were safe from the effect of the floodwaters. Human observers report being able to "hear" an earthquake: I was backpacking on Mammoth Mountain in California in 1980 when the Long Valley caldera directly beneath it—a huge depression created by the collapse of an ancient volcanic cone—was raised by fresh subterranean magma. This lifted the caldera floor by 25 centimeters and created a swarm of magnitude 6+ earthquakes that shook the mountain to its foundation. I sensed the first of these earthquakes for several very long seconds before I felt the ground shaking.

One can marvel at the ability of the human sense organs to detect very faint signals, but it is also possible to marvel at our insensitivity to other important signals in our own internal environment. For example, why is the human body relatively insensitive to the balance of salt and water in the bloodstream? A recent study of runners in the Boston Marathon found that 13% of runners had an abnormally low concentration of sodium in their blood serum and, in nearly 1% of runners, this imbalance had reached a dangerous level.[19] Given that salt imbalance can be fatal, this would seem to reveal an Achilles' heel in the human sensorium.

Nevertheless, one must be impressed at the ability of the human senses to detect and respond to environmental changes that are at or below the threshold of detection of the latest technology. Our sensitivity to the environment is a clear testament to the centrality of simply surviving in a world where the engine of evolution is fueled by the blood of those who failed to survive.

CHAPTER 4

THE "DINOSAUR" BRAIN

T he "dinosaur" brain is the part of our brain concerned with performing automatic functions, so that we need not be consciously aware of them. There is a clear survival value in not having to deliberately draw breath or make your heart beat while, for instance, you are being pursued by a predator. Making such functions as heart rate or blood pressure occur automatically can free the brain to consider issues that have clear survival value in an evolutionary sense, like how to avoid predators in the first place. The dinosaur brain is an informal name for those parts of the brain that were rather well developed in dinosaurs, as shown by careful study of fossil dinosaur skulls. As part of the dinosaur brain, we include the medulla oblongata at the top of the spinal cord, the pons above and in front of the medulla, and the cerebellum above and behind the medulla. All of these structures are shown in an image of the brain of a healthy teenager (fig. 5), obtained by magnetic resonance imaging (MRI).

We note that in biology and medicine there is a persistent bias in the belief that a detailed description of structure will reveal something about function. This bias has resulted in a proliferation of names for structures that often have poorly understood functions. Single struc-

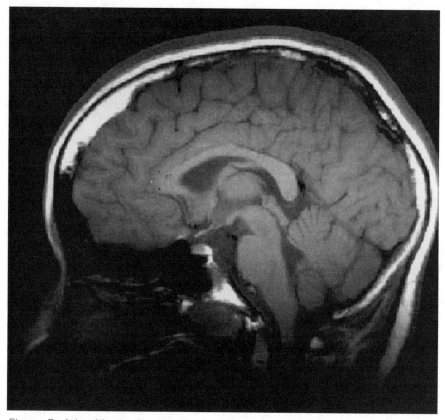

Figure 5. A healthy brain, seen in profile through the midline, with the face to the left. The gyri (wrinkles) of the cerebrum are clearly visible in the mass of gray matter high in the skull and the spinal cord is seen at the bottom of the image. The corpus callosum forms a light dome of tissue at the center of the image; the callosum is a band of tissue that links the right and left hemispheres of the brain. Beneath the callosum is the medulla and the blob-like pons, both at the top of the spinal cord. This is the same subject shown in figure 4.

tures have been given multiple names, and some names correspond to structures that may not really exist as entities. Suffice it to say that every brain function seems to be centered in a particular brain structure (even if we don't yet know the location of that structure) and that every brain structure has been named, often more than once. Yet we will be more concerned with function than structure, and we will resist the temptation to confound naming with knowledge. Here and elsewhere, we will avoid, insofar as possible, getting entangled with anatomic names. This is because names can become confusing and, once memorized, give a false sense of the extent of our knowledge.

THE BASIC ANATOMY OF THE "DINOSAUR" BRAIN

The spinal cord is the traditional starting point for a description of the central nervous system because it is here that individual nerves come together to form a central pathway from body to brain. All of the sensory inputs from the body feed into the spinal cord, and all of the motor outputs that drive the muscles derive from the spinal cord. The spinal cord itself is a structure no wider than about a centimeter at the widest and less than a meter long in even the tallest person. The spinal cord has a segmented appearance because it receives input from a series of nerves that feed into it from the back of the vertebrae. Another series of nerves leave the spinal cord, projecting toward the front; it is these nerves that carry motor impulses to the muscles. Nerves from the front and from the back of the spinal cord coalesce to form "dorsal root ganglia," from which the spinal nerves arise. Because spinal nerves arise at most of the 31 bony vertebrae down the spinal column, the spinal cord, with its spinal nerves, has a segmented, wormlike appearance. In fact, if dissected free of the spinal column, it would look like a clam worm from hell, bait for only the largest of fish.

The spinal cord itself is a bundle of long and incredibly fine nerve axons. In cross section, it has the appearance of a butterfly of gray matter surrounded by a cushion of white matter, which gives it an ovoid shape. The spinal cord has three main functions that are intimately linked to this structure: sensory processing, motor control, and reflex generation. Sensory processing is a function of those sensory neurons that enter the spinal cord at regular intervals up its length; a single sensory neuron may split into 500 or more branches within the

Cross-section of spinal cord

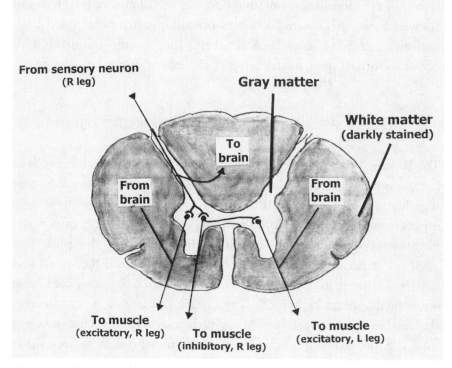

Figure 6. Crossed-extension reflex circuit. This reflex would be invoked by stepping on a tack with the right (R) foot. Pain, sensed by a sensory neuron in the R foot, comes up the R leg and stimulates the foot to withdraw by closing an excitatory loop through the spinal cord. An excitatory action potential also crosses over to the left (L) leg, causing L leg muscles to contract in order to support the weight of the body. Because a muscle on the opposite side of the body is reflexively stimulated, this is called a crossed-extension reflex. At the same time, an inhibitory impulse is sent to the R leg to modulate how far the foot is withdrawn. Because all connections are made through synapses that could be facilitated, long-term potentiation (i.e., learning) may be involved in this reflex (interneurons omitted for clarity).

spinal cord. Some of these branches project up the spinal cord toward the brain, whereas other branches make local connections to neurons in the area. Motor control is a result of motor neurons that drive the muscles to contract; while these neurons may carry signals from the brain, they may also carry signals that arise from local circuits involving the sensory neurons. Finally, reflexes are the result of neural circuits that are largely contained within the spinal cord; sensory neurons are able to connect to motor neurons and induce an involuntary response without requiring input from the brain. These reflex circuits stimulate simple, stereotypical responses such as the knee-jerk reflex, which occurs when the patellar tendon is stretched by a blow to the knee. Were your knee to give way suddenly, the knee-jerk reflex could potentially prevent you from collapsing, but no thought is involved on your part; the reflex occurs without input from the brain (fig. 6).

The spinal cord provides a route for sensations of which we are aware, such as the warmth of the sunshine or the feeling of a chill wind, but it also provides a route for the autonomic nervous system, which regulates those actions that are entirely automatic. The beating of our heart, the rhythm of our breathing, the progress of food through our gut, the secretion of digestive enzymes from our pancreas, the careful control of the acidity of our blood, the routing of blood to active muscles, the action of sweat glands in our skin; all these actions proceed automatically, without conscious input, even in spite of conscious input. This is why it is not possible to commit suicide by holding your breath; all you can do is hold your breath until you lose consciousness, then your autonomic nervous system will reassert itself and reestablish the breathing rhythm. The autonomic nervous system is also responsible for a curious aspect of the nervous system that can have clinical significance. If an internal organ is injured or inflamed, pain may not be localized to that organ but rather referred elsewhere. Coronary artery disease is associated with pain, but the pain is not felt in the heart; rather, referred pain is felt in the upper left part of the torso, and this pain is what we know as angina. This imprecision in pain localization can produce diagnostic enigmas in some cases. Fortunately, physicians are generally well aware of the patterns of referred pain.

At the top of the spinal cord is the medulla oblongata, which is a relay station between the spinal cord and the brain. Nerve tracts that eventually reach the brain must traverse the brainstem, and many of these nerve tracts make synapses within the medulla. Because of the synapses made in the medulla, many integrative functions are possible; control centers for breathing, heart rate, and blood pressure are located here, as are reflexes that control coughing and vomiting. There are even complex integrative functions located in the medulla; the state of consciousness is partially under the purview of a structure called the reticular formation, which is in the brainstem. In addition, all but the first 2 of the 12 major cranial nerves—which control functions as diverse as facial expression, eye movement, and hearing—make connections in the medulla (i.e., the olfactory and optic nerves emerge directly from the brain). Between the medulla and the cerebellum is a fluid-filled space known as the fourth ventricle; in general, ventricles function to enable cerebrospinal fluid to circulate throughout the brain. Cerebrospinal fluid—also called CSF—has a role in regulating the external environment of neurons by providing nutrition and removing wastes. It also serves to cushion the brain from blows to the head.

Adherent to the front of the medulla is the pons, a spheroidal structure about the size and shape of a small plum. The pons contains nerve fibers that traverse from the cerebral cortex of the brain, funnel down through the pons, and then connect to the cerebellum or run down the spinal cord.[1] These pathways are concerned with the control of both gross and fine motor movements. For example, some neuronal pathways control gross motor movements in conjunction with the cerebellum, such as walking on an uneven surface. Other pathways control fine movements, such as buttoning a shirt or playing a guitar. The pons also plays a major role in controlling the rhythm of breathing.

THE STRUCTURE AND FUNCTION OF THE CEREBELLUM

The cerebellum, whose name literally means "little brain," is the rounded mass of tissue that appears to be suspended below the "higher" portion of the brain known as the cerebrum. The cerebellum receives sensory input of all kinds and projects neurons back to the

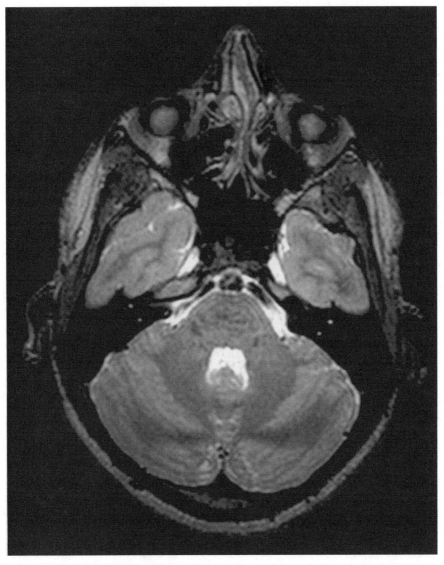

Figure 7. The cerebellum, whose name literally means "little brain," is the rounded mass of tissue at the lower part of the brain in this view. The cerebellum connects directly to the brainstem and indirectly to the cerebrum, from which it receives sensory input of all kinds. The nose is at the top of this image, with the globes of the eyes to either side and the nasal membranes between the eyes.

brainstem and to the cerebrum (fig. 7). Clearly, the cerebellum plays a part in processing sensory information; yet, surprisingly, damage to the cerebellum does not seem to result in sensory deficits. However, the cerebellum does have a central role in controlling motor activity, since cerebellar damage is often associated with abnormalities of gait and equilibrium, of muscle tone and postural control, and of coordination of voluntary movement.[2] Although the function of the cerebellum is not yet well understood, it apparently is involved in learning motor tasks and in assessing error in movement. As a movement is planned and executed, the cerebellum hones the performance of the task by comparing the actual outcome of the movement to the desired outcome. If there is a difference, this represents an error of movement, which is minimized by adjusting motor outputs.

The cerebellum controls large-scale movements, but can also control fine motor movements with an astonishing degree of precision. For example, if a person throws darts at a dartboard, progressive improvement in throwing accuracy is achieved by comparing the target that the dart actually hits to the location of the bull's-eye. Since repetition improves dart-throwing accuracy, a fascinating experiment can be done.[3] If a person throws darts while wearing a pair of prism spectacles that displace the image to one side, throwing accuracy drops sharply. But as the person continues to practice, iterative motor corrections will eventually increase throwing accuracy to approximate the original accuracy attained without the spectacles. Finally, if the spectacles are again removed, throwing accuracy drops again, but the original accuracy is recovered quickly through continued practice. Experiments such as this have convinced scientists that the cerebellum is involved in the learning of complex movements that become more precise and automatic with practice.

COGNITIVE FUNCTIONS OF THE DINOSAUR BRAIN

Evidence is emerging that learning can be mediated by the dinosaur brain. Because it provides a simple, easily studied model of learning, there has been a great deal of study of what is called the "vestibulo-ocular reflex" of the cerebellum.[4] As an object moves across the visual field, our eyes track that object to keep it in focus. Conversely, if the head is moving but the object is stationary, the vestibulo-ocular reflex enables the eyes to move in the opposite direction from the head's motion so that the image will fall on the same part of the retina. The reflex is calibrated quite precisely to compensate for head motion, with every degree of head movement eliciting a compensatory eye movement.

The nature of the vestibulo-ocular reflex is simple, involving a relatively small number of neurons.[5] Without this reflex it would not be possible to read a sign while walking down the street. Because of this reflex, the visual world remains stable and our eyes can easily discern detail. Were connections between the motion-sensing organ of the ear and the motion-conferring muscles of the eye too complicated, it is likely that the vestibulo-ocular reflex could not work fast enough to allow stabilization of an image. Experiments show that the reflex performs accurately at head velocities of up to 360° per second, which is equivalent to 60 revolutions per minute. Visual feedback is not required, as a neuronal connection is apparently made directly between the semicircular canals of the ear, which are exquisitely sensitive to acceleration and angle of the head, and the muscles of the eye. This reflex is so ingrained that it can be retained in a person who is comatose, as if the patient were responding to a visual stimulus.

The fact that the vestibulo-ocular reflex is quite simple gives the scientist an opportunity to manipulate the system experimentally. If an experimental subject wears goggles that change the optics of the eye, then the reflex relationship between semicircular canal and eye will change. In order to keep the image on the same part of the retina, the subject would need to "unlearn" the old reflex relationship and learn a new relationship between head movement and eye movement. Most people can do this with remarkable facility. In the most extreme case, a subject can wear prism goggles that actually reverse the visual field so that head movement in one direction causes a visual image to move in the opposite direction. Even in this case, if a subject wears goggles continuously, he is usually able to adapt over the course of a single day. Of course, when the goggles are removed, the subject has to "unlearn" the new relationship and relearn the original reflexive relationship. In an experimental animal, it can be shown that damage to the cerebellum or the medulla prevents these adaptive changes.

The cerebellum is also involved in regulating complex motion and making subtle adjustments in the amplitude of a movement.[6] Motor learning is a process of improving the smoothness and accuracy of a movement over time through trial and error. There is clearly a need for well-calibrated movement among all organisms; it is obvious that miscalibration of a movement while swinging through trees could be fatal, but it is also true that a tiny miscalculation while running would make an animal likely to become lunch. Evidence suggests that the cerebellum generates an optimal output in response to an input by progressively matching the expected and actual outcome of a movement. The degree of complexity in the cerebellum is a good deal greater than was originally thought; those cerebellar neurons called Purkinje cells may receive synapses from 100,000 other neurons.

There is also evidence that the cerebellum plays a part in learning conditioned responses. If a puff of air is directed at the eye of a subject, the subject will reflexively blink. If the puff of air is paired with the sound of a bell, then the subject will eventually learn to blink in response to the sound of the bell. Such conditioned responses are among the most primitive types of learning, as they can be demonstrated even in an insect nerve cord dissected free of the insect body. In humans, an injury to certain parts of the cerebellum can impair the ability to learn a conditioned stimulus without affecting the reflex itself. Studies in human volunteers, using a method called functional magnetic resonance imaging (or fMRI), have shown that motor learning is acquired in the cerebellar cortex but appears to be maintained elsewhere in the cerebellum.[7]

How Do We Breathe?

Control of human respiration depends upon a complex interplay of different neuronal centers in the medulla and pons. The respiratory rhythm that causes us to breathe in and breathe out, whether asleep or awake, arises from sequential bursts of activity in different neuronal clusters. These clusters together form the central pattern generator. Neuronal activity is associated with inspiration, with the brief delay after drawing breath, and also with expiration, with a different cluster of neurons for each function. The "drive to breathe" is one of the most irresistible impulses in the human body; as anyone who has almost drowned can tell you, the urge to take a deep, gasping breath while still underwater is virtually uncontrollable, and likely overcomes the rational part of the brain in those who actually drown.

Breathing is far more difficult to coordinate than might at first be thought, because breathing must be fitted around other actions that actually interfere with it, such as eating, swallowing, talking, laughing, coughing, or vomiting. Breathing is also closely linked to the heart rate; it would make little sense to breathe more frequently if the heart rate did not also increase so as to circulate the oxygenated blood more rapidly. Sensory inputs to the breathing control center thus include stretch receptors that sense inflation of the lungs.[8] But there are many other sensory cells that have an impact on the breathing rate, including sensors that measure the various types of gas in the bloodstream. Ultimately, the rate and depth of breathing is a function of blood levels of oxygen, carbon dioxide, and pH. Breathing is also influenced by emotion, by stress, by sleep, by changes in light and temperature, by the requirements of speech, and even by thinking about breathing.

Control of breathing is accomplished by three main groups of neurons.[9] Ventral respiratory neurons apparently initiate the respiratory rhythm by stimulating the muscles to fill and empty the lungs; it is noteworthy that expiratory impulses are inhibited during the inspiratory phase, and vice versa. Dorsal respiratory neurons—which receive sensory input from receptors sensitive to blood oxygenation and blood pressure and from stretch receptors in the heart and lungs—serve a key integrative function by coordinating breathing and heart rate. Finally, the pontine respiratory neurons halt inspiration; they are apparently not needed to generate the respiratory rhythm, though they act to stabilize the pattern, to slow the rhythm, and to allow enough time for gas exchange to occur in the lungs.[10] In short, there is not a single ventilatory center: breathing is regulated by different groups of neurons in different parts of the dinosaur brain. A normal breathing rhythm responsive to the needs of the animal can be generated only when the output of the various groups of neurons is integrated in a very sophisticated way.

Not surprisingly, if a person suffers damage to the dinosaur brain, whether by stroke or some other means, there can be disastrous consequences. A brainstem stroke can cause reduced sensitivity to carbon dioxide in the bloodstream or can result in a fixed rhythm of breathing that does not change in response to changing conditions. Stroke in the medulla can impair voluntary breathing without having an effect on automatic breathing. A tumor in the brainstem can cause an irregular breathing rhythm (apnea) and can eventually obliterate the breathing rhythm entirely. Degeneration of neurons as a result of multiple sclerosis (MS) can cause respiratory insufficiency and can eventually result in loss of voluntary control of breathing. Recent evidence even suggests that sudden infant death syndrome (SIDS) is a result of an abnormality in neurons near the pons.[11]

The general level of human consciousness or awareness is controlled in part by the reticular formation, a diffuse assemblage of neurons at the base of the medulla. Because they are so complex and extensive, the pattern of connection of neurons in the reticular formation is quite striking. Since reticular neurons are at a point of convergence of many inputs and outputs, a single neuron may respond to many different stimuli. Though the reticular formation participates in motor control, in sensory control (perhaps including pain perception), in control of the viscera of the body, and in control of the level of consciousness, we will concentrate only on the latter function here.

Brain arousal can consist of waking from sleep in response to sensory stimuli, enhancing arousal in response to novel or important sensory stimuli, and maintaining arousal even during periods of boredom or sensory sameness.[12] Neurons that extend up to the cortex of the brain stimulate activation, whereas neurons that project down to the body maintain muscle tone and a general state of readiness. Neurons that project from the reticular formation to the central gray matter of the brain, particularly to an area called the thalamus, are essential for a normal state of consciousness. If these neurons are damaged by trauma or stroke, the result is coma or a state of profound unconsciousness. It is really rather striking that a brain that remains normal and intact at a level above the dinosaur brain cannot maintain awareness without input from the dinosaur brain. This neuronal circuit, called the "ascending reticular activating system," also plays a key role in maintaining the normal sleep-wake cycle.

LOSS OF CONSCIOUSNESS AND THE PERSISTENT VEGETATIVE STATE

It is now recognized that there are many states between full consciousness and brain death, and that people shift between these states.[13] The most familiar form of unconsciousness is sleep, which is typical of a healthy brain, even though people who are deeply asleep may have little brain activity and may be very hard to rouse. Loss of consciousness more profound than sleep is associated with brain injury, but such loss can form a broad continuum. Like people who are deeply asleep, unconscious patients may move or speak and can rouse from unconsciousness, though they will typically retain few memories of those times. In a deep state of coma, a person is unaware of herself and her surroundings. However, the functions of the dinosaur brain are preserved and there is some hope that the patient can recover function.

There were perhaps 50,000 Americans who suffered prolonged loss of consciousness as a result of traumatic brain injury in 2005 alone.[14]

State	Consciousness	Sleep/wake cycle	Motor function	Auditory function	Visual function	Communication	Emotion
Coma	None	Absent	Reflexes only	None	None	None	None
Vegetative	None	Present	Pain withdrawal random movement	Startle only	Startle only	None	Reflexive or none
Minimally conscious	Partial	Present	Reaches for and holds objects	Localizes sound but inconsistent	Can have sustained visual fixation	Responsive, but inconsistent	Responsive crying/smiling
Locked-in state (complete paralysis)	Full	Present	Paralyzed	Preserved	Preserved	Nonverbal	Preserved

Table 1. Characteristics of various states of compromised consciousness (adapted from Giacino et al., "The Minimally Conscious State: Definition and Diagnostic Criteria," *Neurology* 58 (2002): 349–53. Also included is the "locked-in" state, which can be confused with a loss of consciousness.

Because of this huge problem, an effort has been made to define and describe the various stages of consciousness. The most severe loss of consciousness is referred to as coma; this is basically when the brain has lost all ability to respond to the environment. In coma, there is no response to visual or verbal cues, no evidence of emotion, no sleep-wake cycle, and no evidence that the patient has any brain function at all other than certain reflexes (table 1). The vegetative state is distinguished from coma by the fact that there can be some minimal response to the environment, usually limited to the presence of a startle reflex or very brief orienting to a visual or auditory stimulus. The minimally conscious state is lighter still, distinguished from the vegetative state by a partial preservation of conscious awareness. Evidence of awareness comes from the fact that the patient may respond, albeit in an inconsistent manner, to visual or verbal stimuli. She may also reach for objects and cry, laugh, or make sounds in response to external cues. It can take an extended period of observation to be sure that a patient is responding to the environment rather than merely making random movements. Patients with specific disabilities such as mutism or paralysis are particularly hard to evaluate. Medical imaging methods have now been used to evaluate patients in a minimally conscious state. Evidence suggests that the minimally conscious patient will respond more strongly to auditory stimuli with an emotional valence (the cry of an infant or the sound of the patient's name) than to meaningless noise.[15]

The persistent vegetative state is often described as a loss of function of the cerebral cortex while the function of the brainstem is preserved.[16] A detailed study of brain pathology was done in 49 patients who died after brain injury, all of whom had remained alive in a vegetative state for an extended period (up to 8 years) after injury. Patients in a persistent vegetative state tend to suffer loss of brain volume, with an increase in the volume of fluid spaces that surround the brain. The vegetative state is not associated with injury to the brainstem; in most cases, patients with injury to the brainstem don't survive long enough to become vegetative. Instead, the persistent vegetative state is associated with widespread and bilateral damage to the cerebral cortex, to the thalamus, or to the white matter linking these structures. There was observable injury to the thalamus in 86% of patients in a persistent vegetative state. In general, the potential for recovery from a vegetative state is greater after traumatic injury than after a nontraumatic injury such as a stroke. This is probably because nontraumatic injury is associated with hypoxia, and hypoxia results in rapid and irreversible loss of neurons, whereas traumatic injury may not result in any loss of neurons.

Patients who are reduced to a minimally conscious state but who then recover experience a general decline in level of function.[17] In one small study, the deficit in psychosocial function was mostly severe, although many patients were eventually able to live independently. All patients who become minimally conscious are likely to have some residual impairment or disability after they recover consciousness. For unknown reasons, women are more likely than men to suffer a poor outcome after severe head injury, even when the severity of injury is similar.[18] Women are more likely to have brain swelling after injury, with elevated intracranial pressure and reduced blood delivery to the brain. It is perhaps not a coincidence that all the controversial cases of comatose patients (Karen Ann Quinlan, Nancy Cruzan, and Terri Schiavo) were female.

Brain Death

There is broad agreement that a person is dead when his brain has died. Yet, as the recent political firestorm surrounding the death of Terri Schiavo has shown, a great deal of controversy remains. Advances in our technology have far outstripped advances in our understanding of the legal and moral consequences of that technology.

In the past, brain death was coincident with the cessation of breathing, since the brain will inevitably die without a constant supply of oxygen. Determination of the cessation of breathing is easy;

a physician could simply hold a mirror beneath the nose—condensation on the glass would show respiration. But widespread use of mechanical ventilators means that vital functions can now be maintained long after breathing has stopped and even after the brain has ceased to function. Currently, the determination of brain death is a process that can take more than an hour for a physician to complete.

As brain death occurs, the medulla oblongata is the last part of the brain to shut down.[19] Hours may pass before destruction of the brain stem is complete and, during this period, there can be persistent medullary function (e.g., a normal breathing rhythm). Brain death is defined as a combination of a persistent vegetative state with a cessation of breathing and an absence of all brainstem reflexes. The depth of the vegetative state is assessed by documenting the absence of response to a painful stimulus such as pinching the nail bed of a finger. If there is a persistent vegetative state and if use of a ventilator precludes determining whether there is a cessation of breathing, then it can become very hard to determine whether or not a patient has died. To prove brain death, an examination must document that all of the following reflexes are absent:

- grimacing during pressure at the joint of the jaw
- contractile response of the pupil of the eye to bright light
- withdrawal reflexes of the cornea during a touch to the eye
- response to infusion of cold water into the ear
- gag reflex to something placed at the back of the mouth
- cough reflex in response to tracheal suction
- sucking and rooting reflexes

If all of these reflexes are absent, then there must be a careful test for cessation of breathing. This is accomplished by disconnecting the ventilator and inserting an oxygen catheter, which delivers measured levels of oxygen to the patient. Gradually, the oxygen level delivered to the patient is reduced to determine whether this will eventually stimulate a breathing reflex. If in the face of declining oxygen there is a complete absence of breathing, called "apnea," then the patient can be declared brain dead. It is noteworthy that during an apnea test there can be spontaneous movement of the body, with what appears to be an effort to sit up or to raise the head. Yet misdiagnosis of brain death is unlikely, except in cases of drug intoxication, prolonged hypothermia, or "locked-in" syndrome—a form of total paralysis that results from destruction of the base of the pons. Nevertheless, physicians often do a confirmatory test, such as measurement of the electrical activity of the brain.

The steps for confirming brain death are widely agreed upon among physicians. Yet as one observer has noted, "If one subject in health law and bioethics can be said to be at once well settled and persistently unresolved, it is how to determine that death has occurred."[20] A basic problem may be that a term appropriate for the whole organism ("death") has been used to describe the loss of function in a single organ (the brain). Many people seem to think that "brain death" is a separate type of death that occurs before "real" death. This confusion is compounded if it is said that "life support" has been removed from a patient when that patient was really already dead. People may be suspicious of the motivation of physicians, since there can be an interest in using organs from the newly dead for organ transplantation. Some critics of the brainstem definition of death would rather equate death to the loss of higher cognitive functions, as there is clearly more to life than simply breathing. Because higher cognitive functions are associated with personhood and with the ability to make autonomous decisions, it may seem attractive to define death as the loss of such higher functions. But this is a slippery slope indeed. If death is equated with the loss of higher cognitive functions, patients who are severely demented or retarded could potentially be declared dead.

In the Netherlands, it has been a relatively long tradition both that patients can request to die and that patients in a persistent vegetative state are allowed to die.[21] Dutch courts have focused less upon asking whether it is acceptable to stop treatment and more upon whether it is medically justified to continue treatment. In cases where it would be medically futile to continue treatment, withdrawal of artificial nutrition and hydration from a patient in a persistent vegetative state is not considered euthanasia. Rather, euthanasia is a category that is understood to include only measures taken by a doctor to actively end the life of a person, at his or her explicit request, in the setting of unbearable and hopeless suffering. Since nutrition and hydration of patients in a persistent vegetative state is seen as futile treatment in the Netherlands, the burden is thus placed on physicians to determine when discontinuation is warranted. While no physician would relish such a decision, it is likely that the physician can be more objective and more rational than the family members of a dying person.

An essay written in response to the Terri Schiavo case notes that

"erring on the side of life" . . . often results in violating a person's body and human dignity in a way that few would want for themselves. In such situations, erring on the side of liberty—specifically, the patient's right to decide on treatment—is more consistent with American values and our constitutional traditions.[22]

What Don't We Know?

Although this detailed discussion of coma and consciousness might suggest that we know much of what there is to know, this is not true. A great deal remains to be learned about the nature of consciousness and the injuries that compromise consciousness. For example, it is not known exactly when neurons are damaged during a traumatic brain injury.[23] Does irreversible damage occur immediately or is there time enough to intervene in order to preserve brain function? What determines whether a particular neuron will live or die? Are there genes that can protect neuronal function and make a person more resistant to head injury and, conversely, are there genes that put neurons at greater risk from traumatic brain injury? How can we prevent head injury from progressing to brain death?

How is it possible that Terry Wallis, a mechanic in Arkansas, was able to regain awareness in 2003, more than 18 years after he lost consciousness following a car accident?[24] And how was Sarah Scanlin, a Kansan who was also the victim of a car accident, able to emerge from unconsciousness after 19 years? Will these people fully regain their cognitive abilities or will they always be as impaired as they were immediately after arousal? What accounts for these near-miraculous recoveries? And why was Terri Schiavo unable to recover?

In a broader sense, a sense related less to the clinical outcome of brain injury than to the great remaining questions in neuroscience, there is still much to learn about the dinosaur brain. The cerebellum can be thought of as a paradigm of emergent complexity in that a careful description of cell function does not yet clarify how the cerebellum functions. Though we have a detailed knowledge of how neurons connect to one another in the cerebellum, though we even have data describing how the cerebellum relates to other parts of the brain, we still have a very poor understanding of how the cerebellum helps us to learn and refine a motor task.

Yet it should be abundantly clear that human survival is dependent upon the intact function of the cerebellum and other parts of the dinosaur brain. Absent certain fully automatic functions, which freed the brain from the need to concern itself with such mundane tasks as breathing, it likely would have been impossible to evolve the "higher" functions that we take as indications of our humanity.

CHAPTER 5

THE "HUMAN" BRAIN

I s the human brain as unique as we would like to believe? Or do we show mental abilities that are simply a further evolution of something that is already well established in other, simpler organisms? If we are different, what forces have driven the emergence and elaboration of human behavior? And what forces have constrained development of the human brain? Is the human brain like a computer, or is this a poor analogy? Is brain function at all limited by brain structure? These are some of the questions that will drive us in this chapter and in the chapters to follow. Though many of these questions do not yet have deeply satisfying answers, there is at least a basic answer to each of them.

The "human" brain, also known as the prosencephalon or forebrain, is that part of the brain that makes primates as a group unique, and humans unique among primates. Yet a clear distinction between the "human" brain and the "dinosaur" brain is artificial, for several reasons. As we have seen, humans still have a functional "dinosaur" brain and dinosaurs had a brain structure analogous to the "human" forebrain.[1] While the dinosaur forebrain was no doubt primitive by primate standards, it was nonetheless present. Furthermore, there is no clear anatomic boundary between what constitutes the "human" brain and what constitutes the "dinosaur" brain. Nevertheless, the distinction between "human" and "dinosaur" is serviceable and captures some of the flavor of human uniqueness.

THE BASIC ANATOMY OF THE "HUMAN" BRAIN

Anatomically, the human forebrain is composed of the telencephalon and the diencephalon. The telencephalon is that symmetrical efflorescence of tissue that looks like a cauliflower and forms a rind around the rest of the brain (see fig. 5, chap. 4); the two cerebral hemispheres together form the telencephalon. The diencephalon is made up of the thalamus and certain related tissues, which form a central, symmetrical, globular structure, somewhat like a tulip bulb, around which the telencephalon forms. These arcane and impressive names derive from study of the human brain during development and are important only insofar as they facilitate discussion.

The diencephalon, whose major component is the thalamus, is the more primitive part of the forebrain. When the living brain is visualized in a plane parallel to the eyebrows (fig. 8), the thalamus forms two symmetrical masses at the center of the brain. The thalamus looks as if it forms oval pillars of tissue that support the rest of the telencephalon, which is conceptually not so far from the truth. No sensory information reaches the cerebrum (except for data from the olfactory neurons of the nose) without first being processed in the thalamus. Furthermore, neural circuits involved in generating motor responses often form loops or anatomic pathways that link between cerebrum and cerebellum as they traverse the thalamus. Beneath the thalamus are the remaining parts of the diencephalon (hypothalamus, pineal gland, and subthalamus), which together constitute about 2% of the total weight of the brain.[2]

The telencephalon, which is comprised mostly of the cerebral hemispheres, is what makes the human brain unique. The central por-

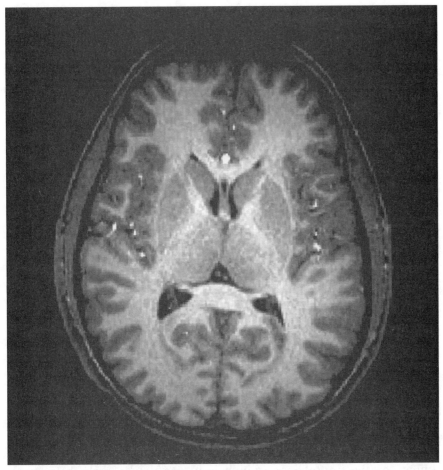

Figure 8. The living brain visualized in a plane parallel to (and slightly above) the eyebrows. The face would be at the top of the image and the thalamus is seen as two symmetrical pillar-like masses at the center of the brain, which support the rest of the telencephalon, with the back of the corpus callosum forming a white band immediately below the dark gray masses of the thalami. White matter forms a complex ramifying mass of tissue with a rind of cortical gray matter between the white matter and the skull.

tion of the telencephalon forms the basal ganglia, which are paired masses of tissue adjacent to the thalami. The thalami grade into the basal ganglia such that the anatomic margin between the two tissues is poorly defined, as can be seen in an image of the living brain (fig. 8).

A great many nerve tracts run between thalami and basal ganglia, and the functional linkage is quite close. The rest of the telencephalon, more commonly known as the cerebrum, spreads in every available direction from the central bulb of the diencephalon. There has been a massive growth of cerebral tissue over evolutionary time so that the volume of the human cerebrum is enormously expanded relative to the rest of the body. As a comparison, the body weight of a rhinoceros is about 30 times greater than the body weight of a human, yet the brain weight of a rhinoceros is only half that of a human, so there is a 60-fold increase in the relative weight of the human brain. Were a similar comparison to be made between a human and a lizard or a human and a fish, the contrast would be far more striking. Massive expansion of the telencephalon is perhaps the key feature that sets the human brain apart from virtually every other brain.

The cerebrum is divided into several separate lobes, with the margins of most lobes defined by deep fissures that score the surface of the brain like wrinkles on a walnut. These fissures, or sulci, increase the surface area of the brain without having much impact on brain volume; it is possible that the ratio of surface area to volume is an important design parameter of the human brain. There are several genetic mutations that result in reduction or loss of the brain sulci and these mutations are generally associated with severe mental retardation. The four separate lobes of the cerebrum, from front to back, are known as the frontal lobe (which is by far the largest), the parietal lobe, the temporal lobe, and the occipital lobe. The division of the brain into four lobes was done early in the history of medicine, and it is possible that the anatomic divisions are more a function of our need to name things than a reflection of true functional differences. It is worth noting that the brain is symmetrical—there are two frontal lobes, two parietal lobes, and so on.

THE FUNCTION OF THE THALAMUS

Clear evidence of the function of the thalamus was provided by a very famous medical case similar to that of Terri Schiavo's, but which began years earlier—the case of Karen Ann Quinlan.[3] At 21 years of age, Ms. Quinlan suffered cardiac arrest after taking a deadly combination of prescription sedatives and alcohol. When she was found, she lacked a pulse and had ceased breathing, she was unresponsive to efforts to revive her, and she had the bluish skin of deep hypoxia. After resuscitation, a pulse returned, but her condition was otherwise unchanged and, since she could not breathe unaided, she was placed on a ventilator. Within the first hour after the accident, she began to breathe spontaneously and her vital signs improved, but she remained unresponsive. Over the first 12 hours, there was an improvement in her condition, with the return of a sluggish pupillary response of her eyes to light, some response to painful stimuli, return of the gag and cough reflexes, spontaneous movement of her limbs, and finally, opening of her eyes in response to sound. Although it is likely that she was irreversibly brain injured at that point, her condition took a marked turn for the worse after aspiration pneumonia set in on the second day. Ms. Quinlan suffered recurrent bouts of hypoxia because her lungs were unable to function well, and her blood oxygenation was dramatically lower than normal. Over the next six months, after the pneumonia had resolved, Ms. Quinlan showed sleep-wake cycles but never showed any sign of cognitive function or awareness of her surroundings. Over time, she developed severe muscle contractures, which made her grimace as if in pain, and her vital signs became increasingly unstable. Ms. Quinlan's movements were not goal directed; instead, they appeared to be stereotypical, as if an automatic behavior pattern had been induced. Her eyes moved randomly and she did not respond to visual stimuli, yet she was not paralyzed; she had sufficient motor competence to respond had she had enough cognitive function. Physicians were thus certain that she was in a persistent vegetative state rather than the paralysis of a locked-in state. She suffered severe weight loss, losing more than 100 pounds from her normal weight, even with hypercaloric tube feeding. She also was treated for multiple infections of lung, bladder, and skin. One year after the accident, her breathing more irregular than before, especially while asleep, she was weaned from the ventilator, yet she did not die. She began to suffer intermittent seizures, with episodic twitching of her limbs, mouth, and eyes, lasting about 30 seconds each time. Medical images of her brain, taken five years after the accident, showed that she had severe atrophy of both her cerebrum and her cerebellum, yet she survived more than ten years before she finally succumbed to an overwhelming infection.

An autopsy was performed soon after death to determine the nature and extent of the injury to her brain. To identify the various sites of injury, the entire brain and spinal cord was sampled. Contrary to expectations at the time, the most severe damage was neither in the cerebrum nor in the dinosaur brain. Instead, careful examination of the brain showed that the most severe damage was to the thalamus, with the extensive scarring characteristic of severe hypoxia visible on both sides (hypoxia is lack of oxygen delivery to the brain, as in stroke). Yet the brainstem, including the regions responsible for control of her heart rate and breathing rhythm, was undamaged. These results made a huge impression at the time, because they suggested that the thalamus was responsible for conscious awareness.

A more recent autopsy study of 49 patients, all of whom had been in a persistent vegetative state, found that damage to the thalamus was present in 86% of cases overall.[4] Thalamic damage was more extensive in patients who suffered a hypoxic event than in patients who suffered brain trauma, and patients who suffered hypoxia were less likely to regain function. These findings show that the vegetative state is a result of damage to the thalamus rather than a result of damage to the brainstem. Prolonged loss of consciousness is usually associated with damage to both the thalamus and the cerebral cortex, with retention of most brainstem reflexes. However, damage to the brainstem is seen in only 14% of patients in a persistent vegetative state, perhaps because damage to the brainstem is likely to be associated with rapid mortality. Another study, which contrasted patients who died after a persistent vegetative state with patients who died after severe head trauma, confirmed that thalamic damage is rarely seen in head trauma patients but is common among patients in a vegetative state.[5]

The thalamus is now thought to be largely the organ of attention.[6] Though we will defer a deeper discussion of consciousness to a later chapter, it may be important to mention some of the salient points here. Perhaps the most difficult aspects of consciousness to explain are "qualia"—the subjective experiences that cannot be measured but that everyone will attest to having. The redness of red, the coldness of cold, the painfulness of pain, even the desolation of heartbreak; all of these are consistent with the human condition, but it cannot be verified that your red has the same degree of "redness" as my red. Rather than dealing with what is probably an unsolvable problem, scientists have focused instead on what is far more solvable: the neural correlates of consciousness. In basic terms, the neural correlates of consciousness are the minimal set of neural events that give rise to a conscious perception. How is it that Picasso could convey motion with a few lines on a sheet of paper? How are we aware that an object is greatly distant when that object is too far away for the physical separation of our eyes to produce true binocular vision? What neural events enable us to have a true and accurate perception of the world?

In essence, the front of the brain, which is associated with the integration of perception, is "looking at" the sensory systems, most of which are located at the back of the brain; it is the thalamus that stands at the intersection, helping us to attend to and make sense of sensory perception. Consciousness, which deals rather slowly with the novel and less stereotyped aspects of sensory input, takes time to decide among appropriate thoughts and responses, and this effort depends upon the thalamus, at least in part.

Consciousness is an ambiguous term that can refer to the waking state, to experience, or to some particular mental state.[7] Self-consciousness is, if anything, even more ambiguous, since it can refer to such widely varied things as a proneness to embarrassment in social settings; or to self-recognition, as of a reflection of yourself in a mirror;

or to the ability to detect sensations and actions that are uniquely your own; or to the awareness of awareness; or even to self-knowledge in the broadest sense, which includes an awareness of your goals and motivations. The thalamus may participate in all of these various processes.

Recent evidence suggests that the thalamus interacts with cortical gray matter of the cerebrum in a complex interplay that, when mildly impaired, can lead to absence seizures.[8] An absence seizure is a momentary clouding of consciousness, as when one loses awareness while staring at a bonfire or a blank wall. Absence seizures are rather common, especially in childhood, and they seem to correlate with a brief but abnormal pattern of neuronal firing that may originate in the intralaminar nuclei of the thalamus. If the intralaminar nuclei of the monkey brain are stimulated with an electrode at a certain rate, a freezing reaction is induced. Whether the genesis of absence seizures is more a function of the thalamus or of the cortical gray matter is an open question, but it is clear that the thalamus plays a key role in consciousness.

Interest in the neural correlates of consciousness is focused, for now, primarily on vision, as this is easiest to study. In addition, there is compelling evidence that awareness plays an important part in visual perception. Yet the study of consciousness is likely to form a focal point for research for many years to come; in the words of Gerald Edelman, a Nobel Prize winner,

> Consciousness is the guarantor of all we hold to be human and precious. Its permanent loss is considered equivalent to death, even if the body persists in its vital signs.[9]

THE FUNCTION OF THE CEREBRUM

Perhaps an adequate way to provide an overview of the cerebrum is to describe what might happen in a monkey brain when the monkey is presented with a visual stimulus.[10] This sequence of events is not unique to a monkey brain and could also be described in a human brain, but the element of time is better understood in a monkey brain because it is possible to measure the rate of flow of information in the monkey brain by using electrodes that are sensitive to electrical currents. If a visual stimulus is presented to the monkey eye, neurons in the retina are activated and send an impulse to the visual cortex of the brain—that gray matter involved in visual perception—within about 30 milliseconds (msec). This nerve impulse passes through the lateral geniculate nucleus—a visual relay center in the thalamus—in about 10 msec (fig. 9). Within 50 msec of first presentation, the nerve impulse arrives in the primary visual cortex of the occipital lobe (at the back of the brain), where neural circuits detect features of the image such as edges, corners, and general shapes. By 60 msec after first presentation, the signal has been forwarded to a nearby visual center, which is concerned with extracting higher-order information from the image such as visual forms and groups of features.

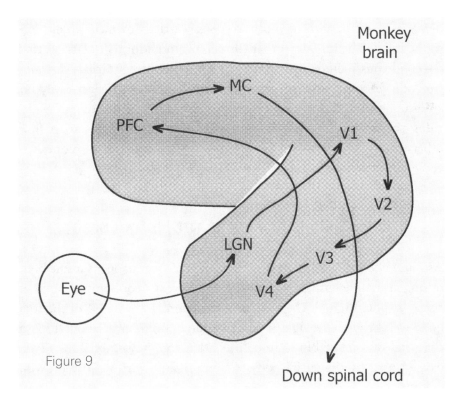

Figure 9

As the neural impulse is shunted from one visual center to another, progressively more sophisticated information is extracted from the visual image until finally there is recognition of a familiar face or a recognizable object in the environment. This whole process takes no more than about 130 msec, at which point integrated visual information begins to reach the frontal cortex where judgments and decisions are made. After a decision has been made in the forebrain, perhaps, for example, to pick up a food item, a message is sent to the primary motor cortex, which generates a neural impulse that is capable of modulating the grasping action of the hand. A motor impulse leaves the cerebrum and begins to travel down the spinal cord within about 140 to 190 msec of first visual presentation. Therefore, the total time required for a monkey to perceive a food item in the visual field and to reach for it may be as little as 180 to 260 msec, or about a quarter of a second. Most of the various functions that have been described are performed in the cerebrum, except for the initial visual integration done in the retina and the motor response carried down the spinal cord to the hand. The functions that the cerebrum performs in far less than a second thus include visual integration, object recognition, formulation of a response, and initiation of a motor action. It is probably not an exaggeration to say that everything we think of as human, everything that makes us unique, everything that sets primates apart from other mammals and mammals apart from other, simpler vertebrates, is centered in the cerebrum.

Energy Consumption by the Human Brain

As should be surmised from the description of what happens in the brain during a simple visual perception, the energy requirement of brain tissue is quite high.[11] The weight of the brain is 2% to 3% of the total body weight of an average person. Yet the brain receives about 15% of the total output of blood from the heart and consumes 20% of the oxygen used by the entire body. Among infants with rapidly growing brains, the brain's resting energy consumption is about 60% of the energy consumption of the entire body. If we base a calculation of brain metabolic rate on the amount of oxygen consumed by the adult brain relative to the rest of the body, the brain metabolic rate is about 8-fold higher than an "average" tissue in the body. Deep anesthesia reduces the energy consumption of the brain by about half, which means that 50% of the brain's energy is needed to drive signals along axons and across synapses.[12] This implies that a substantial amount of energy is needed simply to maintain the gradient of ions across neuronal membranes at rest, without ever firing off an action potential. Thus, the upkeep of a brain is expensive, even if that brain isn't used very much.

The relative cost of the brain can also be determined in other ways. Half of all human genes are expressed uniquely in the human brain. Of the 30,000 genes that are required to build a human, at least 15,000

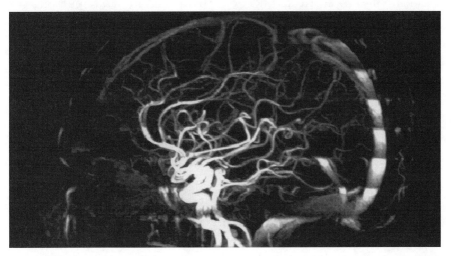

Figure 10. An image showing the arteries that deliver blood to the brain. The head is seen in profile, with the face to the left and the large arteries (and a few of the veins) seen as bright and highly convoluted tubes. Arteries that bring blood to the brain arise from the bottom center of the image. The large venous space that drains the brain is seen as a large band of white to the right of the image (an image artifact that resembles a venetian blind obscures the large vein, in part because blood flow through the vein is relatively slow). Arteries form a beautiful and sophisticated network that penetrates and nourishes the brain.

are needed to build the brain alone. An additional 10,500 genes are expressed in the brain and elsewhere in the body, meaning that roughly 85% of all human genes are expressed in the brain. It has been estimated that the brain contains 100 billion neurons and that these neurons are supported and maintained by a trillion of the nutritive cells known as glia. Together, the neurons of the brain make as many synaptic connections as there are stars, planets, and comets in the known universe. In fact, the human brain is probably the most complex object in the universe. It is hardly poetic to suggest that there are entire universes within our skulls and that we may understand the external universe before we fully understand the internal one.

The energy requirements of the human brain place very severe design constraints upon it. The arteries that deliver blood to the brain and the veins that drain it are extraordinarily complicated, forming a beautiful and sophisticated network that penetrates and nourishes the brain (fig. 10). The major arteries ramify into ever finer vessels, eventually forming vessels that are scarcely larger in cross section than a red blood cell. These tiny vessels, called capillaries, deliver blood to within a short distance of every cell in the brain. The oxygen and glucose that diffuse out of the capillaries are required to nourish each and every cell of the brain. If blood flow through any of these tiny vessels is interrupted for even a few moments, a loss of consciousness can result, and if blood flow is lost for a few minutes, damage to the brain can be irreversible. If blood flow through the major vessels of the brain is blocked by a clot of blood or a bit of tissue, flow to a large portion of the brain may cease. Neurons and other brain cells become distressed within a few seconds and neurons are thought to be irreversibly damaged within about 4 minutes. Even a tiny clot can block blood delivery to a critical brain region; there can be a severe loss of function if even a pea-sized region of brain dies. This is the basic injury of stroke.

DESIGN CONSTRAINTS IN THE HUMAN BRAIN

Neurons receive and deliver signals at an average of 10,000 synapses per cell and they can combine and process synaptic inputs to implement a rich repertoire of responses.[13] Neurons establish new connections and maintain old ones, they vary their signaling properties according to need, and they respond quickly to changing conditions. Neural networks self-assemble, autocalibrate, store information, and adapt to changing circumstances, and they do so for a time span that far exceeds the lifetime of most machines.

Because the energetic cost of a brain is quite high, economy and efficiency must be guiding principles in the design of neural networks. To be clear, even though it is conceptually simple to think of the brain as having design constraints, we are not proposing that there has been "intelligent design." The slow and plodding mechanism of evolutionary experimentation, coupled with the ruthless elimination of all failed evolutionary experiments, has power enough to maximize the efficiency of neural networks without requiring anything like intelligent design. All that is required for the stumbling process of evolution to produce a sophisticated brain is time, and there are many indications that Earth is quite ancient.

The size (and hence the energetic cost) of a brain can be reduced in a limited number of ways without also reducing the computing capacity.[14] Either the total number of neurons required for adequate function could be reduced, the average volume of neurons could be reduced, or neurons could be laid out so as to reduce the length of the connections between them. There have probably been stringent selective pressures for all of these evolutionary solutions, yet the cost of "wiring" the brain has received the most study. Connections between neurons occupy up to 60% of the total volume of gray matter, and these "wires" (axons and dendrites) are expensive to operate because they dissipate energy as they signal.

The "connectivity" between neurons is higher if the neurons are close to each other. In the visual cortex, neighboring parts of the visual field tend to be represented by neighboring neurons, which reduces the wiring problem. The probability that any two neurons are connected to one another is about 1 in 100 for neurons that are less than 1mm apart, but 1 in 1,000,000 for distant neurons.[15] Close connectivity between neurons would enable local information processing to occur. Computer models suggest that allocating about 60% of the

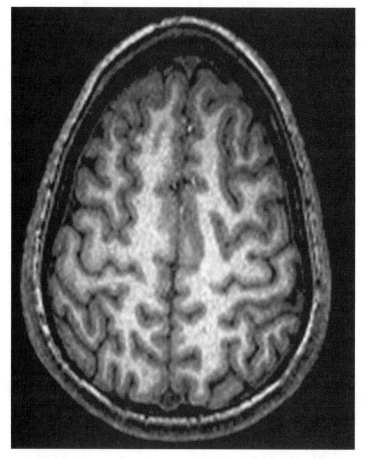

Figure 11. The wrinkles in the surface of the brain are called sulci. Sulci could be a design solution to the wiring problem in the brain; prominent folds in the human brain could enable a large cortical surface area to be packed into the skull, while still allowing white matter tracts to bypass a particular fold. This would minimize the distance that a white matter fiber would need to traverse in order to connect two different cortical regions.

brain to "wiring" can minimize local delays by striking a balance between two opposing needs: high transmission speed and high component density. High transmission speed requires thick wiring, which would limit how close neurons can be to each other. Reducing the diameter of neural wires would reduce the speed at which signals travel, which could slow brain speed. But reducing the diameter of neural wires would also enable neurons to be more densely packed, which would increase brain speed. It is possible that brain sulci are a solution to the wiring problem; prominent folds in the human brain would enable a large cortical surface area to be packed into the skull while still enabling wires to bypass a particular fold, thereby minimizing the length of certain individual wires (fig. 11).

As in any electrical system, "noise" is both commonplace and costly. In the context of the brain, noise is an accident, a neuronal misfire—it may be that noise is simply an unintended consequence of having neurons that are poised to respond. Noise can be characterized as a ratio of the number of action potentials that carry a "true" signal divided by the number of action potentials that carry a "false" or random signal that is not related to the environment. Noise, which is generated if a neuron is stimulated to fire by accidental opening of an ionic channel, can become costly if accidental firing of one neuron induces accidental firing of other neurons down the line.[16] Small-diameter neurons are more prone to fire accidentally, since they have fewer ion channels overall, so random discharge in any individual channel is more likely to generate an action potential. This may set a lower limit on the size of a neuron and it may also be why it is adaptive that neurons sometimes fail to fire when stimulated. The probability that an individual neuron will *fail* to fire in response to an incoming signal is between 50% and 90%. Computer models suggest that, since there are so many neurons that work in tandem, such synaptic failures can reduce the energy consumption of gray matter by 50% without necessarily losing any information. The unreliability of individual neurons strongly suggests that the brain evolved to be efficient and that energy has been in short supply throughout evolutionary time.

It is obvious that energy efficiency could be enhanced by reducing the number of signals in a network, provided that information is not lost.[17] Economy in the number of impulses would also increase the salience of any given impulse. One approach to achieving energy efficiency would be for neurons to "learn" which signals are most salient and which signals can safely be ignored. Like the wise Internet user, neurons could learn to respond to signals from sites that are known to be most informative. This ability of a neuron to select the most relevant signal from a vast panoply of possible signals introduces the concept of synaptic plasticity—the ability of a neuronal synapse to change its response over time (discussed in depth in chapters 7 and 8).

It is expected that, as brains get larger over evolutionary time, there will be a decrease in the degree of connectedness between neurons.[18] This is because if neural connectivity were to be a constant, despite an increasing number of neurons, then the total amount of wiring would have to increase substantially. This would lead to a decrease in the speed of computation, since wires would necessarily become longer. This may explain why there has been an evolutionary trend toward compartmentalization in the human brain, since compartmentalization could enable a neuron to interact preferentially with those neurons that are most salient. In other words, the brain may be composed of a large number of local networks, with each network able to interact with other networks but also capable of functioning rather autonomously.

Compartmentalization and network autonomy would also maximize the potential complexity of the brain. It is obvious that brain complexity would be low if many neurons invariably fired together. Similarly, brain complexity would be low if many neurons rarely fired or if there were few neurons. It is harder to appreciate that the way to maximize complexity is to have a large number of neurons that fire with relative autonomy. But inherent complexity is increased if many neurons are grouped into networks, provided that these networks are able to behave with autonomy. This type of architecture enables a brain to perform disparate functions such as talking and thinking simultaneously and it might also conserve the cost of wiring.

Another key feature of the human brain, which, on one hand, could be considered a design constraint or, on the other, a feature that enabled all other features to emerge, is evolvability.[19] This is simply the capacity of genes to mutate and modify the brain without jeopardizing the fitness of the whole. There has been a great deal of time for evolution to happen but, if the evolutionary fitness of the human organism did not increase over time, there would have been little impetus to elaborate a human brain. Computer modeling suggests that evolvability is favored by the redundancy of neuronal function, by multiple weak linkages among neurons, by the robustness of neurons, and by compartmentalization of neural function. All of these abilities would contribute to flexibility of function and might enable the brain to assume new functions over evolutionary time.

It is possible that redundancy is the single most important feature distinguishing the human brain. Neurons in the insect brain are as profusely branched and as complex as in the human brain, but there are only a million neurons in the bee brain as compared to 100 billion neurons in the human brain.[20] The quantity of neurons in the human brain thus exceeds that in the bee brain by a factor of 100,000, despite the fact that individual neurons in each brain function in more or less the same way. Any notion of design constraints in the human brain must acknowledge the massively parallel pattern of interconnections possible. The strength of these connections can wax and wane over a wide time scale, from milliseconds to months, in behaviorally significant ways, which gives humans a far more flexible repertoire of responses than could be supported by an insect's puny brain.

COMPARING THE HUMAN BRAIN TO A COMPUTER

The maximum potential bandwidth of the human brain—a measure of how much information can be transferred from one place to another per second—is staggering. If we assume that data in the brain can move at a rate of 100 bits per second over each axon in the white

Comparison	Ant Colony	Brain	Computer
Basic unit of construction	Ant	Neuron	Transistor
Basic units per whole	< 10 thousand ants	100 billion neurons	> 100 million transistors
Communication between units	Analog	Analog and digital	Digital
Interactions per unit	Unknown (~10,000 contacts?)	10,000 synapses	~3 gates *
Interactions per whole	10^8 ant contacts	10^{15} synapses	10^8 transistors
Years required for current design	140 million	600 million	50
Total memory	Unknown (<200 Mb?)	~200 Mb	>200 Mb
Longevity of whole	15–20 years	80 years	~5 years
Is self-repair possible?	Yes	Limited	Not yet
Power consumption	Unknown (10 Watts?)	20–40 Watts	50–200 Watts
Interactions per Watt	Unknown (1.0×10^7?)	2.5×10^{13}	5.0×10^5
Memory per Watt	Unknown (<20 Mb/W?)	5 Mb/W	1 Mb/W

Table 2. A crude comparison of a brain to an ant colony and a computer. Where numbers are unknown (as indicated), a guess is tendered to enable further computations to be made; this must necessarily result in order-of-magnitude approximations. Analog responses are graded or modulated, whereas digital responses are all-or-nothing. * In the central processing unit (CPU) of a computer, one gate is connected to 2–3 others. (Interactions per Watt) was calculated by dividing (Interactions per whole) by (Power consumption). (Memory per Watt) was calculated by dividing (Total memory) by (Power consumption). Major references include [12], [14], and Holldobler and Wilson (1990), *The Ants*. (?) indicates a guess. This table should not be construed as anything other than very broadly comparative. It should be emphasized that the power consumption of the human brain is less than that of a dim light bulb.

matter, then the maximum bandwidth of the brain is comparable to the total world backbone capacity of the Internet in 2002.[21] The maximum signaling rate of each wire of the human brain is less than in a computer, but this does not mean that the human brain is inferior to a computer. In fact, the more we learn about the structure and function of the brain, the more we come to appreciate the great precision of the brain's construction and the high efficiency of its operation.

By virtually any measure, the brain is vastly superior to a computer (see table 2). Although a human brain cannot match a computer in logical operations that require pure computational power such as finding the square root of a 17-digit number, a computer cannot match a human brain in seeing, smelling, tasting, moving, or responding to the environment. The fact that we think of these abilities as trivial is convincing evidence of the relative superiority of the human brain, since we have not yet developed a computer that can do the things that are so trivial to us. And how often have humans been hampered by an inability to calculate the square root of a 17-digit number?

Computers deal exclusively with all-or-nothing signals that are inherently digital, whereas the human brain is comfortable with both analog and digital signals. The integration that a neuron is capable of performing is possible only because the neuron analyzes input from a range of sources and responds with a graded depolarization that is inherently analog. This depolarization may reach a threshold sufficient to induce an action potential, at which point the neuron responds in a digital way. If we calculate the total number of interactions that are possible in a brain and in a computer, which may be a very crude correlate of the ability to learn, we see a stark contrast. A brain is composed of about 100 billion neurons, each of which can interact with 10,000 other neurons, for a total of 1,000 trillion interactions per brain. In contrast, a computer is composed of about 100 million transistors, each of which can interact with no more than 2 to 3 other gates in the central processing unit,[22] for a total of 300 million interactions per computer. Thus, a brain enables interactions at a density that is roughly 3,333,333 times higher than a computer.

It is not intuitively obvious that a brain is able to handle more information than a computer, but it is true nonetheless, as can be shown by a simple example. The picture of the brain shown at the beginning of this chapter (fig. 8) is a magnetic resonance imaging (MRI) picture. This image is composed of about 50,000 tiny picture elements (called pixels), which are conceptually equivalent to the dots on your television screen.[23] An experienced radiologist could look at this image of 50,000 pixels and determine that the subject was healthy in just a few seconds. Basically, a radiologist compares the image to a mental database of similar brain images, bearing in mind the age of the subject, the possible clinical problem, and the particular orientation of the subject in the image. The image would be searched for asymmetry, for unexpected anatomy, and for unexplained bright or dark spots, and the radiologist would be able to determine, with a high likelihood of success, whether or not the subject is normal. No computer in the world can do this with the same degree of precision, although it is quite likely that there will be such computers in the future. And this is a static image! Were the image to be moving in three-dimensional space, the computations necessary for a computer to comprehend the image would be totally overwhelming.

This analogy is perhaps not fair to computers, since it is likely that the human brain evolved specifically to perceive and comprehend images. So, let us alter the example to put computers at a relative advantage. In an MRI picture, each of the pixels can assume a limited number of gray scale values between perfectly black and perfectly white. In a typical MRI, the pixels assume any of 256 different gray scale values, and it is the blending of these gray scale values that renders the image. The relative gray scale difference between a lesion and normal anatomy is what enables the radiologist to spot a lesion. The human eye cannot distinguish each of these 256 gradations, but such precision is unnecessary for the human brain to comprehend the image. But what if we were to replace each pixel with a number? Given the ability of computers to deal with a high volume of numbers, this should go to the strength of computers and would befuddle the human eye. In principal, it is possible to replace every pixel in an image with a number between 1 and 256. If 50,000 pixels can assume any of 256 values, simple multiplication shows that the image would be broken down to some combination of 12.8 million different gray scale values. The human eye would then no longer be able to make any sense of the gray scale values. But neither could a computer fathom such data; the eye would lose its relative advantage, but the computer would still be unable to cope.

There is no doubt that computers are evolving at a faster rate than is the human brain; computers are no more than about 50 years old, whereas the brain has been optimized over the course of 600 million years of evolution. Someday, perhaps even someday soon, computers will reach a level of sophistication comparable to what is achieved by the average person, but it remains to be seen whether even the most advanced computer will be capable of creativity.

A brain is a collection of neurons, each of which behaves in a specific way. A mind is what arises when a huge number of neurons are free to perform their functions with a high degree of autonomy. Mind is more than just the sum of neuronal parts; mind is the collection of thoughts, memories, prejudices, and predilections that make us who and what we are. Mind interposes between brain and behavior, especially influencing how the brain responds to novel features in the environment. By this rationale, creativity is evidence of emergent complexity, a property of mind not of brain. The evolution of the human mind may have been inevitable, given the strong selection pressure on neurons to become fast, efficient, autonomous, flexible, and robust. If similar "selection pressures" can be imitated by computer engineers, then the evolution of creative ability, perhaps even the evolution of a type of "mind," may be inevitable in computers.

DOES BRAIN STRUCTURE DETERMINE BRAIN FUNCTION?

It is almost certainly true that brain structure determines brain function. Yet we must be careful to avoid the determinism and dogmatism of a New Age phrenology. Phrenology is the now discredited system of analyzing the psychological strengths and weaknesses of individuals by searching for bony prominences on their skulls.[24] It was once believed that the form of the skull is a fair representation of the function of the underlying brain, such that skull bumps could reveal hidden abilities of the underlying brain. Phrenology had a wide impact on Western medicine, science, and culture in the first half of the nineteenth century but then passed from favor, leaving a residue that still remains. Phrenology represented the first time that brain function was localized to brain structure, which we now know to be essentially true.

There is still a tendency to think that brain function can be revealed by brain structure. For example, there has been a great deal of effort to determine if brain volume is correlated with intelligence. The most common measurement of intellectual capacity is the intelligence quotient (IQ), which is derived from any of several standardized sets of questions. The overall or full-scale IQ correlates with total brain volume, as well as with the volume of certain brain regions such as the cerebral hemispheres, the temporal lobe, and the cerebellum.[25] Gray matter may be more important than white matter in determining IQ, but only 12% to 31% of the total variation in IQ can be accounted for by variation in gray matter volume. Thus, the volume of the brain accounts for a very modest proportion of the variance in human intelligence. This, perhaps, should not be too surprising; men generally have larger brains than women, even after controlling for body size, but men are not smarter than women. Clearly, brain size is to some extent hereditary. But brain size could also be determined by nutrition, by general health, by head injury, or even perhaps by environmental stimulation. For all these reasons, brain size is a poor predictor of intelligence.

Recent work has confirmed that there is a weak but significant relationship between brain size and intelligence.[26] The youngest children typically do not manifest such a relationship, which may mean that the young brain has simply not developed to a point where volume can overcome the many other factors that have an impact upon intelligence. But among older children there is a significant and relatively strong relationship between gray matter volume and intelligence. The correlation between intelligence and brain volume varies from one part of the brain to another, as would be expected given the compartmentalization of the brain. There is a strong relationship between intelligence and the volume of the frontal lobe, specifically in a place called the anterior cingulate, which nestles between the brain hemispheres above the eyes. This makes some sense, as it has long been recognized that the frontal lobe is involved in executive ability, which is the ability to plan and execute higher-order cognitive functions (such as taking an IQ test). But even in the anterior cingulate, where the relation is known to be strongest, brain volume can only explain about 28% of the variation in intelligence. This is statistically significant, but it may not be clinically significant, since it would not enable one to predict a child's success in school based on brain volume alone. What is clear is that intelligence is strongly heritable[27] and that the genes that influence intelligence also have an effect on brain size.[28]

If it is true that brain structure can determine brain function, is it also true that brain structure can limit brain function? There are now a huge number of studies of children in whom brain growth is limited by some factor, and these studies are rather unanimous in showing that brain function can be impaired by abnormal structure. For example, a study of 82 children who were born preterm (at 30 weeks gestation instead of 40 weeks) found that these children suffered an IQ decline by 15 years of age.[29] Children with a higher verbal IQ were more likely to have a large volume of white matter in the frontal and temporal lobes. This is consistent with the idea that the frontal lobe is crucial for higher-order cognition. It was concluded that children with early damage to the brain are unable to maintain a normal rate of knowledge acquisition, which is reflected in declining IQ scores.

DO WE REALLY USE ONLY 10% OF OUR BRAIN?

As we have seen, energy conservation is a major design constraint in the human brain. To help conserve energy in the brain, ways to store information efficiently—called sparse coding schemes—may be crucial. Basically, if there are a great many neurons but only a small number are involved in signaling at any one time, this would require little energy for signaling but could still have a high capacity to represent the environment. This is because there are many ways to distribute a small number of signals among a large number of neurons.[30] Yet having a large number of neurons would be energetically expensive and if the neurons rarely fire it would be wasteful to sustain them. The optimal ratio of active neurons to resting neurons thus depends upon the cost of maintaining a neuron at rest versus the extra cost of sending a signal. If signals are more expensive than neurons, it would be most efficient to distribute a small number of signals among a large number of neurons. But if neurons are more expensive than signals, it would be most efficient to have fewer neurons and involve more of them in signaling. If one estimates the ratio between the energy cost of signaling and the energy cost of constructing a neural network, one can predict the optimal network design. Calculations suggest that, for maximal efficiency, between 1% and 16% of neurons should be active at any given time. Thus, it may be true that we use only 10% of our brain, but this should not be construed as evidence that we are wasting brain power; in fact, we may be conserving it.

There is another way to think about the old saw that we use only 10% of our brain. Epilepsy is a disorder characterized by excessive neuronal discharge, with regions of the brain undergoing what can be thought of as a storm of activity. Epilepsy is a serious affliction, as it is associated with convulsions or seizures and can cause amnesia, hallucinations, loss of muscle tone, and loss of consciousness. During a generalized epileptic seizure, the brain is not only extremely active, but most neurons fire in a synchronous manner. An electroencephalogram (EEG) done while a subject in having an absence seizure shows that neurons distributed over the entire brain all fire together or are all silent together, with these two states alternating about 3 times per second.[31] It is thus possible that epilepsy could be averted if the patient were fortunate enough to use only 10% of his brain.

MIND AS A UNIQUELY HUMAN TRAIT

Clearly, there is more to being human than moving with greater facility than a machine or perceiving with greater acuity than a simple animal. Our brain is a collection of neurons that connect to one another with extraordinary complexity and remarkable resiliency. Our mind is a compilation of concepts, a pool of perceptions, an amalgamation of emotions; mind is a product of brain, but it is also the summation of all that makes us human. It will certainly be possible to understand brain, perhaps in the not-too-distant future. But it may remain effectively impossible to study mind with the rigor that it deserves, since it may never be possible to compare the qualia of one person to those of another with any kind of exactitude.

It should also be noted that great intelligence is a great danger, since it has enabled us to invent the means to self-annihilate.[32] It is possible that mind may be required to ameliorate the enormous capacity for damage that has been created by our brain. Sociality and emotional attachment may be properties of mind, but they have shaped the evolution of human intelligence for millenia. It is not clear whether our intelligence is a by-product of our sociality or whether our sociality grew out of our intelligence, but it is clear that humans differ from most animals in possessing greater intelligence and a greater variety of emotional attachments. It has been argued that the expansion of brain size initiated a pattern of coevolution between social complexity and intelligence that ultimately led to the modern human brain. The emergence of mind could be an indication of the survival value inherent to emotional attachment, given our high intelligence and the attendant dangers of self-annihilation. The fact remains that the human brain is unique in the degree to which it possesses that amalgam of creativity, sociality, and complexity that we call mind.

CHAPTER 6

THE PSYCHOLOGY
OF LEARNING

Memory of even the briefest event can last a lifetime. Yet many details of day-to-day life are forgotten almost as soon as they occur. The human capacity to learn is deeply mysterious; it is something we share with organisms as simple as *Paramecia*, yet the scope of our ability to learn far outstrips every animal, as far as we know.

Learning capacity must have provided a powerful selective advantage during the course of human evolution. The ability to learn a great deal enhanced the survival of our ancestors in many ways: memorizing the location of food sources, learning which berries were edible and which were poisonous, remembering the hunting strategy decided upon by the group, bearing in mind likely sources of danger, recollecting how to build a fire, recalling the nuances of a social network; all of these skills were vital to the survival of our human ancestors. Learning is the crux of what sets humans apart and is one of the key abilities that has made our species so successful.

THE SIMPLEST FORM OF LEARNING

It is possible to learn without the benefit of much in the way of memory; even a single-celled organism can learn by conditioning. Ivan Pavlov's experiments with conditioned learning are still famous. Although he won a Nobel Prize in 1904 for his work on the physiology of digestion in dogs, it is his experiments with "salivating dogs" that are more familiar today. His series of experiments are memorable because they are so easily described yet so important in implication.

While Pavlov was studying the mechanisms that induce secretion of digestive enzymes in dogs, he noticed that his dogs began to salivate before they were fed, as if they knew that food would be forthcoming.[1] Pavlov further noted that salivation could be induced if the dog saw the food container or the attendant who normally did the feeding, or even if the dog heard the sound of the attendant's approaching footsteps. Pavlov called this "psychic" salivation, a wonderfully evocative name.

Pavlov began to pair food with the sound of a bell to see if the dog would eventually salivate at the sound of the bell. In fact, this is exactly what happened; after a training period, the dog would salivate when a bell was rung, even if food was not presented. In explaining these results, Pavlov called food the "unconditioned stimulus" and salivation the "unconditioned response," since the food would evoke salivation without conditioning. Extending this logic, the sound of the bell was a "conditioned stimulus" and salivation at the sound of the bell a "conditioned response," since this response would not exist without conditioning. The psychic link between a conditioned stimulus and a conditioned response represents learning, since there was no such link prior to conditioning and no reason to suppose that a dog should salivate at the sound of a bell.

Pavlov investigated the relationships between stimulus and response in very clever ways. For example, he noted that the temporal sequence was important; it was easier to condition the dog to salivate if food followed the sound of the bell than if the order was reversed or if the food and the bell were presented simultaneously. It was also easier to build a conditioned response if there was a short time interval between the bell and the food. A dog will salivate more if trained with a large piece of food or with a loud bell, but a dog conditioned to salivate to a bell of one tone can quickly be trained to salivate to a bell of a different tone. Finally, if food is no longer given after a conditioned response to the bell has been established, the dog will eventually cease to salivate. It was not known whether the dogs had "unlearned" the conditioned response or simply learned a new conditioned response (nonsalivation). However, if the dogs were rewarded only sporadically during the training period, Pavlov found that it took more time to extinguish the salivation response. Sporadic reinforcement built a more robust conditioned response than was elicited if the dogs were rewarded every time the bell was rung during the training period.

Computer models suggest that extinction of a conditioned response does not really involve "unlearning" the original association.[2] Instead, it is likely that as a behavior is extinguished, it is literally covered up with new, inhibitory learning. Extinction does not reverse learning, since synapses do not return to the native state they occupied before conditioning. There may be layers of learning like sediment in a dried lake; though the older layers may be covered up by newer layers, they remain in place.

It is clear that conditioned learning happens in humans too. The thought of chocolate cake can make us salivate, a smoker may relax before the cigarette is lit, and the arrival of Friday may buoy our spirits merely because we know that we can sleep late the next day. Even the behavior of gamblers can be explained by conditioning; the sporadic reward of a win at the roulette table may keep the gambler coming back through many failed attempts, and gambling behavior is very hard to extinguish because of the sporadic nature of the reward.

Scientists who have followed in Pavlov's footsteps were able to adapt his methods in fascinating ways. In one famous experiment, a dog was trained to associate the sight of a circle with being fed and the sight of an ellipse with not being fed.[3] By small steps, the ellipse was then made to look more and more like a circle until eventually the dog was unable to distinguish between the two shapes. Over time, the dog became more and more agitated and unable to cooperate in the experiment, since it could not distinguish between the two opposing conditioned stimuli. This behavior was described as an experimental neurosis, since it was maladaptive but induced by training. Some dogs were more vulnerable to such neurosis as a consequence of contradictory training, which suggests that a dog's personality can influence learned behavior.

A branch of psychology known as behaviorism grew in response to Pavlov's ideas. The main proponent of behaviorism, B. F. Skinner, argued that human personality develops as a result of the rewards, punishments, sensory inputs, and responses of the developing infant.[4] Skinner believed that the newborn was a blank slate, a *tabula rasa*, upon which anything could be written by parents, by siblings, or by teachers. He grew famous for his experiments with a Skinner box, where he trained pigeons to do various bizarre tasks for a reward. But Skinner also believed that people could be programmed to behave in predetermined ways, in ways that would fit them for a utopian society; Skinner even went so far as to subject his daughter to extended stays in a Skinner box when she was an infant.

FEAR CONDITIONING

Pavlov sought to derive universal laws of learning, and his ideas proved to be very influential over the years. There is still a great deal of research in fear conditioning, which is a learned association between an environmental cue and an aversive event like an electrical shock. This type of learning would have evolutionary value, since it can promote survival in the face of present and future threats. One of the first demonstrations of fear conditioning was in 1920, in an experiment with an infant known as "Little Albert." This unfortunate child was first shown a white rat, which did not evoke any fear but, rather, elicited interest. Later, as the infant was shown the rat, an experimenter made a loud and startling noise by striking a steel bar with a hammer. Not surprisingly, this sound produced a robust fear response in the child, causing him to burst into tears. Later, after the child had been conditioned to see the rat at the same time as the noise was made, the sight of the rat alone could induce the child to cry and crawl away. In this example, crying was the conditioned response to the conditioned stimulus of a harmless white rat. Such Pavlovian conditioning was taken as proof that neurosis can be conditioned, even in a healthy human infant. It is worth noting that this type of experiment could not be done today, under the ethical protections that are routinely afforded to research subjects.

Fear conditioning has attracted a great deal of interest because it stands at the intersection of memory and emotion and may be critical to understanding such troublesome clinical problems as panic disorder and phobia.[5] Early experimental studies with monkeys showed that surgical removal of a part of the temporal lobe known as the amygdala—adjacent to the all-important hippocampus—could completely block fear conditioning. A monkey that once cowered in fear in the presence of people would readily approach his handler after a surgery that damaged his amygdala, since such a lesion prevents the monkey from having an appropriate fear response.

Forms of learning that are motivated by fear require an intact amygdala. An influential recent study showed that humans can recognize fearful expressions in the face of another person only if they have an intact amygdala.[6] Since it would be unethical to surgically remove the amygdala of a person, the role of this structure in human fear was unproven for years. Yet examination of a remarkable patient—one who suffered from a rare disorder (Urbach-Wiethe disease) that resulted in bilateral destruction of her amygdala—confirmed that the amygdala is important for understanding fear in humans too. This woman had average intelligence and was unremarkable except for a history of poor decision making in social situations. To determine if this woman had a normal ability to recognize and respond to fear, she was shown facial pictures of actors feigning fear. The pictures were validated in healthy volunteers with a normal amygdala, so researchers could be sure that these faces actually conveyed fear. Compared to normal subjects, the woman with the damaged amygdala was impaired in her ability to recognize "faces of fear." She was also less able to recognize surprise or anger, even though she could still recognize those facial expressions associated with happiness, disgust, or sadness. By contrast, her ability to recognize the identity of a familiar face was normal. These findings suggest that facial identity and facial emotion are processed in different parts of the brain, with the amygdala needed for emotional insight. Since it is crucial to understand the emotions of another person in order to understand his motivations, these findings could explain why this woman was socially inept.

Recent progress in medical imaging has made it possible to image the intact human brain as it functions without harming patients or exposing them to radiation. This new method, called functional magnetic resonance imaging or fMRI, can be used to characterize how the brain responds to novel stimuli. In an fMRI experiment, a subject is alternately imaged while at rest and while exposed to the stimulus of interest. If images acquired at rest are then subtracted from images acquired during the stimulus, the difference between the images represents the increase in blood perfusion that is induced when the brain is working. In one study, subjects were exposed to a red light that was either presented randomly or that was presented so as to foreshadow a painful electric shock.[7] During acquisition of the fear response, both the amygdala and the hippocampus play a part, with the amygdala particularly involved during the first phase of conditioning. These results confirm that the amygdala is a crucial locus of fear memory and also suggest that the hippocampus is involved in learning new associations.[8]

The amygdala is necessary to encode or extinguish fear memories, which may have relevance in clinical disorders such as panic attack and phobia. Many features of the environment could potentially serve as a conditioned stimulus if a subject is exposed to a strongly aversive stressor.[9] Under conditions where a conditioned stimulus (e.g., a bell or tone) is followed after a long interval by a powerful electric shock, there can be a generalized fear response. This is because the shock is a powerful motivator but it is not necessarily clear to which stimulus it should be linked. A conditioned response could thus be formed to the presence of a lab coat, to the time of day, or even to the room itself, so that a powerful fear response would be evoked by a general stimulus. In essence, if an explicit link between conditioned stimulus and unconditioned stimulus is not learned, then there can be a phobic response to a broad range of stimuli.

This may explain why survivors of a terrorist bomb attack often experience anxiety for months. When an overwhelmingly aversive stimulus is experienced, there can be a fear response formed to a broad range of stimuli such as a loud noise or the sound of people running, even if these stimuli had only a coincidental pairing with the explosion. To some people, a clear September morning in New York City may forever evoke the horror of the World Trade Center attack. An accidental pairing of the attack with a more or less random feature in the environment may be enough to elicit a powerful fear response in the future, which could explain post-traumatic stress disorder.

"ONE-TRY" LEARNING AND SENSITIVE PERIODS

The classic example of "one-try" learning is imprinting, and the most famous example of imprinting is the way in which a young bird learns to recognize its mother.[10] In 1935, Konrad Lorenz showed that young birds go through a critical period during which they "imprint" on their mother. This critical period is essential for the bird to mature properly, since imprinting assures that the adult bird will later mate with a member of its own species. The imprinting process is confined to a very brief period of time and, once accomplished, this imprinting is irreversible. If the young bird sees an object other than its mother during this critical period, it is possible for the bird to become imprinted on that object, which renders the adult bird unable to respond properly to a mating overture from one of its own kind. This bizarre aspect of imprinting is why Lorenz was able to induce young birds to follow him around like tiny shadows. But imprinting is not the only example of one-try learning. It is said that a rat given tainted food will never again eat that type of food, even if the food is offered in unspoiled form and all other food is eliminated.

Whether one-try learning occurs in humans is not known. It is possible for post-traumatic stress disorder (PTSD) to be induced by a single event, but people who suffer from PTSD will typically have suffered more than one event. Even people exposed to the trauma of the attack on the World Trade Center were not exposed to an event at a single point in time but, rather, to an event that unfolded slowly over time and whose cumulative effect can be understood only in the context of the long and stressful period after the attack. It seems unlikely that one-try learning is of much importance in humans, even if it does occur, because so much of what we know is learned by repetition or insight. Given the power of repetition and insight, it seems likely that one-try learning could be overcome by the types of learning that are more characteristic of the human condition. This may be why it is possible to desensitize a person suffering from PTSD so that she is no longer sensitive to the stimuli that once evoked a phobic response.

In humans, it is more appropriate to think of sensitive periods during development. Such periods may be brief, but they are probably not as brief as the period of imprinting in ducklings. And though these periods of imprinting may be powerful, they are likewise probably not as irreversible as imprinting in ducklings.[11] Nevertheless, it is possible that relevant experience during a sensitive period can modify the architecture of a brain circuit in a fundamental way. The most thoroughly studied example of a sensitive period in mammals is when visual experience structures the visual cortex. This has been studied in monkeys, cats, and ferrets, but not in humans, because of the inherent ethical problems of conducting such research on infants. Yet experimental work in animals has shown that visual experience powerfully shapes neural connections in the visual cortex. If one eye is chronically closed, perhaps by being sutured shut, there is a selective elimination of connections with the closed eye and an elaboration of new connections with the open eye. As a result, the visual cortex comes to be dominated by input from the open eye. Even if the sutured-shut eye is then opened, lost connections cannot remake themselves and the visual cortex will continue to be dominated by the one eye. The period during which ocular dominance is established is short enough and strong enough that it is conceptually rather similar to imprinting in a duckling and quite different from the high degree of plasticity that is more characteristic of primates.

Language Learning in Infants

An important example of a sensitive period in humans is the period during which infants learn language. Socially isolated or profoundly deaf children who experience neither oral nor manual language do not acquire language, whereas language learning appears to be an automatic process in infants who are regularly exposed to it.[12] Cross-cultural studies of speech perception show that simply listening to a language induces an infant to learn about the phonetic structure of that language. This learning alters an infant's perceptual system, tuning the infant to the unique properties of the native language long before that language can be spoken. Mothers naturally simplify their own speech production in a way that makes it easier for the infant to learn. As the infant hears speech sounds, the infant brain hones its innate ability to distinguish sound, making new distinctions that are relevant to the particular language while overlooking other irrelevant distinctions. By 6 months of age, an American baby can distinguish the "a" sound in "cat" from the "o" sound in "cot" even though the duration of these sounds is measured in milliseconds. When mothers speak to their infants, their speech is typically at a higher pitch, with exaggerated intonation and a slower cadence, and this is true across cultures. Parents address their infants by producing vowels that are acoustically extreme, resulting in an expanded vowel space that is "stretched" or hyperarticulated relative to adult speech. Infants prefer to listen to speech that is infant directed rather than adult directed, and they respond more to hyperarticulated speech. Language-delayed children sometimes benefit from exposure to spoken language that has been computer altered so as to stretch out the vowel sounds and make the sounds more apparent. To an infant, language exposure provides a rich and detailed source of information that initiates the process of language learning and alters the infant brain in a way that is unique to the specific language. If infants lack this exposure, their later ability to learn a language can be impaired or entirely lost.

Infants learn language with remarkable speed, but how they do it remains a mystery.[13] The infant brain is apparently able to use very sophisticated strategies to pick out the patterns in speech. There is evidence that infants analyze the statistical distribution of speech sounds, and this leads them to the discovery of phonemes, which are the smallest units of sound in speech (e.g., the "m" sound of "mat"). Each child learns how to make sense of an initially unintelligible stream of sound, which is often not clearly broken up into separate words. Infants learn to break sounds into words, based on the particular rhythm of speech; most languages are dominated either by words with stress on the first syllable or words with stress on the last syllable. Each word is also broken down into phonemes; most languages use a set of about 40 phonemes to build different words, but phonemes characteristic of Chinese are quite different from phonemes characteristic of English. Infants are exquisitely sensitive to subtle differences in phonemes, whereas older children lose their ability to distinguish phonemes. The phonemes that are learned early in life seem to bias the brain so that it is more difficult to learn a second than a first language, whereas ever finer distinctions can be learned in the first language. Social interactions are very important in language learning; a child will learn language more rapidly from another human being than from exposure to the same material in a tape recording. It is hypothesized that language learning produces dedicated neural circuits that encode the patterns specific to a given language. As these networks develop, they make it easier to learn new speech elements that are consistent with existing patterns but harder to learn new speech elements that are at odds with existing patterns. This could explain why the sensitive period for language ends; once a certain amount of learning has occurred, the neural circuits are committed, which may interfere with learning a second language. In short, language alters the brain in ways that are irreversible.

PERCEPTUAL LEARNING

Perceptual learning is a lifelong process. We begin by encoding information about the basic structure of the natural world and we spend the rest of our lives augmenting, updating, and correcting those initial impressions.[14] Consider vision, for example: you learn the face of your mother when you are young and when she is young; as you age, she ages, and you must update your memory of how she looks. After leaving home, you may not see your mother for months or years, and return to find her looking different from your mental picture of her. Were we not able to update the memory of our mothers' faces, eventually we would be unable to recognize them. Yet storing a memory or mental representation of an object as complex as a face is still a poorly understood process, although our understanding has grown in recent years.

What is now clear is that the brain must remain plastic, able to alter itself or be altered by the environment, for learning to occur.[15] Such plasticity can be revealed when a human subject learns to make discriminations that could not be made at first. A short period of training can improve the ability of a person to estimate weight, to discern differences in hue, to discriminate the orientation of a line, to distinguish acoustical pitch, to differentiate between two textures by touch, or to detect the precise direction of motion. It is noteworthy that this learning can be quite specific to the type of training, whereas the original motivation behind formal education was the idea that study in a discipline such as Latin could increase a child's ability to learn a "useful" skill. Although there can be poor carryover from one skill to another, there is little doubt that the brain is like a muscle, and that training generally increases the ability of the brain to learn.

The brain is able to identify and encode correlations in the environment. For example, tasty berries may grow on a certain bush at a specific time of year. To make such an association, some neural connections may be strengthened and others weakened during the learning process. There may be a literal competition between neurons, so that neurons that "talk" to each other frequently become more strongly associated and form circuits, whereas neurons that seldom communicate will fail to form such circuits or may form other neuronal circuits. Such change in the relative strength of connection between neurons may be the essence of learning, and it certainly requires a degree of plasticity, even in the adult brain.

Perhaps the clearest demonstration of plasticity in the adult brain comes from studies of how the brain changes after an injury to the body.[16] The sensory cortex of the monkey brain responds to sensations from the hand, and each finger "maps" onto a specific cortical location. If surgery is done to remove a monkey's finger, then the part of the cortex to which that finger normally projects will become idle. Over time, cortical territory that was originally dedicated to the now-gone finger is remapped so that adjoining fingers expand their cortical territory and map onto the idle cortex. This means, in essence, that the cortex is not allowed to lie idle but is recruited for new purposes. Similar effects have been observed in the motor system, the auditory system, and the visual system. The time course of these changes is broad, from changes that occur immediately after a sensory disruption to changes that evolve over weeks and months. Changes can be local or they can occur at a distance; there is often a sprouting of long-range connections during the functional reorganization that follows surgery. It is reasonable to suppose that plasticity of the sensory cortex did not evolve to help a monkey recover from surgery. Rather, plasticity may have evolved to augment recovery from a stroke, and it may even be required for normal learning.

LEARNING THROUGH REPETITION AND INSIGHT

Learning can occur through repetition and insight; this is the sort of learning that we try to foster in schools and that we think of when the topic of learning is raised. Yet it should be clear by now that there are many simpler forms of learning and that life would not be possible if we were limited to learning only by repetition and insight.

Humans are not unique in being able to learn by trial and error. Edward Thorndike, who was a contemporary of Pavlov, became known for his "puzzle-box" experiments with cats, dogs, and chickens. In a typical experiment, a cat might be required to climb down a rope, lift a latch, or do some other simple task to escape from a box and gain access to food; in the first trial, the puzzle would be solved by accident as the cat simply explored its environment. Over time, a cat can learn to get food more quickly by repeating whatever actions were successful in earlier trials. This type of learning tends to occur rather gradually; when Thorndike plotted the time it took for a cat to get the food as a function of the number of separate learning trials, this formed the famous "learning curve." Thorndike believed that trial-and-error learning occurred simply as a result of instinctive movements that were either rewarded and thereby strengthened or not rewarded and thereby extinguished over time. This emphasis on the mechanical aspects of behavior put Thorndike at odds with many of his contemporaries who believed that learning, even in very simple organisms, was a reflection of consciousness.

People can learn by insight, defined as a sudden comprehension of the solution to a problem. Insightful learning was studied in humans by asking them to solve a mathematical puzzle that, though it appeared to be complicated, had a hidden structure that allowed anyone who perceived the structure to solve the puzzle in far less time.[17] With practice but no insight, people were able to solve the puzzle about 17% faster than the rate they achieved at first. However, those people who gained insight into the deeper structure of the puzzle were able to solve the puzzle 70% faster than the rate they first achieved. What is most intriguing about this study is that insight was more likely to occur if a subject was given a chance to sleep between the training period and the test. Of those subjects allowed to sleep, fully 59% were able to achieve insight, whereas only 25% of subjects not allowed to sleep were able to achieve the same insight. Insightful learning can also be proven in monkeys, although not with a mathematical problem. A monkey was placed in a cage with a banana suspended by a string from the ceiling, far out of reach. However, the monkey was given a long stick that would reach the food. Many monkeys had the insight that if they used the stick to start the food swinging, they could reach the banana by climbing up the bars of the cage. This puzzle would seem to require monkeys to show more insight than many people need to use in their day-to-day lives.

WHAT IS LEARNING?

Learning comes in many forms and has many definitions, but learning can be thought of as the retention of explicit and implicit memories.[18] An explicit memory is an image in the mind or something that can be described in words; it is constituted by an object, a place, or an event, and it has a certain concreteness to it. An implicit memory is harder to define, as it may not even enter conscious awareness; it can be a bias or habit of thought, or it can even be an improvement in the ability to discriminate between stimuli that were initially perceived as being identical. In any case, it is clear that learning is not just building a collection of facts: it has more to do with establishing new patterns of thought.

Learning has been defined as behavioral plasticity, although this definition ignores any form of learning that has no effect on behavior. A better definition of learning would acknowledge that learning is memory and that memories persist over multiple time scales (fig. 12). Short-term memory may last only a few seconds; you glance at your watch to answer a question about the time then a moment later you don't know what time it is anymore. If memory lasts longer, say from seconds to hours, this is called working memory. If you take a message for someone, you may forget that message as soon as it has been passed on, even if you remembered the message for hours before conveying it. If memory lasts longer still, from hours to months, this is long-term memory. An example could be your home telephone number, which you are able to recall without effort. A distinction has been made between long-term and long-lasting memory. Long-lasting memory may represent long-term memory that has somehow fossilized; an example could be the telephone number that you remember from your childhood home many years ago. It is possible that working memory, long-term memory, and long-lasting memory are simply different

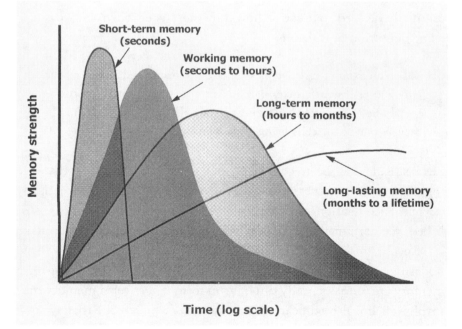

Figure 12. Duration of various types of memory.

phases of long-term memory, although there may be a legitimate distinction among these entities.

More than a century ago, the Spanish anatomist Santiago Ramón y Cajal became famous for his detailed and beautiful line drawings of neurons, which first began to reveal the overwhelming complexity of the brain. On the basis of his experience in drawing hundreds of neurons, Ramón y Cajal proposed that learning is a matter of neurons strengthening connections to one another. This proposal was made at a time when there was wide acceptance of the doctrine that past infancy human neurons do not grow at all. Thus, Ramón y Cajal was led to postulate that learning is a matter of functional changes between neurons that already have physical connections.

The late Donald O. Hebb, a neuropsychologist at McGill University, proposed a definition of learning in 1949 that was roundly criticized then but may provide the best definition of learning now. Many of the original criticisms have proven to be false, based as they were on a flawed understanding of the physiology of neurons. Hebb's Rule was originally derived to describe how a memory is encoded, but it is also a fine definition of learning:

> When an axon of cell A is near enough to excite cell B and repeatedly and persistently takes part in firing it, some growth process or metabolic change takes place in one or both cells such that A's efficacy, as one of the cells firing B, is increased.[19]

Hebb's Rule has been colloquially described as "neurons that fire together, wire together," since a memory may amount to a newly strengthened relationship between neurons. In other words, once established, memory may be the path of least resistance between neurons.

Recent research suggests that Hebb's definition may need updating, as it is likely that short-term memory and long-term memory are fundamentally different in form. Short-term memory may involve chemical modifications at a synapse, which make that synapse temporarily more likely to generate an action potential. In contrast, long-term memory may involve the building of new synapses or the physical alteration of an existing synapse, which makes that synapse permanently more likely to generate an action potential.

BRAIN MECHANISMS OF LEARNING

The human hippocampus is central to learning, as was first shown in a fascinating case study published nearly 50 years ago.[20] A patient known as H. M. submitted to surgery to remove part of his brain at age 27 in an effort to control intractable seizures. The part of the brain that was removed was the medial temporal lobe, which included the hippocampus. After recovery from surgery, it was noted that patient H. M. had certain cognitive problems that were to trouble him for the rest of his life. Although memory of his life prior to surgery was unaffected, H. M. was unable to remember anything that happened or anyone he met after the surgery. He was able to hold new associations in short-term memory for a few minutes, but he was not able to consolidate short-term memory into long-term memory; any task he learned or any fact he heard would have to be relearned the next day. Although there were rather profound changes in intellect, H. M. did not suffer any changes in personality. Yet he was unable to recall the date, where he lived, what was said to him moments before, what he'd had for breakfast, or whom he had just met. In addition, there were subtle and strikingly odd features to his amnesia. H. M. was asked to learn a difficult task that involved using a mirror to draw. Every day he would take up the task afresh, with no memory of having learned the task on the preceding day. Yet his performance steadily improved. Thus, he was learning a motor task without any awareness that he had done so, showing that learning has component parts that are remembered in very different ways. H. M. had no declarative memory of having learned the motor task, yet his body seemed to have learned it nonetheless.

The hippocampus is now understood to bind different sensory stimuli together to form a whole. The sights, sounds, smells, and events that form a memory are first placed in contiguity in the hippocampus. Without these separate elements of memory being placed in an overall context, they could not be bound into a unified whole but might instead rattle around in the brain like loose pebbles. Functional MRI of the brain during memory formation suggests that the hippocampus stores a composite of events rather than specific facts. In ways that are still poorly understood, the hippocampus transfers information to various permanent storage sites distributed throughout the cortex, acting as a staging area for memories that will be stored elsewhere. Damage limited to the hippocampus is sufficient to impair the consolidation of long-term memories, but damage to the hippocampus and certain adjacent tissues can completely block formation of explicit memories.

More recent studies suggest that other parts of the central gray matter of the brain are crucial to other forms of learning. A part of the brain known as the caudate nucleus may be important for trial-and-error learning, as shown by a recent functional MRI study.[21] Learning can be motivated by reward, and it has been proposed that the caudate plays a part whenever there is a contingent relationship between behavior and reward. Functional MRI shows that the caudate is most active when reward is sporadic; in essence, the caudate may provide ongoing motivation in the absence of an immediate reward. As reward becomes more predictable, caudate activation may become less important, perhaps because reward is intrinsic to the task. It is noteworthy that patients with obsessive-compulsive disorder, which could be described as failure to extinguish behavior in the complete absence of reward, often show a dysfunction of the caudate.[22]

LEARNING AS LONG-TERM POTENTIATION

As we have seen, Pavlovian fear conditioning involves the amygdala. Recent evidence suggests that fear learning increases the likelihood of an action potential arising within the amygdala. Activation of neurons in the amygdala is a form of plasticity that has been called long-term potentiation (LTP) because activity of amygdala neurons is increased or potentiated by fear. The process of LTP satisfies Hebb's definition of learning, since fear makes it more likely that a neuron in the amygdala will generate an action potential. Yet the mechanisms by which LTP occurs are not well understood.

Nevertheless, if a drug is used to block the function of neurons in the rat amygdala, then fear conditioning is also blocked.[23] Conversely, if neurons in the amygdala are stimulated, then fear learning is enhanced. Fear conditioning even increases the amplitude of synaptic currents in amygdala neurons that have been excised and placed in cell culture. It has been proposed that the hippocampus and the amygdala both show LTP during fear conditioning and that the tissues interact in very characteristic ways. Elements of the environment are assembled into a unitary perception of context in the hippocampus. The context then causes hippocampal LTP, and this potentiation can directly affect the amygdala. If a particular context is temporally linked to an aversive or fear-inducing stimulus such as an electric shock, the amygdala can form an association between context and fear. LTP in the amygdala can thus explain the origin of a conditioned fear response.

We have discussed LTP as if it occurs only in response to fear or only in the central gray matter of the brain, but this is not true. In fact, LTP was first discovered in the neurons of a sea slug, and LTP has been proposed as a general mechanism by which learning occurs at synapses throughout the brain. It is deeply intriguing that a functional description of LTP is so similar to the general definition of learning that Hebb published in 1949.

CAN WE IMPROVE OUR ABILITY TO LEARN?

Given what we know about learning and memory, is it possible to optimize the human ability to learn? Is learning simply IQ in action or is it a skill that can be honed? Are we limited by our innate abilities or is there a chance that we can intellectually better ourselves?

Recent evidence shows that there are memory-enhancing strategies that rely upon an increased understanding of the way the human brain works. Some strategies are based on common sense and are already in wide use, but some of them are less intuitive. Since there is a multi-million-dollar market for brain-boosting vitamins, seminars about memory enhancement, and special foods meant to increase brain power, it may be worthwhile to review what is actually known.

Perhaps the most important lesson is that it is essential to pay attention.[24] Attention is quite selective; it filters incoming information and allows only relevant information to gain access to working memory. Intense concentration is proven to increase the speed at which a person can learn new facts and ideas, whereas attention has little impact on learning of new motor skills. Explicit memory is therefore more vulnerable to inattention than is implicit memory. Skillful teachers can use this feature of the human mind by helping students to focus on concepts or ideas that are particularly difficult.

When learning new material, it is crucial to use your strengths. For example, most people learn things that they hear more efficiently than things that they see. Yet this is not true for everybody, and it is useful to learn your own strengths and weaknesses. Indeed, there are some students who learn best if they are allowed to touch and play with objects that are related to a lecture; this may be why laboratory classes are so often popular with students. It would be good to figure out your own particular learning style so that you can better utilize your strengths. If you learn best by seeing, make sure that you draw out new ideas or that you find good illustrations for concepts that you find difficult.

It is also helpful to involve as many sensory modalities as possible when learning new information.[25] An oral presentation supplemented with visual material will be understood and remembered better than an oral presentation alone. A lecture that is well illustrated engages two modalities of sensory input, whereas a lecture alone engages only the ears. This strategy is consistent with the idea that words and pictures are processed in different ways by the brain and that engaging both processes makes it more likely that the material will be remembered. In the past, children were often taught to read by listening to a teacher say the sound that each letter makes so that these sounds could be learned and combined. Now, children may be given plastic letters and encouraged to play with them as they hear the sounds. Memory can be augmented if we use motor skills concurrently; this is perhaps why reading is enhanced by note taking. The process of distilling knowledge into a few simple sentences and the act of writing out those sentences can improve comprehension and retention of new ideas.

Information is consolidated from short-term memory into long-term memory by rehearsal.[26] Rehearsal can take the form of reading over class notes (try reading them aloud) or it can take the form of describing significant concepts in new words. The traditional hour-long lecture may effectively fill up working memory so that if no effort is made to consolidate material into long-term memory quickly it may all be forgotten. It is more difficult to hold material in short-term memory, so material that is not moved into long-term memory quickly is more likely to be lost. A good teacher can help students by refraining from pointless digressions; such digressions may be interesting, but they also fill up short-term memory with unrelated facts. Information loss seems to occur when new data actually interferes with data that is already in working memory.

It is better to understand new material than simply to memorize it by rote. If new material is well understood, then a few forgotten details can be derived from the context of what has been remembered. In contrast, if rote-memorized material is forgotten, it's gone. The brain strives to make connections and to make sense of new information. This is why it can be helpful to fit new concepts into the context of previously well-learned concepts. If you already have a related memory, tying new material to that memory will strengthen the old memory and help to consolidate the new material. This may amount to filling in mortar around existing bricks rather than laying an entirely new row of bricks.

A wise teacher will also help students by pointing out some of the interesting associations. This can be done by asking questions or by presenting material in ways that foster connections. A new student of biology, for instance, will probably spend some time defining the properties of a living organism. This can be a fairly dry exercise unless the teacher asks questions to make the material more relevant. One way to make the properties of a living organism more compelling is to ask students to consider whether a virus is alive. Does a virus satisfy the properties of a living organism or are there features that we would all agree define life that are lacking in a virus?

PUZZLES OF LEARNING

An intriguing puzzle in the psychology of learning has been the "calculating geniuses" who are able to perform mental calculations at a level far beyond normal, even though some of these human calculators may be otherwise cognitively impaired.[27] Calculating geniuses can be of several types: children whose ability is precocious but who may not be unusual compared to an adult, learning-disabled people with a special ability that is far superior to the rest of their cognitive capabilities, and adults whose mental ability is far beyond the norm among the rest of the adult population. An example of the last type of savant is the movie character played by Dustin Hoffman in *Rain Man*, an autistic man with an encyclopedic knowledge of airline flight patterns and fatality rates. Although it may be unwise to generalize too much, there are some conclusions about savants that appear to be fairly consistent.

First, most savant abilities are mathematical in nature and are based on deep insight into mathematical structures that may be unperceived by other people. Even extraordinary musical ability may be inherently mathematical, as savants can sense a structure that is imperceptible to most. There are many shortcut methods that enable one to do mental calculations, and some of these shortcuts can be adapted in striking ways. For example, a number divisible by 9 is composed of integers whose sum is 9 (or a multiple of 9); multiplying by 5 is equivalent to adding a zero and dividing by 2; and all perfect squares form a mathematical progression equal to the sum of consecutive odd numbers. By knowing and using such arcane rules, some "mental calculators" can perform mathematical operations faster than people who are using a calculator.

Second, like any ability, mental calculation becomes easier with practice. It is probably safe to say that the ability to calculate mentally does not develop without extensive practice, even if that practice was done before anyone else noticed the ability. There is no evidence that ability develops without a conscious effort to hone it, and ability is known to atrophy with disuse.

Third, mental calculators use different strategies than are taught in school, since school calculation methods are designed to leave a paper trail, whereas mental calculations are not. Savants usually do not depend upon methods that involve carrying of remainders, and they may have mental tricks that generate answers in a left-to-right sequence, rather than the right-to-left sequence generated by paper-and-pencil calculations. This may explain why some mathematical savants never received formal instruction; such instruction could actually be a handicap.

Finally, it can be difficult to discern exactly how savants get their answers, just as it can be difficult to discern how any other intellectual insight is obtained. Mental calculators have been asked to "think out loud," but usually this process slows them down and one cannot be certain that the explicit process is identical to the implicit process. In the case of savants who make their living by entertaining and amazing people, there is a motivation to obscure the process, just as a magician will not reveal the methods of his magic. For example, there are mathematical savants who are able to derive square or cube roots of a number called out by the audience. This ability may not be based on calculation at all, since the savant can probably assume that any number called out is a perfect square or cube. Were it not, the person calling out the number would be unable to verify the correctness of the answer. Therefore, by memorizing a table of squares and cubes or by learning a few simple rules, the savant can quickly calculate the appropriate root.

It is clear that most savant abilities are a result of ability, interest, opportunity, and practice, which makes these abilities seem like a logical extension of normal learning. Yet there are some savants whose abilities so far exceed what can be explained that they seem miraculous. In South Carolina in 1849, a boy named Thomas Bethune was born into slavery.[28] He was blind, cognitively impaired, and perhaps autistic, yet he possessed astounding musical talent. At age 11, "Blind Tom" played at the White House and at age 16 he toured the world giving concerts, even though he was unable to function without a guardian. His speech vocabulary was thought to be fewer than 100 words but his musical vocabulary included over 7,000 pieces, many of them compositions by Beethoven, Bach, Mendelssohn, and Chopin.

A newspaper account that appeared in the *Fayetteville Observer* on May 19, 1862, when Blind Tom was 13 years old, described Tom's wondrous memory and remarkable talents:

> He has never been instructed in music or educated in any way. He learned to play the piano from hearing others, learns airs and tunes from hearing them sung, and can play any piece on first trial as well as the most accomplished performer. . . . One of his most remarkable feats was the performance of three pieces of music at once. He played Fisher's Hornpipe with one hand and Yankee Doodle with the other and sang Dixie all at once. He also played a piece with his back to the piano and his hands inverted. He performs many pieces of his own conception—one, his "Battle of Manassas," may be called picturesque and sublime, a true conception of unaided, blind musical genius.

Memory for music may have relatively little survival value. Yet the human brain is able to learn an astonishing array of things. Savant ability aside, it seems likely that a brain adapted to learn the location of a food source and whether a particular berry is edible may also be adapted to learn how to function in the complex environment of a social network. Our bodies are so very frail compared to those animals that might prey upon us that our brains are probably the only reason we survive as a species. In fact, human physical ability is so paltry that the survival of our ancestors was likely contingent upon their ability to find protection in a social group.

Learning has enormous survival value. But how can memory endure? Given that the duration of an action potential is a few milliseconds, how can we explain those memories that last a lifetime?

CHAPTER 7

THE BIOLOGICAL BASIS
OF MEMORY

Memory is thought to be composed of engrams, which are essentially atoms of memory. *Stedman's Medical Dictionary* defines an engram as "a physical habit or memory trace made . . . by the repetition of stimuli." An engram forms in short-term or working memory where it is transiently held and manipulated but then subsequently lost unless it is moved to longer-term memory. Moving an engram into longer-term memory may require a physical translocation of the locus of memory; working memory is thought to be the purview of cortical gray matter at the front of the brain—in an area called the prefrontal cortex (PFC)—which may not be involved in longer-term memory. After an engram is consolidated into a long-term memory, it may last for hours or a lifetime.

Research in memory currently focuses on working memory simply because it is easier to examine. Studies of long-term memory are hampered by the fact that there would necessarily be hours to weeks between when the memory is made and when the result could be measured. Long-lasting memories, which last for months to years, are essentially inaccessible to research, since it would be prohibitively time consuming to create a memory and then access it so much later.

It might be possible to study long-lasting memory by comparing memories to objective measures of reality, but there are precious few measures of reality that are accurate enough for the purposes of science. Most day-to-day events aren't documented at all, and even events that are recorded in newspapers may be inaccurate, given that newspapers are full of misinformation. These various experimental limitations and technical problems account for why scientists have a good understanding of how new memories form but a rather poor understanding of mechanisms that render a memory more permanent.

The basic problem is this: given that the duration of an action potential is typically a few milliseconds, how can we explain the durability of a memory? We must say at the outset that this question cannot yet be given a satisfying answer, although progress has been very rapid over the last decade. We have a good understanding of action potentials and even of how an action potential can have effects that last longer than the action potential itself, but we still have only a rudimentary knowledge of how a brief event can form an engram that lasts a lifetime.

What Mechanisms Might Explain Working Memory?

What is working memory? If a person is asked to memorize a list of items, recall is nearly perfect if the list has fewer than 7 items, but the success of recall declines for lists that contain more than 7 items.[1] This suggests that there is a type of memory—called working memory—that can hold about 7 items without needing to consolidate those items into longer-term memory. If a subject is given sufficient time to memorize a list of n items and then is asked to recall whether a test item was on that list or not, the time taken by the person to respond is related to the number of items on the original list. Response time increases by about 35 milliseconds per item, which suggests that it takes this long to move individual items in and out of working memory as each test item is compared to each item on the original mental list.

At the turn of the nineteenth century, Ramón y Cajal postulated that memory was somehow a result of changes in the strength of connection between neurons. This idea had little immediate impact because no one could imagine how intricate memories could be reduced to such simple changes or even how such changes could occur. But in 1949, a Spanish neuroanatomist named Rafael Lorente de No, who was a pupil of Ramón y Cajal, described how small neurons formed looplike connections between larger neurons, with the larger neurons tending to connect with distant parts of the brain.[2] This anatomy meant that an enormous number of circuits could be formed within the cortex of the brain. This anatomic description was recognized immediately as having major significance, since it meant that *a circuit could form a loop and have a functional effect upon itself*. If a circuit could reinforce itself, it could form a reverberating trace or echo of the original action potential, which might be a simple analog of short-term memory. Thus, D. O. Hebb was the first to propose that "the persistence or repetition of a reverberatory activity . . . tends to induce lasting cellular changes that add to its stability."[3] This idea was expanded into a hypothesis that long-term memory is the result of a reverberating circuit that somehow becomes permanent:

> Any frequently repeated, particular stimulation will lead to the slow development of a "cell-assembly," a diffuse structure comprised of cells in the cortex . . . capable of acting briefly as a closed system, delivering facilitation to other such systems and usually having a specific motor facilitation. A series of such events constitutes a "phase sequence"—the thought process.[4]

A key element of this conception was the existence of a large number of circuits, each of which would enable "an indefinite reverberation" to occur. The concept of the reverberating circuit was slow to catch on but, ultimately, the Hebbian circuit had an enormous impact because it meant that neurons have a high degree of plasticity. In essence, Hebb had equated learning and memory with the ability of a synapse to change over time, which was highly controversial at that time and for years afterward.

Now, synaptic plasticity is widely agreed to be the basis of learning.[5] It is thought that many computational tasks can be performed simply by adjusting the strength of connection between different neurons at the synapse. Facilitation at the synapse is thought to be the basis of long-term memory; once established, a memory may literally be the path of least resistance. There is also now support for the idea of reverberating circuits.[6] Single-cell recording studies using electrodes in monkeys have shown that neurons involved in object recognition continue to fire after the object has been removed for about as long a period of time as short-term memory lasts. Although this neural activity has been called a cortical buffer, since it can hold items in working memory for a short period of time, it is conceptually indistinct from a reverberating circuit. Thus, reverberatory neural activity may be equivalent to short-term memory.

The brain regions that are most closely tied to working memory are the hippocampus and the prefrontal cortex (PFC). The hippocampus apparently processes information that has not yet formed a memory, an idea supported by observations of H. M., the patient who was unable to consolidate new memories and so was unable to remember anything that happened after surgery to remove his hippocampus. The PFC—at the front of the brain—is also closely tied to working memory. During a working-memory task, the firing rate of individual neurons in the PFC of a monkey remains elevated without external cues, as expected of a reverberating circuit.[7]

THE ROLE OF THE HIPPOCAMPUS IN MEMORY FORMATION

The hippocampus plays a critical role in memory formation, but learning occurs even if the hippocampus has been damaged bilaterally. Apparently, the hippocampus binds the disparate elements of a memory together so as to make a sensory whole. The sights, sounds, and smells of a place may be brought together with information such as the sequence of events that occurred in that place or who else was there. A general hypothesis is that the hippocampus can collate these previously unassociated elements into a cohesive whole, connecting the disparate elements in a unique way.

Binding of stimuli in the hippocampus is crucial to fear conditioning—learning that certain environmental stimuli predict aversive events—but the situation is more complicated than that.[8] The hippocampus is the structure within which the content of a memory is assembled, but the emotion evoked by memory is apparently assigned in the amygdala. Both the hippocampus and the amygdala must therefore be intact for fear conditioning. Interestingly, a rat that learns to fear a shock given at the sound of a bell will also learn to fear the cage in which the shock is given. This has been called "contextual encoding," since the entire context of pain will eventually elicit fear; this may be why some people fear a visit to the hospital even if they are only visiting a sick friend.

The hippocampus is generally involved in episodic memory. This type of memory is defined as the memory of an event or an episode—as opposed to semantic memory, which is the memory of facts or lists. Evidence that the hippocampus is not required for semantic memory came from a study of three children whose hippocampi were bilaterally damaged.[9] These children all suffered brain injury at a young age: one child had been born without a heartbeat and it took 8 minutes to resuscitate her; another premature child had spent 2 months in an incubator and had suffered a series of protracted convulsions at age 4; the last child had received a toxic dose of a drug used to treat asthma, which induced respiratory arrest. All three children were brought to a physician because parents believed that they were unable to remember the events of everyday life. When they were evaluated, the patients ranged in age from 14 to 22 years, but none of them could reliably find their way in familiar surroundings or remember where their belongings were usually kept; none of them were oriented to days and dates, so they could not keep a schedule; and none of them could provide a reliable account of their day's activities. Cognitive testing revealed that their memory was substantially worse than would have been predicted based on their intelligence. All three patients had suffered bilateral damage to the hippocampus, as shown by magnetic resonance imaging (MRI). Considering their limitations, all three patients were able to perform surprisingly well in mainstream schools; all could understand and speak English, all had learned to read and write, and all were normal in verbal intelligence. Despite a pervasive and early onset amnesia—characterized by a disabling loss of episodic memory —their semantic memory was relatively intact. Given that there was an a priori expectation that a person with bilateral damage to the hippocampus would be so severely impaired as to be mentally retarded, these findings are quite striking. More recently, a patient was described with the opposite condition; he had intact episodic memory but severely impaired semantic memory.[10] These findings suggest that episodic and semantic memories are quite different, and that only episodic memory is dependent upon an intact hippocampus.

A great deal of work with animal models has shown that the hippocampus is important for remembering and understanding spatial relationships.[11] When a rat forages for food, there are neurons in the hippocampus that spike in activity as the rat passes through certain familiar places in the environment; these cells are properly known as pyramidal cells, but they have also been called "place" cells. A different place cell may fire for every location with which the rat is familiar, as if the rats' living space was mapped in their hippocampi. Yet the hippocampal map is not a simple point-for-point representation of space; only the more significant locations are mapped at all and place cells fire only when the animal is moving in a specific direction through a location. For example, there may be a place cell that fires only when a rat approaches a preferred foraging location or only when a rat is almost home. When a rat is trained to run through a maze, certain place cells discharge right before a choice point in the maze or right after a correct choice has been made. In this way, place cells can help a rat to navigate through its surroundings without having to densely map every feature of that environment. Hippocampal networks thus encode journeys through space as a sequence of specific locations. This is analogous to navigating through a neighborhood by looking only at street signs and ignoring the houses. This type of hippocampal encoding has been called "memory space" to denote that it is an organized depiction of place.

Rats with hippocampal lesions are impaired in their ability to learn a maze in a way that cannot be explained by differences in motor ability, perception, or motivation.[12] If the activity of place cells is recorded as a rat ranges around a cage, it can be shown that a familiar location or stimulus will evoke activity of a single place cell. However, a novel stimulus will evoke activity of a large number of place cells. Yet within about 30 minutes of exposure to the novel stimulus, the rat will have acclimated, so that activity is evoked from just one place cell. This means, in essence, that the memory of a particular place is encoded in a single place cell. The novel place stimulus has been bound together with other well-known stimuli into a seamless whole. Experiments have shown that as few as 60 place cells can be sufficient to encode the location of a rat in a cage to within a single centimeter. In aged rats, place cells are both less sensitive and more unstable in changing environments. The lack of spatial memory in rats with damaged hippocampi could mean that they are unable to bind the elements of episodic memory into an intact whole. In other words, it is not enough to juxtapose the elements of an episode; elements must be transformed into a permanent memory by binding.

WHERE DOES MEMORY RESIDE?

The locus of memory has been of interest to researchers for many years. Patients who have had surgery to remove a portion of their brain, whether because of tumor or intractable seizures, do not suffer the loss of a subset of their memories. They may suffer the loss of an ability, such as the ability to form new memories, as in the case of H. M., but they do not forget specific periods of their lives. Surgery may generally impair a person's ability to recall memories, but there is never the selective loss of, for example, all memory of 1991. This suggests that memory is somehow distributed throughout the brain.

Another curious aspect of human memory was discovered by Wilder Penfield, perhaps the most famous neurosurgeon of the last century. His observations seem to directly contradict the idea that memory is distributed evenly over the entire brain. Penfield performed surgery on many patients with epilepsy, each time seeking to excise that part of the brain that was a seizure focus. This approach is still used today, as it is sometimes possible to cure intractable seizures by removing only a small part of the brain. Because the brain itself does not sense pain, local anesthetics can be used during brain surgery, allowing the patient to remain awake. This made it possible for Penfield to search for the seizure focus and to identify those parts of the brain that were undamaged by using an electrode to gently stimulate the brain. Penfield found that as he moved the probe over the temporal lobe of a patient, he could stimulate points on the brain surface and evoke specific memories of past events. Some of these memories were so trivial or so long past that the patient lacked a conscious memory of the event. Memories could be evoked in vivid detail, as if the patient were actually reliving the event. Penfield noted that when the same region of the temporal lobe was stimulated, this tended to evoke the same memory. He concluded that everything a person had ever experienced, from a stranger's face to the words of a loved one decades ago, is locked away in the brain permanently. This is why his electrode so often evoked trivial memories rather than important events; in a complete record of all events, the trivial events would far outnumber the significant ones. Such memories may not be subject to recall without the prompting of an electrode, but Penfield concluded that the memories were locked away nonetheless. Of course, we do not know that these memories were true records of events; it is simply not possible to verify the details of a conversation that happened 30 years ago or the color of a dress that was worn to a lawn party in 1949.

It is possible that Penfield's patients were not recalling real events in vivid detail but, rather, were confabulating, or making up details to fill in a story whose outline they vaguely recalled. A recent case report suggests that this type of confabulation can happen, as shown by a 16-year-old girl known as A. K., who was undergoing electrode stimulation to locate the focus of her chronic and intractable seizures.[13] Surgeons identified a small area of her brain where electrode stimulation consistently induced A. K. to laugh. But each time this area was stimulated, A. K. attributed her laughter to a different cause. At various times, A. K. credited her laughter to the object seen during a naming task, to the wording of a paragraph she read during a reading task, or simply to the people standing around in the operating room ("you guys are so funny . . . just standing around"). What is noteworthy is that her laughter was evoked each time the electrode stimulated a specific locus in her brain. Yet each time she attributed her laughter to something different, something that was not objectively funny. In other words, when mechanical laughter was evoked by the electrode, an emotional stimulus was confabulated to give a rational explanation for her laughter. Yet the duration and intensity of laughter was entirely a function of the stimulating current; low levels of current evoked only a smile, whereas higher currents induced a robust and contagious laugh. What this may mean is that laughter normally has motor, emotional, and cognitive components; when an electrode evokes only motor laughter, the brain conjures up the missing emotional and cognitive components. Nevertheless, it seems likely that memories are both diffusely stored and can be evoked by stimulation of a specific brain locus.

The brain may not store whole memories but, rather, may store pieces of information that can be sewn together to create a memory. Different regions of the brain participate in the encoding, storage, and retrieval of engrams of memory, and these different brain regions are activated with near simultaneity so that each region can contribute its part to the whole. Recalling a memory may actually be a creative process, in that disparate shreds of memory are woven together into a seamless whole. This vividly recalls a passage from William Butler Yeats:

The friends that have it I do wrong
Whenever I remake a song
Should know what issue is at stake,
It is myself that I remake.

The prefrontal cortex (PFC) of the brain has been implicated as the locus of most short-term memory. After the hippocampus binds disparate stimuli into a single memory, the PFC has two distinct roles: "active maintenance" is the process of keeping some information immediately available, whereas "executive control" is the process governing which information is encoded to or retrieved from longer-term memory. The PFC is activated whenever verbal, spatial, or object-related information is actively maintained, and tends to be more strongly activated as the amount and complexity of information increases. Information that is remembered well is associated with stronger activation. Executive control is thus a matter of prioritizing, classifying, and storing information as it is being used.

During a learning task, different brain regions participate in a sequential pattern of activity.[14] Studies with monkeys have shown that the PFC is involved at a later stage of learning, as if it is needed for the progressive refinement of behavior. Tasks that involve reward training rapidly activate the basal ganglia and only later involve the PFC. This may mean that the PFC is, in essence, trained by other regions of the brain.

The PFC shows a remarkable degree of specialization in structure, which may facilitate the functions that it serves. Pyramidal cells in the PFC have on average up to 23 times more synapses than similar neurons in the visual region of the brain.[15] Although pyramidal cells are found throughout the cortex, there are regional differences in size, branching pattern, and spine density, such that pyramidal cells in the PFC form denser networks than elsewhere in the brain. Since pyramidal cells are the most common type of neuron in the PFC, the PFC is inherently more complex than most brain regions and the densely connected cells of the PFC may be superbly adapted for storing memories. The PFC also receives diverse inputs from brain areas involved in processing sensory data, so pyramidal cells in the PFC have extraordinarily complicated patterns of connectivity.

It has been proposed that cognitive ability in primates and humans is determined, in part, by the density of connections among pyramidal cells in the PFC. A human pyramidal cell has on average 72% more synapses than similar cells in a macaque monkey and 280% more synapses than such cells in the relatively primitive marmoset. The denseness of interconnection may also make it possible to form circuits with specialized functions, which has been shown by computer modeling to improve the efficiency of brain function. Clearly, one of the trends in evolution of primates may have been a progressive increase in the synaptic density of pyramidal cells in the PFC.

WHAT IS LONG-TERM POTENTIATION (LTP)?

If memories are to become permanent, information must somehow be locked away in the neural network by memory-induced changes in molecules. Long-term potentiation (LTP) is a long-lasting increase in synaptic efficacy that results from a "learned" association.[16] Synaptic efficacy—the ability of a neuron to induce another neuron to fire—is a function of a very limited number of variables: the probability that neurotransmitter is released by a presynaptic neuron, the total amount of neurotransmitter released at any given time, and the probability that a postsynaptic neuron will respond to released neurotransmitter. Evidence is accumulating to suggest that LTP is largely a function of an increased probability that neurotransmitter is released by a presynaptic neuron. But how can the probability of neurotransmitter release be increased?

There is strong evidence that synaptic plasticity is necessary and sufficient for the storage of memory.[17] Synaptic strength is adjusted by a series of events during fear conditioning in the hippocampus. First, the release of neurotransmitter at a synapse induces calcium ions to flow through the synaptic membrane, whether or not an action potential is generated. This influx of calcium can be blocked by drugs that bind calcium; such drugs also block LTP. Conversely, increasing the amount of synaptic calcium tends to facilitate LTP. Calcium ions enter activated neurons through channel proteins in the membrane. These channel proteins actually bind a neurotransmitter called glutamate, and the channel opens only when voltage changes at the membrane co-occur with glutamate binding (see fig 2, chap. 2). This channel protein is called the NMDA receptor because it also binds tightly to a chemical called NMDA (N-methyl-D-aspartate). However, the fact that the receptor binds to NMDA was discovered long before it was realized that the NMDA receptor is involved in LTP.

The NMDA receptor is unusual in that it will not open its calcium channel in response to glutamate until the neuron is depolarized. Thus, the NMDA receptor functions as a coincidence detector. This may help an NMDA receptor to differentiate between "noise" and an information-bearing signal at the synapse. Although NMDA binding alone is insufficient for LTP, blocking NMDA binding blocks LTP. In other words, if the presynaptic and the postsynaptic neurons are active at the same time, and if sufficient calcium enters the neurons, then the strength of the synaptic connection will increase. The NMDA receptor is most abundant in the prefrontal cortex (PFC), which is crucial to LTP. The NMDA receptor makes it easier for one neuron to induce an action potential in another neuron, exactly as Hebb's Rule of learning had predicted.

This model of LTP has been tested in many ways.[18] The temporal aspects of LTP have been thoroughly probed; we now know that two neurons must fire within about 10 milliseconds of each other for LTP to occur. If a drug that blocks the NMDA receptor is injected into a rat hippocampus, the rat cannot learn the pattern of a maze. Such a drug blocks the formation of new memories but does not impair well-established memories. Some of the strongest evidence in support of this model of LTP comes from studies of mutant mice. One study examined mice in which expression of the NMDA receptor had been "knocked out" in cells of the hippocampus. These mutant mice had normal neural transmission but were unable to form long-term memories.

In a fascinating experiment, certain mutant mice were bred to overexpress a subunit of the NMDA receptor in the forebrain.[19] These mutant mice typically allowed calcium to enter the neuron for a longer period of time after a transient depolarization. In essence, this meant that synaptic coincidence would be detected more often such that LTP would happen more easily. These mice showed normal behavior but had enhanced long-term memory. They were quicker to learn fear conditioning, quicker to show fear extinction, quicker to learn their way around a maze, and quicker to show spatial learning. Thus, a small change in a single receptor is able to induce major changes in learning ability, at least in mice. Though similar experiments in humans are unethical, this work may provide a target for drug intervention in people with learning disabilities.

Thus, the majority of synapses that support LTP in the hippocampus and elsewhere require the NMDA receptor to detect a coincident increase in both calcium ions and voltage at the synapse. But nothing is ever simple in the brain, as LTP can apparently occur without NMDA receptors in some brain regions. In those parts of the brain where NMDA receptors are crucial, calcium ions and NMDA receptors are not the only players in the drama. Flow of calcium ions through the NMDA receptor activates a protein called CaMKII. This protein is one of the most abundant synaptic proteins, and it apparently activates a cascade of molecular events. There are more than 30 different forms of CaMKII in the brain, though it is unclear why so many forms of the protein are necessary. Nevertheless, a drug blocking CaMKII function also blocks LTP. In addition to CaMKII, there are at least 8 other molecules that undergo a change during LTP. Some of these changes can occur in minutes, whereas other changes take hours and require protein synthesis.

Very recently, evidence has emerged that LTP can also be mediated by a mechanism that is only partially dependent upon the NMDA receptor.[20] When the NMDA receptor is activated, a gas known as nitric oxide (NO) is released by mechanisms that are poorly understood. This gas can diffuse back to the presynaptic neuron, enhancing the amount of glutamate released, and thereby strengthening the connection between pre- and postsynaptic neurons. When NO release is blocked, LTP is prevented. This does not interfere with memories that are already formed, though it does block new memories. Furthermore, if neurons are exposed to gaseous NO and a strong depolarizing current, LTP is enhanced. This mechanism is intriguing because NO can diffuse very rapidly through tissue, so it may be an effective way to induce LTP. However, NO is a short-lived molecule, so it would have an effect only at the closest presynaptic membrane rather than over a broad general area.

It must be noted that this description of the molecular events that occur during the formation of memory is a gross abridgement of an extremely complicated process, the understanding of which is still emerging. We have said nothing about electrical coupling at synapses (as opposed to neurochemical transmission), yet many neurons are in fact linked electrically to nearby neurons and electrical coupling could be important in getting neurons to fire together.[21] Whether or not electrical coupling is important in memory formation is not known. Nevertheless, we now have what is probably a nearly complete parts list of the molecules that are involved in memory, although we have only a rudimentary understanding of how the parts fit together. Our understanding of LTP may continue to evolve for years as we await definitive proof that LTP is equivalent to memory.

COMPUTER MODELS OF MEMORY

A rapidly burgeoning field in neuroscience has been called "computational neuroscience," which means literally the use of computers to figure out how the brain works. In practice, this means that scientists try to build a two-way bridge between clinical data and computer model building; data guides the formation of the model and the model guides the design of further experiments. The risk, of course, is that the model may blind us to what experiments are most useful and what information is most critical. Yet computational neuroscience has enjoyed some real successes and, at its best, it seems likely to provide insight into brain and mind. In a structure like the brain, where billions of cells make quadrillions of connections, it is simply not possible to do all of the interesting experiments, and it is hard to conceive of progress being made without help from computers.

In 1943, 6 years before Hebb published his seminal book *The Organization of Behavior*, a first attempt was made to describe the function of the brain in terms that would be appropriate for computer modeling.[22] At that time, computers used the binary or "on/off" logic that is still used today, and this paper was an effort to conceive of how the brain might use the same logic. In this first attempt, known as the McCulloch-Pitts model, neurons were assumed to have only two states, active or silent, which we know today to be a false assumption. Yet the model was made starkly simple because anything else would have been computationally impossible at the time. This was a groundbreaking effort, since it suggested that the workings of the brain might be accessible to computer modeling. The McCulloch-Pitts model posited that neurons connect to one another in such a way that a processing or "output neuron" can receive input from several lower-level or "input neurons." The task of the output neuron is to add up the strength of input from all of the various input neurons to determine whether the output neuron should fire. It should be clear that if several lower-level neurons fire simultaneously, an output neuron would be more likely to discharge. A neuronal assembly made in this way can perform fairly sophisticated operations by simply adjusting the strength of connection between neurons. If the threshold for activation of the output neuron is also adjusted, such a network can perform very differently under different circumstances, as would be required of a brain.

There are several weaknesses to this type of model.[23] The most obvious problem is that the strength of the neuronal connections is set by the experimenter and there is no way that a system could know a priori how strong a connection would be needed to perform a specific task. Some type of "hardwiring" of synaptic strength might be adequate for some neuronal tasks, but a hardwired network cannot learn a Pavlovian association. Yet this is not an insurmountable problem, since a modifiable network could perhaps arrive at the right settings by trial and error during a process of training. A greater problem is that such a network, in principle, cannot be trained to do an "OR" operation in which one input neuron or another, but not both together, can induce a discharge in the output neuron. An OR operation is required for many logical tasks, but modelers have been unable to make a simple McCulloch-Pitts circuit that can solve the "exclusive OR" problem. Yet one can solve the exclusive OR problem by adding another layer of neurons to the model, establishing a level of complexity that is certainly no more complex than a human brain. The challenging task for modelers is to conceive a model that can pattern itself as the brain patterns itself during development and that can still reproduce the complex operations of the human brain.

In 1969, a computer model was proposed in which there were many output neurons instead of just one, and which could show a type of associative memory.[24] This network could take a repeated pattern of input activity and eventually, using a Hebbian Rule to adjust the strength of synaptic connections, reproduce a consistent output pattern. Thereafter, presentation of the input pattern would recreate the output pattern very predictably, which may be an analog of memory. Although the new neural network also had difficulty with the "exclusive OR" problem, addition of another layer of neurons to the network solved this difficulty. The hidden layer of neurons is trainable and, given enough neurons in the hidden layer, almost any problem can be solved. Yet it is hard to know a priori how many neurons are sufficient for a particular task. In addition, it is hard to train the neurons, given that the neurons don't know what they are supposed to do. An observer is required to tell the network when it has solved a problem appropriately. In other words, even a good error-correcting rule will work only if the solution to the problem is already known. A hidden-layer neuron knows only that it is supposed to make an output neuron fire correctly; if the hidden-layer neuron knew a priori how to do that, then there would be no need to have a hidden layer. An ingenious way to circumvent this problem is a method called "back-propagation of errors," whereby errors are passed back to the hidden layer as input so that the hidden layer can fine-tune its output. In essence, neuronal output is reviewed in an effort to reduce errors.

If competition between neurons is introduced into a back-propagation network so that input neurons compete with one another to build stronger connections to the hidden layer, the system becomes even more like a brain. This type of network is now called a self-organizing neural network. Such a network can be used as a trainable "expert system" to help scientists understand the rules that an expert uses in making judgments. For example, a self-organizing neural network is capable of finding rules to sort apples from oranges, even if some of the apples are green and sour instead of red and sweet. Some scientists contend that there is little evidence that the brain uses back-propagation of errors, but our discussion of how the cerebellum learns a motor task (chapter 4) seems relevant. It is interesting to note that novel "concepts" or ways of classifying things can emerge from a sufficiently complex self-organizing neural network. Derivation of such concepts is really an emergent property of the system, which could not necessarily have been predicted from a thorough knowledge of the starting point of the system.

Results that have been achieved with neural net programs are rather astonishing, as shown by a program called NETtalk.[25] NETtalk is a computer program that was designed to pronounce English words. It uses written text as input and produces voice sounds as output. At first, having only random connections, the program babbles like a baby as it "reads" a text. Gradually, as the training proceeds, the computer begins to speak more intelligibly. Eventually, when tested with a text that it has never seen before, the computer is able to produce passable English, correctly pronouncing about 90% of the written words. The remaining 10% of words that the computer pronounces incorrectly may be effectively impossible for the computer to learn, since much of

English pronunciation is dependent upon context and the computer knows nothing about the meaning of what it is reading. Thus, the computer can learn rules of English pronunciation not because the rules were written into the computer program but only because *a method to discover the rules was written into the program.* How is this possible? If the hidden layer of neurons is studied to determine what features of text the neurons respond to, it turns out that information about English pronunciation is distributed across all of the neurons.

All information has to pass through the hidden layer, but it does not do so in a random way; instead, different neurons learn different rules (e.g., the difference between consonants and vowels or various ways to pronounce the letter "e"). It turns out that it is important to have the right number of neurons in the hidden layer: too few neurons and the neural network is unable to do the job, too many neurons and the network becomes overspecialized and cannot generalize as well. In the latter case, the network does poorly with a new text that it has not trained on. But with an approximately correct number of hidden neurons, the network can extract meaningful rules and "speak" text that it has never seen before.

This type of neural network has been criticized as a model of human learning because it is not known whether or how neurons could form a back-propagation network. Another issue is that the output of the neural network is judged by an outside observer, usually a human who knows how to read. But if the brain were learning to read in the same way, what would serve as the outside observer? Is this a role taken by some other part of the brain or is this the role taken by the teacher in a classroom? Finally, the back-propagation network becomes very cumbersome if there are multiple hidden layers of neurons. Some neuroscientists have concluded that the brain is unlike a neural network, even though the neural network does provide tantalizing glimpses of how the brain might solve certain computational problems.[26]

THE HOPFIELD NEURAL NETWORK

As John Hopfield noted when he introduced another model of neuronal function:

> Computational properties . . . can emerge as collective properties of systems having a large number of simple equivalent components (or neurons). . . . The ability of large collections of neurons to perform "computational" tasks may in part be a spontaneous collective consequence of having a large number of interacting simple neurons.[27]

The Hopfield model is based on "neurons" that are wired together reciprocally to form a cell assembly, as Hebb postulated. Some of these connections are stimulatory while some are inhibitory, and the strength of the connection between neurons is determined by Hebb-like learning rules. The Hopfield model also introduced several new concepts; energy minimization was incorporated as a way to help the system achieve a stable state and the stable state was assumed to be equivalent to a memory. If a Hopfield network starts near a stable state then, over time, it gravitates toward that stable state as a way to minimize energy in the system; this is analogous to the recovery of a memory.

An intuitive way to understand this is to visualize a network as having multiple stable states that correspond to hills and dales, like the piedmont of North Carolina, with dales representing low-energy states. If a huge ball were set rolling in the piedmont, it would come to rest in a low spot somewhere simply because it lacked the energy to climb the next hill. If the ball were again set rolling from the original spot, the position of the ball in the low spot might be the same. If the low spot is equivalent to a memory, this implies that memory can be recovered in its entirety by starting with some shred of that memory. In other words, if the network is given part of the input pattern, it can regenerate the entire pattern. This has been called "content-addressable memory" because any part of the content of a memory could potentially be used to re-create the entire memory. Clearly, a problem would arise if many memories were stored; this would be analogous to a region with many hollows and dales, so that a false minimum ("false memory") might be found. This problem would grow in severity, as more and more memories were stored, so that it might eventually be impossible to recover the original memory. Yet the Hopfield model reflects several emergent properties of the brain, including a capacity for categorization, for recognizing the familiar, for error correction, and for retention of time-sequence information.

The original Hopfield model of neuronal function was criticized for many reasons.[28] First, real neurons behave continuously, sometimes integrating input without generating any output, whereas the Hopfield model assumed that neurons were either on or off, with no state between. Second, real neurons have time delays and cannot respond to a signal instantaneously, whereas the model assumed that neurons would suddenly and randomly change states from passive to active. To deal with these criticisms, Hopfield introduced a more mature model that did away with both of these assumptions so as to better reflect the behavior of real neurons. Unlike the original model, the new model could generate graded responses and could do so in a specific time frame. Interestingly, the desirable properties in the original model were retained; the graded-response model was capable of content-addressable memory and was also more resistant to the effects of random noise. Hopfield claimed that a computer could be built to exactly reflect the properties of his computer model.

Plasticity in connections between neurons allows learning to occur, but it also allows noise to degrade the function of a network. In a neuronal network, noise is defined as an action potential that occurs by accident without being stimulated by a real event. In the most recent Hopfield model, rules were incorporated into the system to minimize the impact of noise and to allow the system to self-repair.[29] These rules regulate neuronal plasticity and enable the system to learn new tasks without supervision. With these rules, the system also became more robust to the impact of noise, since the system can self-repair. The problem that the repair rule must solve is that it must correctly identify a presynaptic impulse as being an "appropriate" one to connect functionally to the postsynaptic neuron. In essence, the overall pattern of neuronal discharge is used to structure the network and to repair the network if it becomes degraded. A network that can learn to recognize a specific pattern is, in the end, functionally equivalent to a network designed specifically to recognize that pattern.

Computer models have shown us how sustained activity can be stable in the presence of noise, how different channel proteins contribute to persistent neural activity, how modulation can affect the robustness of a system, and how novel items are categorized.[30] It is now clear that distributed and localized models of memory are not mutually exclusive; content-addressable memory can recover an entire memory from a shard of memory. A memory may be distributed over the entire brain, but stimulation of a specific locus can kindle recovery of that memory.

REMAINING PUZZLES OF MEMORY

Despite this discussion of how memory might work, the basic problem remains: Given that the duration of an action potential is typically a few milliseconds, how do we explain the durability of a memory? This problem is one of the most difficult and resistant problems facing scientists today. But it is not the only puzzle related to memory.

We still do not understand what impact the diverse morphology of neurons has on brain function.[31] It is probably true that neurons themselves are powerful nonlinear computing devices, equipped as they are with a great diversity of shapes, a large repertoire of inputs, a wide variation in cable properties, and a broad range of possible outputs, including inhibition of future responses. But is it true that the morphology of certain neurons is essential to working memory? If so, what neuronal morphology is necessary for long-term memory?

Unlike the simple summing device envisioned by McCulloch and Pitts, the neuron is a very sophisticated self-contained computer. Yet the brain is still subject to mysterious failures, which may turn out to be clues as to how the brain stores memory. How can the brain conjure up what seem to be memories but which turn out to be fictions or "false memories"? How is it that a false memory can seem as real as a genuine memory?[32] Why is eyewitness testimony so often worthless? Why are memories so malleable?[33] Why is it sometimes possible to recall shards of a memory but not the entire memory? Why, once remembered, are memories forgotten at all?

Perhaps the greatest puzzle of all is the brain's ability to wire itself to store memories. How can the brain generate patterns among neurons so that, later, these patterns are interpreted as memory? How can the phenomenally complex patterning of the brain occur without external guidance?

Clearly, there is some guidance in wiring the brain, in the form of human DNA, which has guided construction of the human brain for millennia. The DNA molecule is subject to stringent selection, since the human organism bearing that DNA is subject to illness and predation. DNA that did not adequately define the structure of the brain, that did not enable a brain to wire itself, would have been selected against eons ago. Yet it is not possible, at present, to understand how a series of base pairs in the DNA molecule can provide instructions that are sufficiently detailed and sufficiently flexible to pattern the human brain. Even if we assume that synaptic connections are initially made at random and are strengthened or weakened as a function of use,[34] it is still astonishing that any one of us has a functional brain.

CHAPTER 8

BRAIN PLASTICITY AND NEURAL STEM CELLS

earning is a form of behavioral flexibility, a plasticity of mind. The flexibility of behavior is, in turn, supported by a plasticity of the brain itself. As it grows, the brain is able to wire itself as a function of the relevant genes. As it learns, the brain is able to rewire itself as a function of the relevant environment. The brain constantly changes, and it is clear that plasticity of the brain is essential for plasticity of the mind. In other words, our ability to modify our own behavior is a function of the brain's ability to modify its own structure.

It is hard now to remember how deeply held was the view that neurons in the mammalian brain simply cannot divide, that they lose the ability to undergo cell division within a short time after birth. Until recently, the idea that neurons could be born in the adult brain was considered to be a heresy. This view was so ingrained that textbooks didn't bother to state it, since everyone knew it already. There were a very few reports of dividing cells found in the brain of adolescent or mature animals but, for the most part, these reports were ignored or discounted.[1] Less than 20 years ago, this dogma began to change. The revolution in thinking since then has been so rapid that the scientific literature is still rife with contradictory claims. But there is now a new dogma: neurons in certain brain regions, even in humans, are capable of cell division well into adulthood. And an even more revolutionary idea appears to be emerging—that cell division may be necessary to support new learning.

But a fair-minded consideration of brain plasticity will unavoidably project us into one of the most controversial topics of the past few years: stem cell research. Thorough appraisal of neural stem cells will necessarily involve a discussion of how exciting medical advances have been stymied by crass political considerations, how a scientific revolution has been hijacked by a religious movement that is fundamentally ignorant of and hostile to science, and how an empty argument has trumped the thinking of some of the best scientific minds in the country. But this discussion must start with something that could not seem more inconsequential: the song of male birds looking for a mate.

THE PROBLEM OF BIRDSONG

Song behavior plays a crucial role in bird reproduction, both for defending territory and for attracting a mate. Male canaries that reach sexual maturity learn a characteristic song with which to woo females of their species.[2] There are two regions of the canary brain that are involved in this process: one region for learning and remembering song and one region for producing song. Together, the two brain regions form the "song control system." The first interesting observation to emerge from a study of canary song was that male and female brains are profoundly different in that females have much less brain tissue devoted to the song control system. What was even more striking was that the male brain changed over the course of a year; the song control system was rather small in the fall, when males were not trying to find females, and increased in volume markedly in the spring. The two brain regions are 99% and 76% larger in spring, when males are producing song, than in fall, after several months of not singing. It was hypothesized that this increase in brain volume was related to the ability of the birds to learn new songs in the spring. What was most striking about this observation was that because this was a cyclical change that could apparently happen year after year, it implied that new neurons were produced in the spring, and that these neurons were responsible for both the increase in brain volume and the seasonal ability to learn song. In other words, neurons in the adult canary brain are able to divide!

Another study from the same laboratory a few years later showed unequivocally that neurons that contributed to the control of song production in canaries and zebra finches continued to grow throughout adulthood.[3] Production of new neurons occurred prior to the acquisition of those motor skills needed for song making. This suggested that newly produced neurons might be necessary for learning to occur.

We now know that neurons are constantly added to the forebrain of songbirds.[4] New neurons replace older neurons that have died, and peaks in cell replacement rate coincide with peaks in information acquisition. The new neurons mostly arise from a location in the wall of the lateral ventricle, a fluid-filled space in the brain. Neurons arise from precursor cells—called stem cells—which are able to give rise to many new neurons. These stem cells are not simple-looking undifferentiated cells, as expected, but, rather, appear to be well differentiated. Life expectancy of the newly produced neurons ranges from weeks to months, and neuronal survival is a function of hormones, the need for new cells, and song production by the bird.

Why are new neurons born in the healthy adult bird brain? Clearly, it would be useful to have a fresh supply of neurons if older neurons die (due to stroke or trauma), but the role for fresh neurons in the absence of neuronal death is less clear-cut. It has been proposed that new neurons may be necessary to maintain learning potential over the life span. This idea is very intriguing, since it implies that changes in synaptic number and efficacy are not an adequate mechanism for long-term storage of memory. In other words, learning may require more than just a change at the synapse: new learning may require new neurons.

Neurogenesis in the Mammalian Brain

The idea that neurons can divide and grow in the adult brain was startling enough when the brain under consideration was that of a bird, but recent evidence indicates that neurons can also divide and grow in adult mammals.[5] Certain cells can be isolated from a mouse brain, which are able to grow and differentiate into mature-seeming neurons in culture. This confirmed that neural stem cells exist in adult mammals. Another recent study showed that mouse neural stem cells arise from the wall of the lateral ventricle, as in birds, and that these cells can form many cell types, not just neurons.[6] Cultured rat stem cells grafted into the rat hippocampus develop into cells appropriate to the hippocampus, and stem cells grafted into one region of the brain can migrate over fairly long distances to repopulate neurons in another region of the brain.[7]

Technically, the term "stem cell" means a cell that is not in the final stage of differentiation. Stem cells may be able to divide throughout the life of the animal and can have progeny that either continue as stem cells or differentiate into a mature form that is no longer capable of cell division.[8] In short, neural stem cells are capable of both self-renewal and renewal of mature neurons and other cell types.

Evidence for the existence of neural stem cells in rodents is now overwhelming. Neural stem cells can be extracted from the mouse brain and cultured, and can be coaxed into generating new neurons in culture or forming new neurons when injected back into the mouse.[9] Neural stem cells injected into the mouse brain can migrate for long distances to take up residence within the olfactory bulb of the brain; as the cells move through the brain, they form characteristic chainlike structures of distinctive appearance.[10] If bone marrow stem cells are injected into the bloodstream of a mouse, they can find their way to the brain, where they can express proteins characteristic of neurons. Thus, stem cells in mice show a capacity to change from bone marrow to brain.[11] Some researchers have reported that neural stem cells

Table 3. Factors reported to affect growth of new neurons in mammals. Neural growth is called neurogenesis.

Increased neurogenesis	Decreased neurogenesis
Exercise	Increasing age
Environmental enrichment	Stress
Stroke or ischemic injury	Opiate use
Traumatic brain injury	Methamphetamine use
Increased serotonin	Decreased serotonin
Seizure activity	
Various growth factors	
Estrogen	
Electroconvulsive shock	
Restriction of food intake	

cannot change in this way,[12] but evidence is fairly strong that up to 0.3% of mouse neurons can be derived from bone marrow after only 12 weeks.[13] Thus, bone marrow stem cells experiencing the environment of the brain may be able to undergo transformation into a completely different cell type. Production of mouse neurons from neural stem cells is depressed by aversive experiences and stimulated by pleasant experiences.[14] Newly generated cells in the mouse hippocampus are able to generate action potentials and form functional synapses.[15] There is some controversy as to how this happens. Some investigators claim that bone marrow stem cells cannot actually change into neurons; rather, it has been proposed that these stem cells fuse with troubled neurons and thereby rescue those neurons from death.[16] But there is essentially no doubt that neural stem cells exist and can produce new neurons in the rodent brain in response to a variety of stimuli (table 3).

But is there evidence of neural stem cells in the advanced brain of the primate? It is known that the motor cortex of the monkey can reorganize itself after amputation of a hand, but it is not clear whether this reorganization involves the growth of new neurons or simply the elaboration of existing neuronal connections.[17] Four macaque monkeys were examined 1 to 10 years after surgery to remove an injured hand. Since there was no longer a hand to control, the area of the motor cortex that normally controlled the hand could remain idle or could control something else. In all 4 monkeys, that part of the cortex that normally controlled the hand had been remapped so that it could perform some other function; the monkey that had surgery 10 years previously displayed the largest shift, since part of the face was now controlled by that cortex which would ordinarily have controlled the hand. While this work does not prove that neural stem cells are present in the monkey brain, it does prove that there can be neuronal growth and the making of new neuronal connections in the adult monkey brain.[18] More recent work suggests that in adult macaque monkeys neurogenesis—the growth of new neurons—does occur, and that new neurons can be added in several brain regions. Critics have suggested that the primate brain can add new neurons only in the hippocampus and olfactory bulb and not in the cortical gray matter, but it is no longer controversial that neural stem cells are present in the primate brain.[19]

Evidence for Neural Stem Cells in the Human Brain

Do neural stem cells also exist in human beings? A groundbreaking study suggests that the human hippocampus retains stem cells that are capable of neuronal renewal well into old age.[20] This striking result was obtained by studying five patients, all of whom were diagnosed with squamous cell cancer. An important prognostic factor in such patients is the rate of growth of cancer cells, and this is measured by injecting a drug (called bromodeoxyuridine or BrdU) into the patient. BrdU labels the DNA in a newly produced cell—a cancer cell that is dividing rapidly will incorporate a lot of BrdU. This is usually interpreted as predicting a poor prognosis. As it happened, all five of these patients died within 2 years after the injection of BrdU and, with the permission of the families, brain tissue was sampled to see whether there was any evidence of growth specifically of brain cells. A key issue was to exclude a possibility that growing cells in the brain were metastatic cancer cells. But this type of cancer does not usually spread to other parts of the body and there was no evidence of metastasis in these patients. Therefore, neurons labeled with BrdU must be new cells—the progeny of neural stem cells—since they contain newly synthesized DNA. A thorough search found new neurons in a region of the hippocampus known as the dentate gyrus; in one patient who died only 16 days after BrdU injection, there were more than 200 labeled cells per cubic millimeter. In patients who survived for a longer time after injection of BrdU, there were fewer labeled cells, which might be expected if some of the newly generated cells had died over time. The average fraction of labeled cells in part of the dentate gyrus was 22%, which is an extraordinarily high proportion of new cells if we consider that the time between BrdU injection and death of the patient averaged less than a year. The implication was that 22% of hippocampal cells were renewed within a single year. This result is especially impressive given that the average age of patients was 68 years and the patient with the greatest number of newly produced cells was 72 years old.[21]

This astonishing result has been confirmed by several other studies. Growing cells were found in adult human dentate gyrus tissue surgically removed from the hippocampus of 18 patients with intractable epilepsy.[22] Although less than 1% of neurons seemed to be newly produced, this estimate could be wrong, since growing neurons were identified without the benefit of BrdU. Several studies found that neural stem cells could be isolated and grown from brain tissue surgically removed from adult patients with brain tumor or intractable epilepsy; neural stem cells were found in both white matter[23] and hippocampus,[24] and both brain sites produced stem cells that differentiated into mature neurons (fig. 13). Finally, neural stem cells were found in surgical samples of the dentate gyrus removed from children with intractable epilepsy.[25]

Are neural stem cells common in the human brain? A very clever new method suggests that many—perhaps most—neurons don't divide over the life span of a person.[26] This method relies upon the fact that aboveground nuclear weapons testing in the late 1950s and early 1960s resulted in the release of a large amount of radioactive carbon into the atmosphere, which quickly spread over the entire globe. The release of

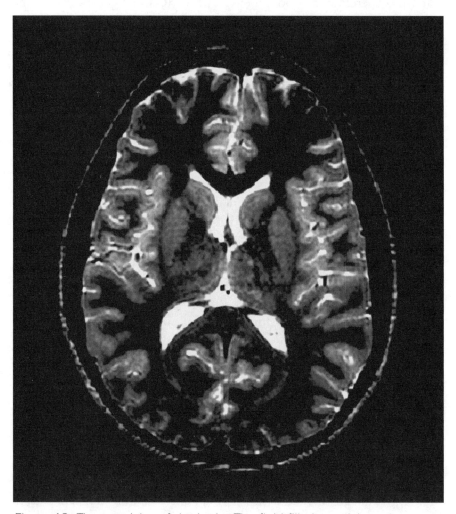

Figure 13. The ventricles of the brain. The fluid-filled ventricles are seen as bright spaces forming an X across the center of the brain. New neurons mostly arise from tissue in the walls of the large ventricles shown here.

radioactive carbon ended with the test ban treaty in 1963, and atmospheric radioactivity has declined greatly since then. Yet radioactive carbon was incorporated into the DNA of cells that were growing prior to 1963, so the amount of radioactivity in cells yields an estimate of the average date when cells in a tissue were born. Single cell resolution is impossible with this method, as only about one radioactive carbon is incorporated into DNA for every 15 cells. Furthermore, the half-life of radioactive carbon is 5,730 years, so the temporal resolution of the method is poor. Yet, if a large amount of tissue is sampled, the method can be fairly accurate. The average age of cells in gray matter of the human cerebellum was almost as old as the individual, suggesting that these neurons hardly divided at all. In contrast, neurons from the gray matter of the occipital lobe were substantially younger. This suggests that there was more neuronal growth in the cortex than in the cerebellum. Yet neural growth, even in the cortex, was not extensive. This does not rule out a possibility that there was a far higher growth rate in the dentate gyrus of the hippocampus—since this area of the brain was never measured—nor does it reveal whether neuronal growth could be stimulated by injury. A final weakness of the method is that it is not clear what would happen if a labeled cell died and released its store of radioactive carbon; it is possible that this carbon would be taken up by the surrounding cells and used during DNA repair. If that happened, the cell that took up the carbon might appear to be much older than it actually is. Overall, this method cannot be taken as proof that neural stem cells cease to grow, but the study does suggest that neural stem cells are not abundant in the human brain.

The human brain contains a germinal region—called the subependymal (or subventricular) zone (see fig. 13)—where neural stem cells reside that can proliferate in the brain and grow in culture.[27] A layer of growing cells, presumably neural stem cells, was seen in the walls of the lateral ventricles in every one of 68 specimens studied, and these cells were present in both men and women, and in patients ranging in age from 19 to 68 years. These cells were able to differentiate into cell colonies that had neuronal properties in culture. Yet thorough microscopic examination could find no evidence that newly produced cells moved out of the subependymal zone to form the structure that is typical of migrating neuronal stem cells in both primates and rodents. Thus, there was no proof that human stem cells could move away from their home in the subependymal zone. What does this mean? Does it imply that the human brain is able to produce and incorporate new neurons only in the dentate gyrus?[28] Or does it mean that the subependymal zone supplies neural cells only in the event of some catastrophe, such as a stroke or brain injury? The answers to these questions are not known, but this highlights the great importance of studying stem cells from the human brain, since the behavior of human stem cells appears to be at least somewhat different from the behavior of stem cells in rodents or primates.

Thus, we have seen that the growth of new neurons continues in the adult human brain through adulthood, although the dentate gyrus of the hippocampus may be the only place with spontaneous creation of neurons.[29] Cells produced in the subependymal germinal zone in the walls of the lateral ventrical do not seem to replace neurons spontaneously, although there could be replacement of neurons that die due to injury. It may be possible for subependymal stem cells to produce daughter cells that then migrate over long distances to fill in where needed. Yet it is noteworthy that there have not yet been successful clinical trials of human brain repair strategies based on mouse or rat research. This may mean that even though the human brain can produce new neurons, it is simply more resistant to accepting immigrant neurons than is a rodent brain. If the human brain really is more resistant to accepting immigrant neurons into a mature neuronal network, this may represent an evolutionary adaptation. If immigrant neurons were not accepted, this might help to stabilize a neuronal population, with its accumulated experience and specific interconnections, for an entire lifetime. Yet, it is still possible that immigrant cells would be accepted in the event of stroke or brain injury. If subependymal stem cells are part of a normal cellular defense mechanism, then it may become possible to harness these cells to supplement the natural response to brain injury. Transplantation of neural stem cells to motor areas of the brain might eventually be used as a therapy for a person paralyzed as a result of stroke. Understanding the molecular and cellular mechanisms that prevent adult neural stem cells from integrating into a functional neural network would be a huge achievement with far-reaching clinical significance for patients with stroke, spinal cord injury, or even dementia.

IS NEUROGENESIS NECESSARY FOR LEARNING?

There have been several recent studies that suggest that neurogenesis is necessary for the formation of memory. Clearly, such experiments cannot be done in humans; we are therefore forced to interpret the results of animal studies, despite the caveat that neurogenesis in animals may be distinctly different from neurogenesis in humans. Yet the results from animal studies are clear-cut and seem to suggest that human neurogenesis is necessary for at least some types of learning.

In a landmark study, adult rats were given the drug BrdU, which labels dividing neural stem cells, as the animals were randomly allocated to receive one of four different behavior-training regimens.[30] Two of the regimens involved tasks thought to require hippocampal learning and two involved tasks thought to be independent of a hippocampal role in learning. Learning either of the hippocampus-dependent tasks roughly doubled the number of newly produced neuronal cells in the dentate gyrus. The increase in neurons was specific to the dentate gyrus, as there was no increase in neurons in the subependymal zone. Nor was there an increase in neurons among those rats that received training that did not involve the hippocampus. These results show that neuronal growth or survival in the dentate gyrus is enhanced only by tasks that require hippocampal learning. The most reasonable conclusion is that hippocampal learning facilitates the integration of new neurons into existing circuitry in the hippocampus, thereby insuring the survival of the new neurons.

Another research group was also able to show that exposure to an enriched environment could increase neurogenesis in the adult mouse dentate gyrus.[31] Because "environmental enrichment" is a vague term that can include expanded learning opportunities, increased social interaction, more physical activity, or even more living space, an effort was made to determine which of these factors was most critical in stimulating neurogenesis. Careful parsing of the various possibilities suggested that voluntary exercise, in the form of wheel running, was more crucial than any other factor. Mice typically make heavy use of an exercise wheel if one is provided, running between 20,000 and 40,000 revolutions per day. Such voluntary exercise acts to increase neuronal proliferation, cell survival, and resultant neurogenesis. In fact, running roughly doubles the number of surviving newborn neurons. There was also evidence that the greater the cognitive demand on the hippocampus, the greater the survival-promoting effect.[32] This is consistent with evidence that exercise improves a patient's recovery after a stroke and that it can enhance cognitive function even in people who are well. It may be noteworthy that what is called "enriched" under laboratory conditions is arguably impoverished in comparison to the natural habitat of a rodent.[33] This may mean that neurogenesis is artificially depressed by the stark environment most rodents experience in the laboratory setting. Still, new data suggest that environmental enrichment and physical activity in a laboratory result in a 5-fold increase in the rate of neurogenesis and that this is not a transient effect; the stimulation is maintained for months.[34]

These early results have been confirmed by studies showing that rodents can produce several thousand new hippocampal neurons every day. While most of these cells die within a few weeks, learning enhances the survival of new neurons. Treatments that block neurogenesis also seem to block learning. If the number of newborn neurons in rats is reduced by a treatment that blocks neurogenesis in the dentate, then hippocampal-dependent learning is also blocked;[35] and treatment-induced reduction in hippocampal neurogenesis even impairs conditioned learning.[36] Conversely, treatments that stimulate learning also stimulate neurogenesis; exposure to an environment that is rich in new smells specifically increases the number of newborn neurons in the olfactory bulb of mice.[37] Natural variation in the rate of production of new hippocampal neurons could explain natural variation in the spatial memory of different rats.[38] Mice treated with low levels of radiation to the hippocampus, a treatment that blocks neurogenesis, are unable to learn spatial tasks as well as mice with a normal hippocampus.[39] Extremely low levels of radiation to the hippocampus can impair spatial memory in mice and rats.[40] Remarkably, the life span of a new neuron is roughly equivalent to the duration of the memory it supports.[41]

PROSPECTS FOR HUMAN DISEASE TREATMENT USING NEURAL STEM CELLS

Brain injury, whether due to trauma, stroke, or degenerative brain diseases like Alzheimer's and Parkinson's, has long been thought to result in permanent loss of neurons, with no possibility for renewal or regeneration.[42] But this pessimistic viewpoint has begun to change as a result of the truly revolutionary work going on in experimental neurogenesis. Brain cells are now thought to have a surprising degree of plasticity. Neural stem cells have been proven to be relatively abundant, easy to work with, capable of forming a wide range of cell types, able to engraft in the brain with a high degree of specificity, and perhaps even capable of reestablishing neuronal connections that were damaged or lost. Stem cells may actually be attracted to sites of neuronal damage as though they are specifically recruited to areas where they are most needed. It may become possible to augment such natural regenerative processes by supplying critical growth factors or hormones. This could open up an entirely new field of medicine—regenerative neurobiology.

However, many problems remain before regenerative neurobiology can become a legitimate clinical specialty.[43] We still do not know the safest, most practical, or most effective methods for exploiting stem cell biology even though a rational research program for the future has begun to emerge. It will be important to study each disease in depth, since each disease will probably have a unique treatment strategy. It will probably be fruitless to dump neural stem cells into a diseased or injured brain in the hope that stem cells will "know what to do." For many diseases, we do not even know the site of disease, let alone what type of stem cell is needed or how that stem cell could reconstitute lost function. It could be that brain injury is fundamentally different from brain disease in that brain injury damages the environment to which a stem cell homes, whereas brain disease may kill the cell but leave the cell environment intact. If this is true, then brain disease may be more amenable to stem cell treatment than brain injury.

It will also be crucial to better understand the biology of neural stem cells before we attempt to use them therapeutically.[44] For example, it might be pointless to use patient-derived stem cells to treat Parkinson's or Huntington's disease, since the stem cell may have the same genetic lesion that produced the disease in the first place. It will be crucial to learn if a given stem cell can produce a particular cell type; if the needed cell type is not in the repertoire of a transplanted stem cell, then transplantation is unlikely to be beneficial. Inducing neural stem cells to yield neurons of the right type in the right place at the right time, and inducing those neurons to make the right connections with the right partners would be a tour de force of modern medicine.

It will also be imperative to understand how neural stem cells might affect the biology of the host into which they are introduced.[45] An injected neural stem cell could give rise to a brain tumor, since stem cells might express genes that stimulate the brain to grow or remodel when it should not do so. It is also possible that an injected stem cell would simply stimulate an immune response, which could ultimately cause rejection of the transplant and injury to the host. However, an intriguing possibility is that transplanted neural stem cells might be able to "rescue" host cells without replacing them. Stem cells may be able to alter the environment that host cells experience. Experimental work suggests that neural stem cells can reduce scarring, necrosis, inflammation, and host cell death in certain animal models of brain disease.

An authoritative review of neural stem cells in potential treatment of human brain disease has identified several targets that may be ripe for stem cell therapy.[46] The "low hanging fruit" of stem cell therapy could include diseases in which cell rescue or cell protection is critical (e.g., seizure, ischemic stroke, Parkinson's, Huntington's, amyotrophic lateral sclerosis), metabolic diseases of childhood (e.g., mucopolysaccharidosis, Canavan's, other metabolic storage diseases), and brain tumor (e.g., medulloblastoma, glioma, glioblastoma). While deferring a discussion of seizure for a later chapter, we will briefly review some experimental data related to disease treatment.

Stroke, which results from the interruption of blood supply to a part of the brain, is associated with neuronal death. Yet, in a rat model of stroke, interruption of the blood supply stimulates a marked increase in cell proliferation in the subependymal zone.[47] Newly produced cells migrate away from the germinal zone to take up residence in the stroked region, coming to resemble mature neurons. In other words, stroke stimulates the production of cells to replace lost neurons as if this is part of the normal repertoire of the rat brain. Stroke was associated with a 31-fold increase in the rate of neurogenesis! Two weeks after stroke, there were more than 6,000 newborn neurons per cubic millimeter of tissue in the damaged region of the brain. Yet only about 0.2% of dead neurons were replaced by newly produced neurons, and up to 80% of the newly produced neurons died within a few weeks. Poor survival of neural stem cells may mean that the brain "environment" is hostile to new cells in the stroked region. If a method could be found to make this environment more supportive of neuronal growth, perhaps by treatment with growth factors, the neurogenic response following stroke could potentially be amplified.[48] It is not known whether such a therapy will ever be possible in humans, although the data showing that human neural stem cells do not move away from their point of origin in the subependymal zone is perhaps not encouraging.[49]

On a completely different front, brain imaging studies in patients (using magnetic resonance imaging or MRI) have shown that the hippocampus is often reduced in volume in patients with depression or severe stress. Recent evidence suggests that clinical depression may be related to a loss of hippocampal neurons or to a depression in the rate of hippocampal neurogenesis.[50] Long-term treatment of rats with drugs that are used to treat depression in humans stimulates the rate of neurogenesis. Short-term treatment has no effect, consistent with the fact that it can take weeks for antidepressant medication to be effective. Another study suggests that the behavioral effect of antidepressant medication in mice is caused by a stimulation of hippocampal neurogenesis.[51] Whereas exploratory behavior of mice was stimulated by long-term antidepressant medication, blocking neurogenesis also blocked exploration. This is exciting in two ways: it may mean that clinical depression is caused by a reduced rate of neurogenesis and it suggests that the hippocampus plays a role in emotional response, as well as in learning and memory.

CAN ALZHEIMER'S DISEASE BE TREATED WITH NEURAL STEM CELLS?

Alzheimer's disease is the most common form of late-life dementia. Alzheimer's may already affect 20 to 30 million people worldwide, and it is increasing in prevalence as the population of elderly people expands in developed countries. This disease once epitomized the mechanistic ignorance and therapeutic nihilism surrounding brain disorders, but it is now the object of a great deal of research.[52] Alzheimer's may be a disease of synaptic failure since, in its earliest clinical stage, there is a characteristic and very pure impairment of memory. Symptoms of amnesia often occur without any other symptoms, and the signs of amnesia can be subtle and intermittent. This suggests that amnesia may be associated with alterations of synaptic efficacy in the hippocampus. Synaptic loss may be the next stage of the disease, with loss of neurons perhaps not occurring until well after symptoms begin. Synaptic density is generally decreased by about 25% in the cortex of people with mild Alzheimer's, compared to age-similar people free of dementia.

Synaptic loss explains cognitive deficits better than either the "plaques" or "tangles" that have been described in the brain of patients with end-stage Alzheimer's.[53] The protein that forms plaques or clumps in brain tissue—known as amyloid β-protein (Aβ)—may be unique to patients with Alzheimer's. Small molecules of Aβ protein gradually accumulate to form larger protein clumps, which then induce the other signs of disease, including eventual death of neurons. There have been efforts to treat Alzheimer's, either by inhibiting an immune response to these protein clumps or by stimulating the immune system to clear the protein clumps.

However, an innovative new therapy suggests that it may be possible to treat Alzheimer's with neural stem cell grafts.[54] It has been known for years that a specific subclass of neurons is lost in patients with Alzheimer's; cholinergic neurons, or those neurons that use acetylcholine as a neurotransmitter, seem to be most vulnerable. A protein known as nerve growth factor (NGF) stimulates cholinergic neuron function through mechanisms that are not presently understood. Yet it may be possible to use NGF stimulation of cholinergic neurons as therapy for Alzheimer's. There are formidable problems with such therapy: NGF does not cross the blood-brain barrier, which is the physical barrier between brain and bloodstream that is meant to insulate the brain from toxins in the bloodstream. If NGF is injected directly into the ventricles of the brain, it can cause intolerable side effects, including pain, severe weight loss, and damage to cells in the spinal cord and medulla. The issue, then, is how can NGF be delivered to the brain precisely where it is most needed, without getting into parts of the brain where it will be damaging?

In a human clinical trial of NGF as a treatment for Alzheimer's disease, NGF was provided where it could have the greatest effect on cholinergic neurons by injecting cells—genetically modified to make

NGF—directly into the brain.[55] The cells used for this purpose were harvested from skin biopsies in people with mild Alzheimer's disease. The cells were then modified and injected back into the same person. In the months after cell injection, the viability of the cell graft was assessed and the rate of cognitive decline in the subjects was characterized. Overall, results were fairly encouraging; brain scans using positron emission tomography (PET) showed a broad increase in the metabolic rate of the brain at the site of injection, suggesting that injected cells were viable or that the host brain was more active. Cognitive decline was arrested in 5 of 6 subjects, whereas cognitive loss would normally have increased over the 2 years of follow-up. An autopsy of one patient showed that injected cells were viable and that the brain was responding to stimulatory or "trophic" factors released by the transplanted cells. Nevertheless, there were problems with the study; it was an "open-label trial," not a double-blind placebo-controlled trial. Thus, both patients and physicians knew that the patient was getting an experimental drug and may have expected a benefit from treatment. Such expectation can actually create a therapeutic benefit through a process called the placebo effect. The placebo effect is likely to have been powerful in this study; if patients with only mild Alzheimer's consent to be treated in such an aggressive manner, then the patients must have been strongly motivated, perhaps by family tragedy. Such motivation could make them more vulnerable to a placebo effect. Furthermore, 2 of the 8 patients suffered substantial and damaging side effects from treatment; both were partially paralyzed during the injection procedure, and one of these patients later died. Finally, the sample size in the study was very small.

Would the treatment of Alzheimer's disease by cell injection have been more successful if neural stem cells were used rather than skin cells? This is an important question because it will have an impact on the ease of treatment in the future. Is it more feasible to work with a limited number of stem cells that have been altered to make them appropriate for all patients with Alzheimer's or will it be more practical to derive new stem cells for each individual patient, as was done in this case?[56] If there are "universal neural stem cells," this would be preferable, since such cells could become an off-the-shelf treatment, available in large quantities and with rigorous quality control. If, however, stem cells must be derived anew for each patient, there will be a tremendous cost associated with treatment. If neural stem cells must be created afresh for each patient, it will be necessary to establish each cell line in culture, prove that the cells are free of contaminants, characterize the cells to verify that they will do as they are supposed to do, and determine whether growth factors are needed to help the cells survive. Such "individuation" of therapy would be expensive and might take so long that a critical window of opportunity for treatment could conceivably be lost.

Even if neural stem cell lines are derived, many clinical problems will remain.[57] What is the best method to expand cell numbers to meet the demands of treatment? How should stem cells be given to the patient? How much of a problem will immune barriers be, and can such barriers be overcome? Is it true that stem cells can "home" to the site where they are most needed? Can stem cells be used in therapy without creating a risk of tumor growth? The greatest challenge may be that there are relatively few stem cells in the adult brain, and they may not all be equally useful in transplantation.[58] Most physicians would like to have such problems, as they would herald the advent of a new therapy for Alzheimer's. Unfortunately, physicians who hope to treat patients with stem cell therapy have an entirely different set of problems foisted on them by ambitious politicians who are serving a small but strident segment of the electorate.

THE POLITICS OF STEM CELLS

On April 10, 2002, President George W. Bush announced that he believed "all human cloning is wrong" and both reproductive cloning and research in stem cell cloning "ought to be banned." The three reasons he gave for his position were that research cloning "would require the destruction of nascent human life"; anything other than a complete ban would result in "embryo farms" that would inevitably result in "the birth of cloned babies"; and the "benefits of research cloning are highly speculative."[59] As we have seen, this last claim is completely false. But by failing to separate reproductive cloning, which virtually no one supports, from research cloning, which would benefit a huge number of patients, Bush has done an enormous harm to patients. The Parkinson's Action Network, the Juvenile Diabetes Research Foundation, the Alliance for Aging Research, the American Infertility Association, and the American Liver Foundation have all come out against the administration's position. Because the debate about cloning has been formulated as a debate about the moral status of human embryos, the issue of patient benefit has been swept aside, replaced with a specious argument about abortion politics.[60] President Bush is now in the perverse position of valuing the life of an ailing human being less than that of a tiny clump of cells no larger than the period at the end of this sentence.[61] Because of Bush's foolish stem cell policy, 400,000 blastocysts, which are produced by nuclear transplantation preparatory to in vitro fertilization, are thrown away every year in fertility clinics rather than being used for research that could benefit hundreds of thousands of people.

All of the lifesaving therapies that may derive from stem cells are stymied by crass political calculus. Will our politicians permit stem cell research to be done in the future or will they continue to bow to a vocal minority who are ill informed and hostile to science? On June 6, 2005, *Newsweek* reported that 10 states currently have banned or are considering a ban on embryonic stem cell research, another 8 states have placed severe restrictions on research, and only California and New Jersey have approved embryonic stem cell research. The result of this situation is likely to be that scientists who do stem cell research will flock to California and New Jersey, private dollars will follow in the form of grants, bequests, and patients willing to pay out-of-pocket, business will boom in these 2 states, and future patients will have to make pilgrimages to these 2 states to receive adequate therapy.

More than 25 years ago, there was a call to ban all research involving recombinant human DNA, but there can be no doubt that hundreds of thousands of people are alive today because of such research.[62] Recombinant DNA technology has been used to make many recombinant human proteins that have had an enormous impact on human health:

- insulin for diabetics;
- interferon for patients with hepatitis, cirrhosis, SARS, or various forms of cancer;
- erythropoietin for patients with myocardial infarct, chronic kidney disease, end-stage renal disease, and cancer-related anemia;
- etanercept and infliximab for arthritis, Crohn's disease, inflammatory bowel disease, and psoriasis;
- erythropoietin and human cell cytokines for cancer patients taking chemotherapy;
- insulin-like growth factor I for diabetics and children of short stature;
- brain natriuretic peptide for patients with congestive heart failure;
- various blood coagulation factors for hemophiliacs and trauma patients;
- growth hormone for AIDS patients and growth-impaired children;
- antithrombin III for patients needing cardiopulmonary bypass;
- basiliximab and antibodies against the IL-2 receptor for transplant patients;
- epoetin α for ventricular hypertrophy;
- tissue plasminogen activator for acute stroke patients;
- chorionic gonadotropin for infertile women who wish to bear children;
- hematopoietic stem cell factor for severe anemia;
- parathyroid hormone for osteoporosis;
- activated protein C for sepsis;
- neutrophil elastase protein inhibitors for chronic obstructive pulmonary disease, cystic fibrosis, and asthma;

- keratinocyte growth factor for cancer patients with treatment side effects;
- tumor necrosis factor for breast cancer patients;
- granulocyte colony stimulating factor for lung and colorectal cancer patients;
- ErB-2 (Her-2) proteins for experimental breast cancer vaccines; and
- countless human proteins used in various diagnostic tests.

If stem cell research is allowed to go forward, there can be little doubt that new treatments will emerge for Alzheimer's, Parkinson's, multiple sclerosis, heart disease, diabetes, cancer, and other diseases.[63] Yet such research is stymied by a focus on "human dignity." This focus is a smoke screen that hides another conflict: a mother's dignity and right to have authority over her own body is in direct conflict with the supposed rights of a fertilized egg. A fertilized egg (or even an early-stage embryo) is incapable of independent life under any imagined circumstances. It is irrational to imagine that a fetus has any legal rights that are independent of the mother.

But the Bush administration has proven itself to be hostile to science time after time. Sound science is denigrated as "preliminary" and consensus is dismissed as a minority view. More than 20 Nobel laureates and another 40 leading research scientists signed a letter criticizing the Bush administration for the practice of "misrepresenting and suppressing scientific knowledge."[64] A prominent cell biologist who had been asked to serve on the President's Council on Bioethics was summarily dismissed from the council for opposing the administration's view on stem cells.[65] This scientist noted that there is a growing sense in the scientific community that research is being manipulated for political ends by stacking the membership of advisory bodies and by delaying and misrepresenting the reports of such committees. When prominent scientists must fear that their work will be distorted or misused by the government, something is deeply wrong.

CHAPTER 9

CONSCIOUSNESS

For many years, consciousness was an embarrassment to neuro-science.[1] It may be an essential aspect of our brain, it may even be vital to our survival as a species, but it is also an elusive entity. It is very challenging to study patients with impaired consciousness, since they cannot cooperate in research. It is not yet possible to designate a particular part of the brain as the locus where consciousness resides. It is hard to get research funding to study something so difficult when many simpler questions still lack an answer. It is even problematic to define consciousness.

Because rigorous research in consciousness is difficult, the topic tended to appear in research journals only in the form of philosophical musings written by eminent neuroscientists nearing the end of their careers. An atmosphere developed in which it seemed that scientists might remain silent about one of the most salient features of human existence. The view of many in the neuroscience community was expressed by the Nobel laureate John Eccles in 1964:

> We can, in principle, explain all input-output performance in terms
> of activity of neuronal circuits; and consequently, consciousness

seems to be absolutely un-necessary! [A]s neurophysiologists we simply have no use for consciousness in our attempts to explain how the nervous system works.[2]

But this viewpoint has changed. In July 2005, *Science* compiled a list of the 100 most interesting and compelling questions for the next century of research.[3] Second on that list was, "What is the biological basis of consciousness?" This question followed the question, "What is the Universe made of?"

There has been a flurry of hypotheses relating to consciousness, some so vague and arcane as to highlight the fact that all of them are exploratory and tentative approaches to a very difficult problem. Part of the problem may be that consciousness is so hard to define and many studies of consciousness never adequately define it. As an example, one scientist (who should remain nameless) wrote that "one may state that neural states cause states of consciousness only in the broad sense of explanatory causality with regard to the observation of a law-like psycho-neural relation." This is clear evidence of a state of confusion.

WHAT IS CONSCIOUSNESS?

In a solipsistic sense, consciousness subsumes time and the physical dimensions; without consciousness, we could not be aware of the four dimensions. Consciousness could be thought of as the "fifth dimension" of the physical world. In his *Nicomachean Ethics,* Aristotle wrote, "To be conscious that we are perceiving or thinking is to be conscious of our own existence." Written thousands of years ago, this definition remains pertinent, though not as detailed as we would like. The *American Heritage Dictionary* defines consciousness as "a critical awareness of one's own identity and situation," which seems true, but also vague. In contrast, *Stedman's Medical Dictionary* defines consciousness as "the state of being aware, or perceiving physical facts or mental concepts; a state of general wakefulness and responsiveness to the environment."

This definition invokes both attention and awareness as elements of consciousness, which seems apt, yet still incomplete.

We propose that consciousness is by nature multifaceted. Clearly, attention is needed, as attention is the arbiter of awareness, the gatekeeper to the senses. Perception must then arise from attention; perception is a key to the kingdom of the mind. Though attention and perception enable us to be sensitive to our environment, they don't enable us to be conscious of it, unless we also have a template to which new perceptions can be compared. Thus, memory is an integral part of consciousness. The final and most ineffable element of consciousness is awareness; this can be understood as one part of the brain monitoring another part, watching it work. Thus, we propose that consciousness arises only when a subject shows a combination of attention, perception, memory, and awareness.

Even with this definition, it may not be possible to determine "What is consciousness?" in satisfying detail. Francis Crick, the Nobel Prize winner celebrated for elucidating the structure of DNA, pointed out that this may be the wrong formulation for the question. If one were to find a mysterious machine in one's garden shed, the first question might be "What is it?" Yet this question is fundamentally unanswerable without penetrating the mind of the inventor. Instead, it may be better to ask of the mysterious machine, "What does it do?" since that question can be answered. A short answer to that question might be that consciousness controls behavior, which would seem to conflate consciousness with mind, since mind is defined as what interposes between brain and behavior. Certainly, consciousness is an element of mind and may have a decisive role in determining what mind can do, but mind is more than simply consciousness. Therefore, a better formulation of the consciousness question is, "What neural events correlate with consciousness?" Such correlates can be thought of as the minimal set of neuronal actions that give rise to consciousness.[4] Because we have defined consciousness as a product of attention, perception, memory, and awareness, we should therefore examine the neural correlates of each of these phenomena.

ATTENTION: THE ARBITER OF AWARENESS

Visual scenes typically contain many stimuli, with each stimulus vying for control over behavior; attention biases this competition in favor of the most relevant stimulus. The ability to focus attention is critical and impairment of attention is potentially fatal, whether for a modern human driving a car or an early human hunting and being hunted on the veldt. Impaired attention can result from brain injury and certain brain diseases that impair consciousness (e.g., multiple sclerosis, Parkinson's, encephalitis, narcolepsy). Attentional problems are common if the midbrain is injured, or if the connection between the thalamus and forebrain is weakened by injury or disease.[5] Many patients with severe fatigue find that it interferes with their ability to focus and devote attention to a subject, with the result that consciousness is impaired. Patients with narcolepsy suffer an irresistible daytime sleepiness and may also have a range of unusual symptoms, including cataplexy (abrupt loss of muscle tone induced by laughter or emotion), vivid hallucinations while falling asleep, disturbed sleep, and sleep paralysis. Narcoleptic patients suffer brief but unpredictable impairments of consciousness, such that it can be unsafe for them to drive.

Attention acts to bring unruly neurons into synchrony so that neurons firing in a rather haphazard manner become much more regimented in their behavior.[6] As we shall see, synchronization of neurons is thought to play a critical role in consciousness. Donald Hebb postulated many years ago that a single neuron could be involved in forming many different cell assemblies over time. Scientists have hypothesized that neurons that form a single Hebbian cell assembly— the coalition of cells hypothesized to hold a memory—should give evidence of their relationship to each other by firing in synchrony. To test this hypothesis, neuronal activity was measured in the motor cortex of macaque monkeys trained to perform a specific motor task. The task comprised two visual cues presented in sequence; one cue indicated the target to which the monkey should move his hand and was also a signal to "get ready," while the second cue, presented a few moments later, was the "go" signal. The time interval between the first and second cue was variable, so that the "go" signal could appear up to 1.5 seconds after the "get ready" signal. The time delay between cues varied randomly from one trial to the next so that the monkey did not know what to expect but had to await the "go" signal. Scientists found that neurons in the motor area of the monkey increased their synchrony when the animal expected the "go" signal to appear. This means that as the monkey was preparing to make a hand movement, motor neurons controlling that movement became synchronized. The degree of synchronization was really quite impressive: neurons in the same cell assembly could fire within as little as 1 to 2 milliseconds of one another. Some neurons did not synchronize at all, suggesting that they were not in the same cell assembly, and the neurons that did synchronize with each other changed over time as new cell assemblies formed and reformed.

Attention has also been found to modulate synchronization of neurons in the somatosensory cortex—where sensing of physical texture occurs.[7] Monkeys were trained to switch attention between a visual task and a tactile discrimination task. Neurons in the somatosensory cortex are specialized for the tactile task, so these neurons responded more intensely and became more synchronized during that task. When the monkey switched attention, 78% of the neurons under study changed their firing rate, and there was a high degree of synchronization in these neurons. The degree of synchrony was affected by the monkey's state of attention, with greater synchronization during the most difficult task. There is also fresh evidence that attention can synchronize the firing rate of neurons in the visual system.[8] If neurons fire together, this may increase the overall strength of a neuronal response and help to control behavior. Neural synchronization thus seems to be a reliable indicator of increased attention, and so may be a neural correlate of consciousness.

But what evidence is there that attention is needed for human consciousness? Clearly, it is not possible to do experiments in humans that mirror those done in macaque monkeys. Yet we know that attention is necessary for the consolidation of memory,[9] and we have postulated that memory is a component of consciousness. We also know that human infants can differ in the way they allocate attention to a visual task and that these attentional differences predict visual memory. More mature infants typically shift their gaze often and tend not to fixate on a particular visual feature, whereas less mature infants tend to fixate on a specific visual feature for long periods of time. Infants born prematurely, prenatally exposed to alcohol, or fed formula with a low concentration of omega-3 fatty acids all tend to fixate attention for longer intervals on a visual stimulus and also tend to be impaired in their ability to form memories.

Attention is crucial for memory in humans. If memory is indeed required for consciousness, then attention must also be necessary for human consciousness. Activation of specific clusters of neurons is associated with increased attention. If a person is unable to recruit and synchronize neurons, he will be unable to modulate attention, to form memories, and to remain fully conscious of his surroundings.

PERCEPTION: THE KEYS TO THE KINGDOM

To understand perception, we will use visual perception as a paradigm of all perception and we will use the word "perception" in the broadest sense possible. Thus, by visual perception we do not simply mean the sensitivity of the retina to light; rather, we mean the sensitivity of the visual apparatus, including the visual cortex of the brain, to specific features of the visual environment. Perception thus has an integrative function, combining the parts of a perception into a synergistic whole. In other words, how do we know that a face is a face and not just a random collection of bloblike features? How do we know that a particular face is the face of a loved one and not the face of a stranger? Briefly, we will consider how it happens that the way things look can enable us to make reliable judgments about the way things are.

Perhaps the first person to consider perception in detail was George

Berkeley, an eighteenth-century Irish philosopher who contended that decisions about the size, distance, and position of objects are not made on the basis of a knowledge of optics, since most people have no such knowledge.[10] Instead, we interpret the visual environment using rules learned through extensive visual experience. Yet these rules are ultimately rather arbitrary; my "red" has the same name as your "red," but we cannot be sure that both of us really see the same color. Here, Berkeley may have gone further than current readers would be comfortable with. He contended that "language and knowledge are all about ideas, words stand for nothing else." Berkeley believed that the world was composed entirely of ideas, that the apparent unity of experience for different observers is an act of God, that the objects of vision are purely visual ideas in the mind, and that the material world does not exist, except in the mind. Although it is tempting to dismiss his philosophy as a sophisticated form of solipsism, it is worth noting that we still cannot be sure that your "red" is the same as my "red."

Perception forms the keys to the kingdom: nothing can enter awareness but through the senses. Perception is a filter between the individual and the environment, a barrier against the world. Still, perception also compensates for shortcomings in our ability to sense the world. Though the human retina has a blind spot where the optic nerve leaves the eye, though it has essentially no color receptors outside the central fovea, though it lacks dense receptors for light and dark within the fovea, and though the retinal image is blurred by incessant movements of the eye, we still perceive the world with remarkable acuity.[11]

Clever experiments have begun to reveal the neural correlates of perception. Binocular rivalry is an experimental approach in which a subject—whether monkey or human—has different images projected simultaneously onto the retina of each eye.[12] The observer does not see the two images superimposed, except very briefly, because this makes no sense to the brain. Instead, the observer sees first one image, then the other, then the first again, alternating every few seconds. Essentially, the images are in rivalry with one another. A key question becomes: Do visual neurons in the brain monitor the image (which is constant) or the perception of the image (which is rapidly changing)? Studies of monkeys show that neurons in the visual cortex, at a point close to the sensory input from the eyes, monitor the image and fire constantly. In contrast, neurons at the top of the hierarchy—in that part of the brain that integrates visual stimuli—fire discontinuously, as if they are monitoring the perception. Up to 95% of the visual integration neurons fire when the monkey perceives one stimulus and are silent as it perceives the other stimulus, even though both stimuli are always present. What this means is that the activity of integrative neurons conforms to the perception, not to the image—these neurons form a neural correlate of consciousness.

Neurons across broad regions of the human brain can synchronize during a binocular rivalry task.[13] This synchrony is similar to what happens in a task that requires focused attention, but there are interesting subtleties. An electrical recording method, called magnetoencephalography (MEG), uses up to 148 electrodes pressed to the scalp, to record electrical activity over broad areas of the human brain. Competing stimuli were projected onto each retina, with each image flickering in intensity at a different frequency. Since neurons respond to the rate of flickering, "frequency tags" could be detected at each sensor. Thus, it was possible to know to which stimulus a given brain region was responding. The power detected at each electrode varied as a function of which perception was dominant. Neurons in a specific brain location, which were presumably all involved in one perception, became synchronized. There could also be synchrony between widely separated brain regions, which means that these regions were recruited to fire together. Conscious perception was associated with broad changes in the firing rate of neurons, with synchronization extended far beyond the visual cortex.

Thus, certain neurons in the human brain respond not to a stimulus but to a perception.[14] Studies of brain surgery patients have shown that the activity of individual neurons track stimuli with astonishing specificity. In order to surgically treat epilepsy, surgeons identify the seizure focus by implanting electrodes in the brain for up to two weeks. After implantation, these electrodes continually record information until the seizure focus is identified. During the period when electrodes are implanted, patients can be tested to see whether visual training has an effect on the activity of those neurons involved in processing visual information. Researchers have found that neurons often respond not to the visual environment alone but to cognitive features of the visual environment. For example, a certain neuron may respond to a picture of an animal but not to pictures of cars or human faces. A neuron that responds to animals may respond to a range of different animals, including tigers, dolphins, elk, rabbits, and even scorpions. Some visual neurons will even respond when a subject is asked to imagine an animal in the absence of retinal input. Among neurons that fired in response to an image of a dolphin, roughly 85% also responded when the subject was asked to imagine a dolphin. It is as if some neurons are tuned to fire for the idea and not the image of "dolphin."

Neurons can be tuned to ideas or "percepts" with an astonishing degree of specificity.[15] In one patient awaiting epilepsy surgery, the activity of neurons in the amygdala was recorded as the subject was shown images of then president Clinton. A specific neuron was identified that could respond to a line drawing of Clinton's face and to a photograph of Clinton in a crowded room. Yet this neuron did not respond to 50 other images, including photographs of three other presidents. Does this mean that scientists had found a "President Clinton neuron," which fires only in response to the perception of Clinton? While this seems too astonishing to believe, the experiment is hard to interpret in any other way.

Can there be awareness without visual perception? Can the blind "see"? Blindsight is the ability of some patients who are clinically blind (due to destruction of the visual cortex in the brain rather than to destruction of the retina) to detect a visual stimulus.[16] This is a highly controversial subject, in part because it would seem to be impossible, in part because it may be vulnerable to hoax by an unscrupulous person seeking attention. Nevertheless, there is a

growing consensus that blindsight is real and that there can be awareness without perception. Patient GY is a 46-year-old man who sustained a head injury in a car accident at age 8. Hemorrhage after the
accident caused a unilateral degeneration of the left occipital lobe of
his brain, as confirmed by magnetic resonance imaging (MRI). GY is
thus clinically blind in that part of the visual field that would report
to the left occipital lobe. Because of the unusual connections between
eye and brain in every human being, the visual loss in this subject was
in the right visual field of both eyes. This condition of blindness in
half of both visual fields is given the alliterative name of "homony-
mous hemianopsia."

Patient GY has an intact retina, so stimuli to the right side of the
retina are passed on to the brain. Nevertheless, because the brain
region that would ordinarily process data from the right visual field
was destroyed, most of this visual information is never processed. Yet
GY can perform some tasks that would seem to require vision. The
patient was asked to fixate visually on something in the center of the
visual field. Then images were flashed randomly to the right and left
side of the retina and the subject was asked to note precisely when an
object was presented to the blind side of the visual field. It was proven
that the image fell on the retina where the subject claimed he could
not "see." Nevertheless, GY was able to report, at levels much better
than random chance, when an object was presented to his blind side.
Does this mean that the subject was lying or perhaps that he had hys-
terical blindness that precluded him from "seeing" what he could see?
This is, of course, the crux of the controversy. However, sophisticated
scientists believe that due to the damage to his visual cortex GY really
is blind, but that he can detect visual signals in ways that do not reach
consciousness. There may be many pathways that a nerve impulse can
follow from eye to brain, so it is possible that information can traverse
a part of the brain that is inaccessible to conscious perception. In short,
unperceived stimuli can potentially have an effect on behavior. Nevertheless, the fact that some patients are visually aware without perception cannot be taken as proof that consciousness can happen without

perception. Patient GY has senses other than vision, and these are all normal, just as most blind people have otherwise normal sensoria.

Sensory information undergoes extensive elaboration and alteration as it is incorporated into the texture of cognition. For a subject to be aware of a visual stimulus, there has to be a group of neurons that fire in response to that object.[17] We are probably unaware of activity among neurons in the primary visual cortex; people cannot tell which eye they are seeing with unless they close or obstruct one eye. Yet we eventually become aware of activity among neurons as a result of "higher" parts of the brain, which integrate visual stimuli into a coherent whole.[18]

MEMORY: THE SACRED STOREHOUSE

It may be of interest to describe a familiar situation in which there can be memory without awareness or consciousness.[19] Imagine that you are reading at home and you suddenly become aware that the house is silent; perhaps the refrigerator motor has switched off. Thinking back, you may come to believe that you had been hearing the noise of the motor all along but that the noise never reached conscious perception. Still, you cannot be sure, because you cannot explicitly remember the motor being on or even, for that matter, the motor switching off. Nevertheless, had you spoken to someone before, you might have spoken in a louder voice to compensate for the noise of the refrigerator without consciously being aware that you were doing so. It is the free access to information in memory that defines consciousness; this is why simple awareness of the environment has been called "the remembered present."[20]

A patient with epilepsy can occasionally perform a complicated behavior during a partial complex seizure without having any recollection of that behavior.[21] Such actions are a form of automatism. One epileptic patient periodically had a seizure while riding his bicycle to work. This patient might "awake" to find himself at home without ever having reached work, without any recollection of the intervening time, without any idea of how he came to be at home again. Apparently, in the midst of a seizure, the patient was able to turn his bicycle around and ride home safely through traffic without any memory after the fact. This state of partial awareness, without memory or consciousness, has been called "zombie mode." It is possible that zombie mode is involved in much moment-to-moment behavior—it requires no consciousness to reach for a door handle or to grasp an object. Zombie mode may simply be a behavioral module that is induced during a conscious behavior but that does not need conscious control.

Is it possible to learn without awareness, to be unconscious of memory? While there are no familiar examples of unconscious memory, intriguing evidence suggests that memory without awareness is not only possible but relatively common.[22] Such "unconscious memory" is called priming, which is defined as an increase in the speed or accuracy of processing a stimulus because of prior exposure to a similar stimulus. Priming is an example of implicit memory in that prior experience has an impact on behavior even if the prior experience is never explicitly remembered. If patients with amnesia are asked to learn a list of words and then are shown a list of word "stems" (the first three letters of the learned words), they do very poorly when asked to use the word stems to recall the original list of words. Yet if amnesia patients are shown a list of word stems and asked to complete the stems with the first word that comes to mind, they do as well as healthy people in recalling the list of words. This shows that amnesia patients have intact priming even if their explicit memory is poor. Among healthy people, priming can cause a huge change in response, even if the word stem is not explicitly remembered. In a study that offered 3-letter word stems, each of which could be completed in at least 10 different ways, people tended to complete the stem in the same way as the primed word 70% of the time. In contrast, without priming, the stem was completed with a desired word less than 10% of the time. Interestingly, priming can be associated with a decrease in neural activity, as if the brain found it easier or more efficient to fall into familiar patterns.

Awareness: Master of the Servant Brain

The sprinter starts to run before he "hears" the gun, before the sound of the starting pistol has entered conscious awareness. That there are relatively few false starts at the highest levels of competition shows that runners are responding to the gun appropriately, although their response is visceral rather than cognitive. Thus, awareness is not necessary for human behavior to occur, especially if a behavior is endlessly trained to become automatic. One consequence of learning is the release of behavioral performance from the control of attention, the liberation of response from awareness so that a task can become fully automated or "unconscious."

There is substantial support for the idea that simple behavior doesn't require awareness. Classical conditioning—for instance, the eye-blink response—has been studied in depth to determine the extent to which learning can be acquired without awareness.[23] The eye-blink response occurs when a conditioned stimulus, such as a sound or tone, is presented before or coincident with an unconditioned stimulus, such as a mild puff of air to the eye. In a simple experiment like this, the subject quickly learns to blink in response to the tone, thereby establishing a Pavlovian or automatic reflex. In experiments with human subjects, it is often true that people become aware of the reflex at about the same time that the reflex is established, so the reflex may be dependent upon awareness. But the simple eye-blink experiment can easily be made complex in order to probe whether awareness is required for conditioning. In one experiment, subjects were asked to watch a movie and told that they would later be tested on their memory of the movie, but this was not the actual goal of testing. While subjects were distracted by the movie, they also underwent a conditioning regimen in which a tone was paired with an air puff. Afterward, the subjects were asked to fill out a questionnaire that tested their understanding of the experiment they had just participated in. It was found that awareness was not required for the establishment of a conditioned response; subjects without any understanding of the experiment had established an eye-blink response that was just as robust as subjects who understood the experiment.

Human learning can happen in a trial-and-error fashion, without awareness and independent of declarative memory.[24] Two men, each of whom had extensive damage to the hippocampus, were studied; both men were so badly amnesic that they lacked any recollection of what had happened to them during the study. Nevertheless, these men were able to learn a complex task that required them to discern a desired object from among 8 pairs of objects. Healthy subjects usually master this task quickly and can easily describe the rules that govern rewards, yet the amnesic men were unable to describe the task. Nevertheless, over the course of many weeks of practice, the men learned to perform the task with an accuracy comparable to healthy subjects. However, amnesic patients took 28 to 36 trials to reach the same level of competence that a healthy subject could achieve in only 3 trials.

If behavior can occur without awareness, this suggests that behavior cannot be taken as proof of consciousness. This is a compelling conclusion, given that simple behavior has been offered as proof of consciousness in certain patients—like Terri Schiavo—who were in a persistent vegetative state.

If we could define the neural correlates of consciousness, this might make it possible to have a diagnostic tool to assess the level of a patient's coma. In an ingenious experiment, thirst was induced in human subjects by the injection of a strong saline solution into the bloodstream.[25] Subjects were evaluated by functional magnetic resonance imaging (fMRI) and positron emission tomography (PET) as they became conscious of their sense of thirst. Since both methods can highlight regions of the brain that are activated by a stimulus, both can be used to establish neural correlates of the perception of thirst. Both the fMRI and the PET studies showed that multiple areas of the brain are activated as a person first becomes aware of thirst. Such activations are consistent with sensing thirst and planning a behavioral response. But it may be crucial that multiple brain regions were activated as the subject became conscious of thirst, suggesting that awareness and consciousness itself may require the involvement of a large number of brain regions.

Awareness can perhaps best be understood as one part of the brain monitoring another part, watching it work. Neuroscientists concerned with artificial intelligence have designed neural networks to learn more about how the brain categorizes things. To validate that a network has developed appropriate rules, the output of the neural network must be assessed by an expert.[26] If the brain works in the same way as a neural network, then a part of the brain must be acting as an "expert." That part of the brain must be aware of the output, aware of the template to which output can be compared, and able to make a judgment as to whether an output matches the template to an acceptable degree. This is why awareness may be necessary for conscious behavior; awareness serves the function of the expert.

SYNCHRONY AND OSCILLATION: NEURAL CORRELATES OF CONSCIOUSNESS?

If it is true that consciousness can arise only when a subject simultaneously is able to attend to and perceive stimuli, remember the stimulus in relation to a template, and be aware of the whole process, this suggests that different parts of the brain are activated simultaneously when a person is conscious. If various parts of the brain are recruited to work together, one would necessarily expect there to be some degree of synchrony between the parts.

If a person were fleeing from a predator, what neurons might be involved? Sensory neurons embody features of the environment, motor neurons embody actions committed or contemplated, prefrontal neurons embody plans being made, and the activity of all of these neurons must be coordinated for an effective escape response. All these different neurons with their diverse functions must cooperate to generate behavior; it would be of no survival value to be aware of a predator but unable to respond or to respond without a plan of action. The fact that different brain regions are required for each component of a response suggests that each brain region must subsume itself to a single overarching goal. In fact, survival may be contingent upon the ability to harness the parts to generate a coherent response. Consciousness may therefore require an increase in coherence between different brain regions.

The strength of individual synapses can change over time because of synaptic plasticity. This enables the brain to learn, but it also may make the brain somewhat vulnerable to instability. If synaptic strength is not modulated in some way across many synapses, the level of activity in an individual circuit could potentially increase or decrease in an uncontrolled manner.[27] However, the brain has developed processes that appear to regulate the overall activity in neural circuits. These regulatory processes adjust the threshold of individual neurons so that in an active circuit it becomes more difficult to induce long-term potentiation (LTP). It is noteworthy that the process of stabilizing synapses could ironically incite neural circuits to compete with one another. Competition arises because a neuron will, over time, become less responsive to some inputs and more responsive to others such that the former input will lose out to the latter. There is a prediction that as synaptic weight within groups of neurons gradually increases to build a cell assembly, stable persistent states will appear abruptly rather than emerging gradually.[28] A correlate of this prediction could be that different brain regions may be recruited suddenly to become part of a network of brain regions involved in maintaining consciousness.

In computer simulations, as neurons share more and more inputs, they tend to fire with greater synchrony, and there are larger and more regular oscillations over time.[29] Neurons that fire in synchrony may be more likely to have an effect, in terms of generating a certain behavior. Yet correlation of activity among neurons is not necessarily meaningful: two neurons can fire because both are stimulated by another set of neurons or because they are directly connected to one another. Synchrony between visual neurons could reflect nothing more than an overlap in their visual fields. Similarly, neuronal oscillations could arise due to an intrinsic property of neurons as well as from one neuron directly exciting another. Yet reproducible *changes* in how the activity of one neuron correlates with the activity of another neuron may reflect functional connection. This means that the pattern of change in a neural circuit can be a probe of how that circuit works.

Oscillations or modulations in synchrony between populations of neurons have been linked to changes in attention, expectation, and binocular rivalry.[30] Such oscillations can link neurons into cell assemblies, can facilitate synaptic plasticity, and may even be involved in learning or LTP. It has been argued that oscillations are important in maintaining consciousness, or that cortical oscillations are responsible for binding various parts of a perception into a coherent whole. Computer simulation suggests that complex systems tend to find a frequency for self-sustained oscillations, even if the systems are not initially correlated. This process remains mysterious, and it is not yet known whether spontaneous synchronization can occur in the human brain.

As various neurons synchronize with one another, they form complex electrical rhythms that can be detected by a method called EEG or electroencephalography. But these patterns are very hard to interpret.[31] Taking an EEG of the functioning brain is rather like putting a huge microphone over a football stadium and recording crowd noises during a game; in the absence of a clear understanding of what is happening on the field, the noises alone would be very hard to interpret. Yet the complex composite rhythm of the brain can be broken down into component rhythms, and these rhythms can be studied separately from one another. Patterns can be picked out of the noise, and if such patterns are correlated with an independent measure of activity, the complex EEG can be understood. Each electrode of the EEG detects a signal from a different population of neurons, so it should come as no surprise that many rhythms are detected. Alpha rhythms, which occur at a rate of 10 Hertz or 10 cycles per second, occur across broad regions of the brain. Gamma rhythms, which occur at a rate of 25 to 90 Hertz, occur in working memory tasks and may mean that neurons are forming reverberating circuits. Mu rhythms, which occur at yet another frequency, may be present in the motor cortex that controls the hand but abruptly cease as the hand completes a motion. EEG may eventually enable scientists to understand the flow of information from one brain region to another as attention is modulated, as perception occurs, as mental templates are checked, and as the brain maintains awareness.

Consciousness may be equivalent to neuronal synchronization.[32] If so, does anything that increases neural synchronization also increase consciousness? Recent evidence shows that Buddhist monks are able to synchronize large brain areas as they meditate. This is consistent with the focused attention and increased consciousness that practitioners claim as a benefit of meditation.[33] Equating synchronization to consciousness makes a fairly simple hypothesis, with the obvious appeal that it would be easy to test and easy to prove false.

SPLIT BRAIN EXPERIMENTS AND THE NATURE OF CONSCIOUSNESS

How can a brain be aware of itself? Some insight into this question can be gained from a fascinating series of experiments with patients who underwent a type of brain surgery that left them with a "split brain." About 40 years ago, a patient was described who had suffered *grand mal* seizures for at least 15 years after sustaining a head wound in World War II.[34] These seizures had been resistant to treatment and varied in frequency from once a week up to 10 times a day. In an effort to surgically control epilepsy, a neurosurgeon severed the corpus callosum—a thick bundle of white matter that forms the main connection between right and left hemispheres in the human brain (see fig. 5, chap. 4). Radical as this surgery may seem, this patient enjoyed a substantial reduction in severity and frequency of seizures, with no change in either intellect or temperament. Nevertheless, there were strange changes in ability that only became evident with careful testing. If the patient was blindfolded and given objects to handle, such as a pencil, a cigarette, a ring, or a hat, the subject could name the objects he touched with his right hand but not the objects he touched with his left hand. This curious inability to find words arises from the fact that language is controlled by the left side of the brain in most people. The left side of the brain controls the right side of the body, so objects in the right hand could be named but objects in the left hand were sensed by a part of the brain that lacked any language ability. It was as if this patient had two brains in a single skull, with one of the brains being mute.

The bizarre changes shown by this patient were not limited to language.[35] He was unable to draw objects that appeared in his left visual field and unable to read text that fell on the left side of his retina. When tapped lightly on either hand, the patient could replicate the number of taps with that hand but not with the hand on the opposite side. Simple jigsaw puzzles could be put together correctly with either hand alone but not when cooperation between both hands was required. Whenever stimulus and response were both confined to the same brain hemisphere, performance was normal, but if the patient was required to integrate the performance of the two hemispheres, he found it to be impossible. If he learned a motor skill with one hand, there was no carryover to the opposite hand; it was as if the opposite hand had to learn the skill from nothing. When he was at home, his wife would sometimes see him pick up the evening paper with his right hand and put it down abruptly with the left hand, only to pick it up again with the right hand. In short, each hemisphere seemed to be completely unaware of what was happening in the opposite hemisphere.

We now know that each cerebral hemisphere has its own set of skill and abilities, with the left hemisphere specialized for language and problem solving and the right hemisphere devoted to more prosaic tasks like facial recognition and focusing attention.[36] We normally have the subjective sense that our thoughts and actions are the result of a single mind, but this is only possible if all the normal neuronal pathways are intact. In an unbroken brain, the left hemisphere houses what has been called an "interpreter" whose

> job is to interpret our responses . . . to what we encounter in our environment. The interpreter sustains a running narrative of our actions, emotions, thoughts, and dreams. The interpreter is the glue

that keeps our story unified and creates our sense of being a coherent, rational agent. To our bag of individual instincts, it brings theories about our life. These narratives of our past behavior sink into our awareness and give us an autobiography.[37]

Do Dogs Show Consciousness?

We have defined consciousness as including attention, perception, memory, and awareness, and dogs certainly show at least the first three of these attributes. But do dogs show awareness? And, if so, should they be considered conscious? Most scientists are reluctant to consider that animals have consciousness, perhaps because it is essentially impossible to prove conclusively that animals have self-awareness.[38] Yet it is equally difficult to prove unequivocally that other people have self-awareness, so this seems to be a trivial objection. What we know is that if there are structures in the human brain that are required for consciousness, there are at least rudiments of all these same structures in a dog brain. Dogs show versatile behavior that is suggestive of thinking, they are aware of objects and events, they understand social relationships, they have clear memories, and their ability to anticipate when they are likely to be fed indicates some sense of time. Even though dogs are clearly not as intelligent as human beings, human-level intelligence is not required for consciousness.

Dogs show a high degree of awareness specifically of humans; in fact, dogs are more aware of human cues than are higher primates.[39] In a common experiment to probe awareness, a scientist hides a piece of food in one of two opaque containers without showing the test subject where the food is hidden. Before giving the subject a choice, the scientist then gives a clue as to food location by looking at or pointing to the right container. A chimpanzee, given such cues, will perform slightly better than chance in finding the food, whereas a dog will usually make the correct choice. This is not simply a matter of the dog having a better sense of smell, since results are the same if smell is eliminated as a possible clue. Dogs are more skillful in using human cues than are wolves, and puppies only a few weeks old can use human cues even if they have been raised with little human contact. We can thus conclude that dogs did not inherit their social skills from wolves and do not require human contact for manifestation of these skills; instead, it seems that dogs have an inherent ability to understand humans even without training. Clearly, this is a skill that has been selected for over the 100,000-year history of dog domestication, yet it would be wrong to conclude that such social skills are trivial.

In order to follow social cues, dogs must understand the viewpoint of humans. The ability to understand the viewpoint of another individual is called "theory of mind" by psychologists, since one individual must develop a theory that another individual has a similar mind. If there were not an inherent similarity in thought processes between individuals, it would be impossible to follow social cues.[40] In fact, dogs are not only able to understand the human mind, they are also able to make humans understand the dog mind. Dogs can show their masters where an object has been hidden using only their gaze, showing that the dog has a "theory of mind" that relates to people. Thus, the "Lassie effect" is not at all trivial.

Another recent study has documented that dogs are able to learn human language in a way that is reminiscent of how an infant learns.[41] The rate at which infants learn language is astounding; by the age of 2 years, typical English-speaking children incorporate roughly 10 new words per day into their vocabulary until they have a total vocabulary of about 60,000 words by the time they graduate from high school. This would not be possible without specific strategies to learn the meaning of words. As children acquire language, they form a tentative hypothesis about what a new word means after just a single exposure in a process that has been called "fast mapping." In essence, fast mapping is the process of quickly attaching a meaning to a word.

Scientists tested the ability of a border collie named "Rico" to learn new words by fast mapping.[42] Rico was reported by his master to know more than 200 words, most of which were the names of toys that Rico liked to fetch. These 200 objects were split into 20 sets of 10 different items and Rico was asked to fetch one of the 10 objects from each set by name. In 40 trials, Rico was able to fetch the right object 37 times, even though his master was not there to help. Thus, Rico's vocabulary was comparable to language-trained apes, dolphins, sea lions, and parrots. To make this task yet more difficult, a novel object was mixed with 7 familiar objects and Rico was asked to fetch the novel object by name. In order to identify this object, Rico had to recognize that the new word did not refer to any of the familiar 7 objects. Thus, Rico was forced to generate a hypothesis that the new word referred to the new item, which is essentially a fast-mapping task. Rico was correct in 7 of 10 trials, a level far beyond what could be expected by chance alone. Four weeks after the first training session, Rico's retention of the novel names was tested and he was able to remember correctly on half the trials. What this means is that Rico understood that objects have names, that new names must relate to new items, and that new names should be remembered. This astounding level of linguistic ability had never before been demonstrated in an animal.

Clearly, there has been a selective pressure for dogs to understand the social cues of humans, since dogs have depended upon humans for food and companionship for at least 100,000 years. Clearly, this relationship would have been facilitated if dogs and humans shared certain abilities. When hunting together, it would have been of great selective advantage if dogs and humans could attend, perceive, remember, and be aware of the same environmental cues. No doubt, a dog does not have the same level of attention, perception, memory, and awareness as a human, but a dog does have the rudiments. Perhaps consciousness in dogs is an unanticipated outcome of the selection pressure to be a hunting partner for humans. Though it is unlikely that consciousness is limited to dogs, it is not yet known whether all animals share some level of consciousness.

EMERGENT COMPLEXITY AND CONSCIOUSNESS

Does consciousness perhaps provide a general-purpose mode of perception,[43] which can cope with complicated situations while also enabling the brain to handle simple things automatically? Clearly, there would be a survival advantage if an organism could operate in two different modes simultaneously. One mode might provide a limited number of rapid, stereotyped, unconscious responses appropriate for recurrent situations while another mode might provide slower, more varied, conscious responses appropriate for novel situations. One could imagine that different parts of the brain might engage in developing alternative responses to a given situation. Thus, an unconscious competition could arise between different brain regions as each region strives to gain control of the behavioral output. This could be a neuronal analogy for what it means to be "of two minds."

At the core of many current hypotheses about consciousness is the idea that consciousness results from competing coalitions of neurons.[44] There may be low-level cell assemblies that detect simple correlations among sensory stimuli and other cell assemblies that, in a hierarchical fashion, detect correlations among correlations across broad regions of the brain. This could potentially give rise to the neuronal oscillations that have been observed in the human brain. The enormous number of neurons in the human brain, their ability to form temporary coalitions, and the ability of these coalitions to learn or to induce action makes the human brain arguably the most complex object in the universe.

It is possible that consciousness is simply an inadvertent function of the parallel processing that is characteristic of the human brain. We are used to evaluating a problem in two or more ways at the same time to determine what would be the better solution—this is a form of parallel processing. A brain that is capable of such parallelism could be aware that it is working to solve a problem, could be aware that there are alternative approaches to that problem, and could even be able to allocate effort to each of the different approaches to a problem. This type of parallel thinking could explain the origin of consciousness, since a brain capable of parallelism might also be capable of holding a specific attitude about a problem or about life in general. If an attitude is defined as a viewpoint held in the background while the brain works on a problem in the foreground, this begins to seem like a simple analog of personality. If such parallelism is equivalent to consciousness, perhaps consciousness is simply an awareness of self.

But what is the function of consciousness? Is consciousness even necessary? Would we be recognizably human without it? Is consciousness a precondition for the adaptations that set us apart from other organisms? Is language or social adaptation even possible without consciousness? Or is consciousness needed only if language and sociality evolve to the level of sophistication seen in humans? Could we make do and be human with less consciousness? Or is consciousness the minimal qualification for the human condition? In our struggle to understand consciousness, we must confront the possibility that the brain may be unable to understand the mind and that we may always be a bit mystified by our own abilities and capacities.

It seems likely that consciousness is a correlate of complexity, an emergent property of the brain. The nineteenth-century American philosopher Charles Sanders Peirce said,

> Our whole past experience is continually in our consciousness, though most of it sunk to a great depth of dimness. I think of consciousness as a bottomless lake, whose waters are transparent, yet into which we can clearly see but a little way.

CHAPTER 10

ALTERED STATES

Why do we sleep? Sleep is surely one of the most mysterious processes involving the brain. The evolutionary pressures against sleep would seem to be enormous, since an organism is fatally vulnerable to predation while asleep. To make matters worse, sleep is an energetically expensive process. Rapid eye movement (REM) sleep is associated with intense neural activity, combined with loss of muscular control and an often profound loss of consciousness. It is hard to imagine a process that would be potentially more dangerous for a human on the African veldt, where prey is scarce and predators are rather too abundant. In short, there are many reasons not to sleep and no clearly established reasons why sleep is essential. Yet sleep deprivation itself can be fatal.

To understand sleep—which amounts to a periodic and profound loss of consciousness—it will be necessary to explore other altered states of consciousness. What does an unconscious or comatose person perceive? Is consciousness a minimal requirement for the human state or are we still human without it? Is it possible to be aware or to feel pain without being conscious?

We have defined consciousness as a state characterized by attention, perception, memory, and awareness. If this is true, and if one of the components of consciousness is missing or attenuated, then a person would have an altered state of consciousness. If just one component was impaired, consciousness might only be diminished. Yet some patients experience outright loss of all these components, in which case consciousness may be entirely gone.

THE MINIMALLY CONSCIOUS STATE

We have already discussed the persistent vegetative state in great depth (see chap. 4), but there has recently been an impetus to define another state of reduced consciousness, far deeper than sleep but less profound than a persistent vegetative state. The "minimally conscious state" is thought to differ from the persistent vegetative state in showing inconsistent but discernible evidence of consciousness. A definition of the minimally conscious state requires that a patient show behavioral evidence of self-awareness or awareness of her surroundings. This is controversial because some patients in a persistent vegetative state seem as if they are able to respond to the environment. To probe the minimally conscious state (MCS), radioactively labeled water was infused into the bloodstream of five MCS patients, fifteen patients in a persistent vegetative state, and eighteen healthy people. A positron emission tomography (PET) machine was used to detect radioactivity in the brain as a way to measure brain blood flow. Since perfusion increases if the brain is actively consuming oxygen, increased radioactivity in the brain is taken as evidence that the brain is metabolically more active.

In order to stimulate the brain of subjects in this study, a simple series of clicks was played into each ear individually as the subjects lay within the PET machine.[1] Not surprisingly, the sounds activated the auditory cortex of the brain in healthy subjects as well as the association cortex at the front of the brain. Also, not surprisingly, among vegetative patients there was activation of the auditory cortex but no activation of the association cortex. This shows that while a portion of the brain was still responding to the auditory signal in a very simple manner, there were no higher thought processes, no associations being made, no evidence that the subject was able to "hear" the sound in any but a purely sensory way. However, the MCS patients showed a condition that was essentially intermediate between full consciousness and the persistent vegetative state. Patients who were minimally conscious showed activation of the auditory cortex in response to sound, but they also showed some activation of the association cortex. The MCS patients, whose consciousness was still strikingly impaired, thus had a stronger functional connection between auditory cortex and association cortex than did the vegetative patients. This implies that some "thought" was happening in the MCS patients. Nevertheless, it would be easy to make too much of this finding; among the MCS patients studied, only one was able to recover enough to live independently. One patient died and two patients were markedly dependent, so the minimally conscious state is a long way from being fully functional. Furthermore, hospitals that have MCS patients do not necessarily have PET machines, so it will not be possible to assess all patients in this way. Finally, it is not clear if PET offers any insight that would not be available from a careful neurological examination. In fact, all of the MCS patients in this study showed some level of function on the Glasgow Coma Scale, which is the most common way to assess consciousness and does not require a PET machine. Nevertheless, this study seems to validate the concept that the minimally conscious state is distinct from the persistent vegetative state.

ANESTHESIA AND LOSS OF CONSCIOUSNESS

Patients undergoing painful medical procedures may be offered an anesthetic, which can reduce the perception of pain. Anesthetic drugs induce a reversible loss of consciousness similar to deep sleep but different from sleep in a profound way. During sleep, the sleeper has impaired consciousness, but consciousness can be recovered; the sleeper can become fully conscious again within seconds and without help. During sedation, the sleeper has impaired consciousness that is externally controlled and the sleeper can become fully conscious again only after a long period of recovery or with the intervention of another person. During sedation, the locus of control of consciousness shifts from the sleeper to an observer.

The perfect sedative would produce a transient reduction in the level of consciousness, so that a patient can be sedated and tranquil during a painful episode. However, the perfect sedative would do more than that; it would also produce a predictable pattern of deep sleep and would never result in tolerance, dependence, or addiction.[2] There are, of course, no perfect sedatives. Most sedatives are to some degree toxic, associated with abuse or dependence, and prone to induce tolerance or addiction.

There are several types of drugs used in anesthesia: analgesics, which diminish the perception of pain and lessen the reflexive response of the body to a painful stimulus; relaxants or paralytic agents, which diminish or block the muscular response to pain; anxiolytics, which reduce the anxiety that is associated with pain or the anticipation of pain; and sedatives or hypnotics, which can diminish consciousness to an extent that varies from mild sedation to general anesthesia. There is also at least one dissociative agent, which is an analgesic that diminishes the emotional urgency of pain and induces amnesia.

Ketamine—a dissociative agent, which is also a drug of abuse (Special K)—is an interesting form of anesthesia. After intravenous administration, the patient passes into a trancelike state during which his eyes may remain open, though he cannot respond to pain.[3] This trancelike state of catalepsy is characterized as a dissociation between the brain and outside stimuli such that a state of "sensory isolation" is induced. Rather than having a typical "dose-response" effect, with higher drug levels inducing more sedation, ketamine dissociation is either present or absent, with an abrupt transition. There is a profound analgesia with amnesia, but with normal breathing, normal heart rate, and normal reflexes to protect the airway. Because the amnesia induced by ketamine is usually complete, it is possible that the patient still feels some level of pain but has no emotional response to it and retains no memory of it. Ketamine—which is a street drug—is associated with an increased risk of psychosis, although it is also possible that a person prone to psychosis may be more prone to abuse ketamine.

Virtually all anesthetics share the property of dissolving well in fats and oils. The efficacy of an anesthetic can be predicted with remarkable accuracy by simply measuring how much of the anesthetic will dissolve into solution in olive oil under specified conditions. Fat solubility is important because anesthetics appear to work by dissolving into cell membranes and such membranes are composed of fatty substances. Recall that neurons are normally able to maintain an ionic gradient across a cell membrane. After an anesthetic dissolves in a neuronal membrane, that membrane may become more permeable so that the function of the neuron is disrupted.

Inadequate sedation during surgery, so that a patient retains awareness and a sensation of pain, is a source of tremendous anxiety for patients about to undergo surgery. However, the field of anesthesiology has progressed very rapidly and inadequate sedation is less a risk now than it was even a decade ago. New anesthetic agents have been introduced, and new ways to monitor the depth of sedation are routinely used. Nevertheless, the molecular mechanisms that underlie general anesthesia remain unknown, even though most scientists now think that anesthetics work by binding to specific neuronal receptors in the brain.[4]

It is possible to use anesthetic agents in clever ways to probe the nature of consciousness. For example, a large group of healthy volunteers was given a low dose of one of several different anesthetics at the same time as they were asked to perform a memory task.[5] This experiment revealed that amnesia and sedation are distinctly different: certain drugs at a low dose can block memory but cannot induce sedation. Furthermore, some anesthetic agents increase the risk of seizure, suggesting that they work by a mechanism like an epileptic loss of consciousness.[6]

Consciousness, Hallucinations, and Déjà Vu

Hallucinations are sensory perceptions in the absence of sensory stimulation. They represent an altered state in that consciousness cannot discern between reality and illusion. Hallucinations can be visual, tactile, aural, olfactory, or even gustatory, but visual hallucinations are the most common and can be most alarming. Hallucinations may simply be thoughts or memory images that are experienced as if they were perceptions. This conflation of imagery and perception is accepted as true even if it conflicts with other sensory input or is inconsistent with what others perceive.[7]

Hallucinations tend to be disowned, in that a hallucinating person may feel that she has no control over the imagery, as if the hallucinations were an objective record of a reality that cannot be altered. This is essentially a failure of consciousness to test reality. Were a person to be fully conscious but hallucinating, the falsity of perception could be determined either by using other sensory modalities or by using social clues. For example, if you perceive a dead body in the room but no one else is upset, then the body is probably a hallucination.

This discussion highlights a key element of perception: in most people and at virtually all times, perception is objectively tested before it is accepted. If you hear a faint noise, you turn your head to determine if there is a change in the strength or character of the sound that conforms to your expectations. If you see something indistinctly, you turn on a light, reach out to touch it, or ask another person what they see. As fatigue begins to set in, people may become less able to test reality and less vigilant as to the boundary between reality and perception.

People who have been deprived of sensation for a long time, who are put in a darkened and soundproofed room, may be unable to test reality rigorously because of a lack of sensory input. Tortures have been developed based on sensory deprivation, using the principle that the absence of environmental stimuli causes sufficient mental anguish that physical pain is almost irrelevant.[8] The solitary confinement or "boxcar" cells at the maximum security federal penitentiary in Marion,

Illinois, which are used for the most incorrigibly violent of prisoners, are small steel cubicles without natural light. The cells are insulated so that sound cannot penetrate, ventilation is poor, and the only light is provided by a single low-watt light bulb. A prisoner who becomes ill has no way of alerting his captors because the cells are so isolated from the rest of the prison. These cells have been involved in several court cases in which inmates claimed that confinement was driving them insane. Because of the extent of sensory deprivation, incarceration in these cells may indeed make it hard for the inmate to test perception against reality.

Many medical conditions are associated with visual hallucinations, and it is noteworthy that similar hallucinations can result from very different disease processes.[9] Most conditions that produce hallucinations are also associated with some impairment of consciousness (table 4). Sometimes hallucinations result directly from a lesion in the visual cortex that causes defective visual processing, but many hallucinating patients have a normal visual cortex. Some patients retain normal consciousness and insight, so that visual hallucinations are experienced with an awareness that they aren't real. Other patients lack insight and thus experience hallucinations as if they were real. In some cases, it is not at all clear why a patient experiences hallucinations. For example, Parkinson's disease patients may begin to suffer hallucinations, but only after prolonged treatment with levodopa (L-dopa). The severity of hallucination is usually a function of the dose of L-dopa. Such patients usually also lack insight and have sleep disturbances, vivid dreams, and episodes of reduced awareness while awake. It is unknown why symptoms arise late in the disease course after a fairly long period of successful disease treatment.

There may only be three mechanisms that produce hallucinations: an irritative process in the association cortex that is falsely interpreted as visual input; a release phenomenon in which defective visual input causes abnormal stimulation of the cortex; or faulty visual processing in which inputs are normal but lesions cause an inappropriate pattern of cortical response.[10] The capacity to generate hallucinations appar-

Table 4. Characteristics of different conditions associated with visual hallucinations

Disease	Type of hallucinations	Duration	Consciousness	Insight	Cause
Schizophrenia	Elaborate, often paranoid	Nearly continuous	Can be impaired	Can be very poor	Genetic lesion?
Bipolar disease	Can be elaborate	Can be frequent	Can be impaired	Can be poor	Genetic lesion?
Drug-induced hallucinations	Can be elaborate	Minutes to hours	Can be impaired	Can be poor	Drug intoxication
Fever	Can be elaborate	Minutes to hours	Can be impaired	Can be poor	Unknown
Focal epilepsy	Brief, stereotyped	Usually seconds	Often impaired	Usually preserved	Temporoparietal lesions
Delirium tremens	Highly variable	Often prolonged	Agitated, confused	Usually reduced	Alcohol dependence
Parkinson/Lewy body dementia	Often in the evening	Minutes	Drowsy, impaired	Usually preserved	Cortex or brainstem lesion
Infarction or stroke	Abnormal visual field	Days to weeks	Can be impaired	Usually preserved	Visual pathway lesion
Visual field disturbance	In disturbed visual field	Prolonged	Normal	Preserved	Visual pathway lesion
Hypnagogic hallucinations	Upon falling asleep	Seconds to minutes	Drowsy	Usually preserved	Brainstem lesion

adapted from M. Manford and F. Andermann, "Complex Visual Hallucinations: Clinical and Neurobiological Insights," *Brain* 121 (1998): 1819–40. Used by permission of Oxford University Press.

ently resides in the association cortex, and hallucinations occur when this cortex is released from a normally inhibited state. Visual hallucinations may be analogous to the "phantom limb" sensation that can trouble amputees; traumatic loss of a limb can release the sensory cortex from inhibition so that there may be a sense of pain from the absent limb. Similarly, if the visual cortex is released from "real" visual sensations, that cortex may become spontaneously active and may interpret random activity as a visual stimulus. This could explain why relatively small lesions can lead to florid hallucinations. In addition, there may also be an involvement of the thalamus in hallucinations, which could explain why hallucinations are often associated with an alteration of sleep or awareness.

Another form of altered perception that may be familiar to some is déjà vu, a vivid sense that something was "already seen," that something has happened before.[11] Déjà vu can be very powerful, and the compelling strangeness of the experience has provoked many explanations. Déjà vu experiences have been taken as evidence of precognition, or of having lived a prior life, or of having seen a place through the eyes of another person. At issue is an assumption that the déjà vu memory is an accurate recollection of an event that really happened. Yet we know that memory is not reliable: eye witnesses are often wrong and false memories can be generated in a laboratory situation. Thus, it is possible that déjà vu is simply a false recollection. The question is then no longer, "Why does this person fail to remember the original event?" but rather, "Why does this person think that the present recollection is real?"

It may be that a déjà vu is simply a neuronal misfire.[12] Information traveling from the senses to the brain can follow multiple pathways; ordinarily these separate sensations are bound together into a single perception in the hippocampus. If part of a perception is slowed by an accidental misfire, it could arrive at the hippocampus a moment too late to be perceived as a part of the same whole. Perhaps the brain is unable to tell whether the other parts of a perception arrived milliseconds or months sooner. Yet the delayed percept could still seem familiar, since other parts of the perception have already made a trace in the brain. Thus, instead of one unified perception, the brain could have several fragmentary perceptions, all of which seem equally true, all of which are equally familiar. The idea that the hippocampus plays a role in the origin of déjà vu is supported by an observation that some epileptic patients who have a known seizure focus in the hippocampus have a frequent sense of déjà vu as a part of their illness.

It is also possible that déjà vu arises when one part of a perception resonates with something else in memory's attic. A lamp in a doctor's office may look like a lamp you saw at the store last week; a phrase uttered by a friend may be the same as something you heard on the radio recently; the mannerisms of someone you just met may remind you of your third-grade teacher. We may not be conscious of these similarities, but they may confer a strange familiarity onto a scene that cannot have been perceived before. There is some experimental evidence to support the idea that déjà vu is simply flawed recollection.[13] In one experiment, people were shown pictures of various places that they had never been, but their attention was diverted by asking them to find a small cross hidden somewhere in the image. A week later, these subjects were shown a set of pictures that included some of the images from the week before and they were asked if they had ever been to the place in the picture. Some subjects falsely claimed that they had been to the place that they had seen only in pictures the week before. This false familiarity has much in common with déjà vu, and it provides an insight into the alarming subjectivity of memory.

EPILEPSY AND IMPAIRED CONSCIOUSNESS

Epilepsy is one of the most common neurologic problems worldwide. About 2 million people in the United States have epilepsy, and roughly 3% of all people will have an episode of epilepsy at some point in their lives.[14] Epilepsy is characterized by short periods of altered consciousness, which can be accompanied by seizure and a loss of muscular control. Most forms of epilepsy last no more than a few minutes, although seizure activity can recur many times over the course of a day. Yet some seizures last longer than half an hour or recur so often that a person may not regain full consciousness between seizures. Epilepsy is associated with abnormal electrical discharges—like "storms" in the brain—that arise from a small seizure focus.

Epilepsy is classified into various types based on the type of seizure, the presence of other symptoms, and the form of electrical activity revealed by electroencephalography (EEG). During an absence seizure, a patient may simply stare into space for a few seconds then return to normal, without losing muscular control and without retaining a clear memory of the event. Absence seizures can occur tens or hundreds of times in a day, yet patients may be unaware of them or may be diagnosed with attention deficit disorder or "daydreaming." During a partial seizure, a patient may experience brief hallucinations followed by a loss of awareness, and the patient may stare blankly, speak unintelligibly, or show various automatic behaviors such as lip smacking or picking at clothing. During a generalized seizure, a patient may become completely comatose for moments to minutes, losing muscular control, twitching violently, and returning to consciousness very slowly after muscular control returns.

The electrical storm that gives rise to seizure activity can be caused by many mechanisms.[15] In absence seizures, there may be an alteration of the circuitry that normally connects the thalamus to the cerebral cortex. Since the thalamus is thought to be responsible for awareness, impairment of this circuit would tend to impair consciousness. In partial seizures, there may be damage to a small part of the brain such that a seizure focus is formed; hippocampal sclerosis, a form of neuronal loss and tissue shrinkage, is relatively common in people with partial seizures. In generalized seizures, there may be a mutation in a protein that forms an ion channel in neurons. Since ion channels are everywhere in the brain, this may explain why generalized seizures are so overwhelming.

There are many puzzling questions related to epilepsy.[16] For example, how can a chronic brain alteration lead to sporadic seizure?

A person with a stable and persistent brain abnormality might be expected to show stable and persistent symptoms, but epilepsy is the prototype of a disease that has sporadic symptoms. Does this mean that the environment has a role in evoking the symptoms of epilepsy? It is also not known why epilepsy has an age-dependent onset, with greater severity in childhood and old age, or why it can remit spontaneously. The fact that epilepsy is often quiescent suggests that drug therapy could prevent seizures. Seizure prevention medication is already available, but none of the current medications are completely effective and some of them have alarming side effects. A better understanding of the mechanism of seizure generation could have a major impact on treating epilepsy.

There is new evidence that neurogenesis—the birth of new neurons—can be stimulated by a seizure.[17] In patients with hippocampal sclerosis, there is often evidence of newly sprouted neurons, which can form abnormal connections in the hippocampus. If rats are given a chemical that induces epilepsy, there can be a dramatic increase in the rate of growth of hippocampal cells, and the vast majority of dividing cells may differentiate into neurons. Yet these fresh neurons tend to form abnormal connections and integrate into abnormal circuits within the hippocampus. This suggests that epilepsy is associated with a disorder of neuronal proliferation, neuronal migration, or cortical organization. Injury-induced neurogenesis after stroke is apparently both common and reparative, but seizure-induced neurogenesis after epilepsy may not compensate for the effects of disease.[18] In fact, some data even suggest that newly produced neurons in the epileptic brain can contribute to seizure vulnerability. It has also been suggested that neurogenesis is stimulated in patients with recent-onset epilepsy but depressed in patients with chronic epilepsy.[19] This could occur if a pool of neuronal stem cells gradually became depleted as a result of an abnormal growth stimulus associated with epilepsy. In order to treat patients with epilepsy effectively, a compelling need exists for a better understanding of neurogenesis in the adult brain and of how the process of neurogenesis can be controlled.

SLEEP: THE FORGOTTEN COUNTRY

There is a balance between the drive for sleep and the need for wakefulness so that most people are able to maintain a regular circadian rhythm of sleep and wakefulness.[20] Certain neurons induce sleepiness and other neurons are involved in arousal; the interplay between neurons determines our state of awareness over the course of the day. If the relationship between neurons is disturbed, one consequence can be narcolepsy, which is an irresistible urge to sleep in the middle of the day. Among healthy subjects, the drive to sleep increases at night so that eventually it is very hard to stay awake. After we fall asleep, there is a period of rather light sleep, followed by a deepening, so that the proportion of sleep given over to rapid eye movement (REM) sleep increases as the night passes. Blood flow to the brain is somewhat less while asleep than while awake, except in REM sleep, when energy demand is roughly comparable to being awake. After a night of sleep (or when the alarm clock rings), we begin to awaken in a process that has been studied by Positron Emission Tomography (PET).[21] There is a rapid reestablishment of consciousness followed by a relatively slow reestablishment of full alertness, which can take more than 30 minutes. Blood flow to the brain first increases in the brainstem and thalamus, consistent with what we know of their role in the conscious state. Thereafter, the rate of blood flow to the rest of the brain increases.

Since there is a rather high demand for energy during sleep, and specifically a high energy demand during REM sleep, it is a mistake to think that sleep is simply a diminution in normal metabolic activity.[22] The increase in energy demand is distributed broadly across the brain, and the pattern of activation in the sleeping brain is rather similar to the pattern of activation in the awake brain before sleep.[23] This does not imply that sleeping is like being awake; rather, the sleeping brain may continue to produce a pattern of activity that was first established while awake. Scientists interpret this to mean that one function of sleep may be to stabilize the pattern of neural activity that was established while awake when the brain was storing new memories.

REM sleep is associated with activation specifically in the visual association cortex, with a simultaneous reduction of activity in the primary visual cortex.[24] It is as if the brain is still processing—yet not perceiving—visual stimuli. This may mean that the brain is briefly resistant to new visual input. A recent study suggests that thinking and hallucinating alternate during sleep.[25] As people go from waking through sleep onset to non-REM and finally to REM sleep, they report that visual imagery is progressively more hallucinoid, that they lose conscious control over the thought process. Roughly 82% of REM dreams are hallucinoid, so it is not true that the sleeping brain is simply rehashing visual stimuli from when it was awake. REM sleep is associated with reduced insight into time, place, and state, and there can be a lack of insight into perceptual anomalies. It may be quite literally true that REM sleep is like being insane. It may also be noteworthy that if people are forcibly kept awake, their sleep-deprived brain will begin to hallucinate, and visual images will intrude more and more aggressively into the conscious brain.

Sleep seems to keep madness at bay in healthy people, whereas people who are mentally ill often have great difficulty sleeping. Patients with schizophrenia can have disturbed patterns of sleep, even if the effects of antipsychotic medication are factored out. Schizophrenia patients take longer to fall asleep, spend less time asleep, spend less time in deep sleep, and tend to spend more of their night wide awake.[26] These trends are worse in patients who have been treated but then stopped taking medication. But even schizophrenic patients who have never been treated have problems with sleep. Antipsychotic medication may indeed disturb sleep, but sleep disorder seems to be a part of the schizophrenic disease process rather than a result of disease treatment.

Hypnagogic dreams—those fragments of dreams that occur while falling asleep—are linked to memories from the waking period before sleep, but REM sleep is freed from such constraints.[27] Instead, REM may be activated when the mechanism for storage of episodic memory is inactive. REM dreams occur during periods of fairly chaotic brain activity stimulated by the brainstem. During REM, the mind may simply be making the best sense possible of a stream of images that make little inherent sense. Disturbances of REM have been reported in depressed patients, in people who are bereaved, and in patients suffering from severe anxiety or posttraumatic stress disorder.

The purpose of sleep, especially REM sleep, is deeply mysterious.[28] REM sleep varies greatly from one species to another, but the amount of REM does not seem to correlate with intelligence in any obvious way. Humans do not exhibit an unusually high amount of REM sleep compared to other animals; in fact, some of the most primitive mammals have far more REM sleep than do humans. Whales and dolphins, which have some of the largest brains on Earth, typically have very little REM sleep. Even if we limit our attention to humans, there is no correlation between intelligence quotient (IQ) and REM sleep.

Why Do We Sleep?

Because we sleep, despite enormous selection pressure against it, sleep must be important. But what does sleep actually do? The oldest idea is that sleep is restorative, that the enforced rest of deep sleep repairs the tired body and brain. This seems obviously true for the body; as we lay in bed, very nearly motionless, the body can recover from the rigors of the day. But if sleep were merely restorative for the brain, we would expect that the brain should be as inactive as the body. Yet sleep is a very active process for the brain. What could be the point of imprisoning an active brain in an inactive body for so long every day? This question becomes particularly compelling when put in an evolutionary context. For millennia, our ancestors must have been vulnerable to predation while asleep. However close our ancestors huddled to the fire, the flames eventually died down and predators crept closer. Were the mind alert while the body was resting, this might not have been a problem; yet the protohuman brain was probably similar to our own, so sleep was probably associated with some loss of consciousness and some degree of vulnerability to predation.

Why must we sleep? A fascinating theory proposes that sleep is necessary to consolidate memory.[29] At the onset of sleep, there are episodes of neuronal synchronization across broad expanses of the forebrain. This type of neural synchrony is called spindling. Spindling gradually deepens into a state of slow-wave sleep, which sporadically alternates with episodes of the rapid neural activity known as REM. It has been proposed that spindling opens a molecular gateway to plasticity. As spindling stimulates the frontal cortex, depolarizing pulses of calcium ions enter the cortical neurons, and this may begin a process of long-term potentiation. Coherent patterns of excitation could regulate gene expression in neurons, helping to adjust the strength of synapses and foster the consolidation of new memories. Slow-wave sleep could be an iterative process of "recalling" and "storing" new information that was initially primed into neural circuits during waking hours. In short, brain activity during sleep could be used to reorganize cortical networks after learning occurs in the awake state.

This theory makes some fascinating predictions that can be directly tested. If sleep is indeed associated with memory consolidation by the mechanisms described, then one would expect that patterns of neural activation during sleep might mirror what happens during the learning process, that performance on a motor task would improve after sleep, that recall of information would improve after sleep compared to no sleep, that sleep could reorganize as well as consolidate memory, and that sleep would be associated with changes in the regulation of different genes. Each of these predictions has been tested and most have been confirmed.

During sleep spindles, synchronized action potentials arise in the thalamus, and these bursts of depolarization interact with neurons in the cortex.[30] However, the interaction with cortical neurons is such that the neuron is depolarized yet prevented from generating an action potential. Cortical neurons are thus simultaneously stimulated to fire and inhibited from firing. This creates a condition that is ideal for calcium entry into neurons. Because spindling is coordinated across large expanses of cortex, a great many neurons may be stimulated to allow the entry of calcium ions. Calcium entry into neurons is, of course, the first step in long-term potentiation (LTP). Experiments with neurons in culture also suggest that calcium entry induces the synthesis of certain proteins that are required for long-lasting synaptic change. Spindling may thus activate gene expression and result in memory consolidation.

The fact that cortical activation is widespread may be quite important. Animal studies have shown that the hippocampus is needed for memory storage, but only at first. As time passes, the role of the hippocampus in memory diminishes as memories are broadly distributed across the cortex. For example, a memory of how to play the violin is broken into several parts: a memory of the shape of a violin is stored in the visual cortex, a memory of the sounds that a violin makes is stored in the auditory cortex, a memory of how to move your fingers while playing the violin is stored in the motor cortex, and a memory of the music itself is stored at yet another site. The hippocampus is responsible for "binding" these separate elements into a complete memory of playing the violin, but the hippocampus does not actually store the memories.

During memory storage, the brain solves two related problems.[31] First, where should changes be made in order to create new memories or to create new links between existing memories? And second, how can new memories be made compatible with previous memories so that older memories are not lost? Computer modeling suggests that it would be dangerous to change the strength of synapses while the system is active, especially if the brain may be required to make snap judgments unrelated to the memory task. Instead, it is safer to wait until the cortex is no longer processing information before making any permanent changes that could interfere with sensing the environment. In essence, this may be like taking a computer offline before making a software upgrade for fear of bringing the entire network down if there is a problem. Network reorganizations must be made carefully, especially if it is not possible to undo a change. Sleep may be a means of taking the brain offline so that sensory input cannot interfere with forming new connections and so that there is less risk of significant "glitches" developing. This may be why the brain becomes progressively less responsive to external inputs and less concerned with actively gathering information during the transition from wakefulness to sleep. During sleep, the brain is no longer attending to the outside world; it is attending only to itself.

IS SLEEP ASSOCIATED WITH MEMORY FORMATION?

It is well known that sleep deprivation interferes with learning and memory in both animals and humans.[32] Yet sleep deprivation studies are potentially flawed in ways that can interfere with memory; sleep deprivation studies in animals are almost always associated with stress, and stress can interfere with memory formation. Nevertheless, it has been shown in rats that the pattern of neural activation during sleep mirrors what happened during a learning task that preceded sleep. It is as if the brain was rehearsing while asleep.

Clear evidence has begun to emerge that motor skill learning in people improves after sleep. In a particularly strong study, 62 people were trained to perform a motor task, which involved pressing computer keys in a specific sequence as quickly and as accurately as possible.[33] Some subjects had a training period that was at 10 AM in the morning, while other subjects had a training period at 10 PM at night. There was no difference in skill or learning rate between the subjects who learned in the morning and those who learned in the evening, so there was no evidence that subjects were fatigued while they were learning at night. People who had learned the motor task in the morning were tested again in the evening; they showed no significant change in ability over the course of the day. Yet those subjects who had learned the task in the evening showed a significant improvement in their ability after a night of sleep. In fact, a night of sleep produced a 20% increase in motor speed without any loss in accuracy, whereas an equivalent period of wakefulness provided no benefit. This result could not be explained as a result of hand fatigue, since some subjects who trained in the morning wore mittens over the intervening day. To probe whether this result could be explained by something other than sleep, subjects who were trained in the morning and tested at night were tested again the following morning. Even though their evening performance had not been any better than when they were trained, their performance after a night of sleep was much better. There was even a correlation, albeit a weak one, between the amount of non-REM sleep and the improvement in motor ability. This finding is consistent with the idea that sleep spindles, which occur during non-REM sleep, act to facilitate memory consolidation. Thus, anyone concerned with learning a complex motor task might do well to take a nap shortly after practicing.[34]

Several people who learned this finger-tapping task were then studied using the method of functional magnetic resonance imaging (fMRI), which highlights those regions of the brain that are involved in learning a specific task.[35] When subjects were trained in the evening and then tested the next morning, there was a substantial change in the way the brain performed during the retest. Certain parts of the brain that may be able to support faster and more precise motor output were fully engaged, whereas other parts of the brain—needed for conscious monitoring of the task—were less engaged. This implies that the motor task became more automatic after a single night of sleep. This study was weakened somewhat because findings in control subjects—who learned the task but had no opportunity to sleep— were not reported in detail. Nevertheless, the study suggests that sleep restructures the way that a motor memory is stored in the brain.

Sleep-dependent neural plasticity appears to play a role in learning and memory across a broad range of species.[36] Studies show that daytime naps may be able to restore performance decrements that result from fatigue or "overpractice," and sleep is necessary to optimize motor performance. In humans, declarative memory—the ability to recall facts and figures—does not depend upon REM sleep,[37] but REM plays a crucial role in the consolidation of procedural learning—the ability to perform a new task. If people are deprived of REM sleep, they are less able to learn the rules of a complex logic game, less able to acquire a foreign language, and less able to retain memories after an intensive period of study. While the evidence for memory consolidation in sleeping animals may be weak,[38] this is not a compelling criticism, since experiments in humans show that sleep can profoundly reorganize memory.[39]

It may be noteworthy that patients with schizophrenia, who have disturbed REM sleep, also tend to have poor memory.[40] One study found no difference in the rate at which patients and healthy controls could learn a finger-tapping task, but schizophrenic patients failed to show any improvement after a night of sleep. In fact, patients performed 4% worse after a period of sleep, whereas healthy people performed 11% better after sleep. This study confirms that REM sleep is essential for learning, but it may also give some insight into the nature of schizophrenia.

GENE REGULATION ACROSS THE SLEEP CYCLE

The theory that sleep is necessary to consolidate memory requires that sleep result in systematic changes in gene expression.[41] If there were no changes in gene expression, there could be no long-term changes in protein synthesis. If there were no changes in protein synthesis, this would mean that synapses could not be strengthened by long-term potentiation (LTP).

Gene expression profiling has been used to study changes in gene expression in the rat brain as a function of the sleep/wake cycle.[42] Gene expression profiling is used to characterize gene expression levels in a tissue sample. After a small piece of tissue is harvested, cellular RNA—a chemical relative of DNA—is extracted and then amplified and plated onto a microarray plate. A microarray is like a computer chip that identifies specific gene sequences in RNA. Because RNA is needed to direct the synthesis of new protein, the microarray method can reveal which proteins the cell is about to make. With recent improvements in technology, the expression of up to 40,000 proteins can be assessed simultaneously.

Spontaneously awake rats were compared to spontaneously asleep rats and to rats that were sleep deprived by gentle handling for 8 hours.[43] Pieces of rat cortex were removed from the brain by surgery and RNA was chemically extracted from this tissue. Then RNA levels were compared between the groups to see whether the sleep/wake cycle had an impact on gene expression. In comparing asleep rats to awake rats, more than 10% of genes were differentially expressed. Roughly half of these genes were expressed as a reflection of a circadian rhythm—the 24-hour cycle of physiological change that is keyed to day length. The other half of the gene changes were directly related to sleeping and waking, no matter what the time of day. Hundreds of genes changed in expression between sleeping and waking. The proteins related to these genes serve a broad range of purposes. During wakefulness, genes are turned on that facilitate synaptic plasticity (especially LTP), that control energy metabolism, that are involved in the response to stress, that moderate membrane depolarization, and that are used to synthesize the neurotransmitter glutamate. During sleep, genes are turned on that block LTP, that moderate membrane hyperpolarization, and that synthesize the neurotransmitter GABA. Thus, the sleep/wake cycle is associated with neural plasticity as well as with changes in energy metabolism and the stress response.

When muscle tissues were assessed, there were no significant changes among the most important sleep-related genes between sleeping and waking.[44] Thus, sleep-related changes in gene expression are limited to the brain. Although sleep is a state of behavioral inactivity, many genes increase their expression in the sleeping brain. In fact, roughly as many genes are turned on by sleep as are turned on by wakefulness. In short, sleep is associated with the expression of specific proteins, consistent with the theory that sleep acts to consolidate memory.[45]

SLEEP AND CREATIVITY

For centuries, there has been an assumption that sleep and creativity are somehow linked. The experimental physiologist and Nobel Prize winner Otto Loewi wrote in an autobiographic note that on the night before Easter in 1921

I awoke, turned on the light, and jotted down a few notes on a tiny slip of thin paper. Then I fell asleep again. It occurred to me at six o'clock in the morning that during the night I had written down something most important, but I was unable to decipher the scrawl. The next night, at three o'clock, the idea returned. It was the design of an experiment to determine whether or not the hypothesis of chemical transmission that I had uttered 17 years ago was correct. I got up immediately, went to the laboratory, and performed a simple experiment on a frog heart according to the nocturnal design.[46]

Loewi claimed that while sleeping he had begun to link new ideas together in a pattern that he had not conceived while awake. Similarly,

the German chemist August Kekule had the insight that the benzene molecule might be shaped like a ring after he fell asleep while watching sparks in the fireplace make circles in the air. If he had stayed awake, Kekule would have presumably rejected as ridiculous the thought that there might be a connection between the sparks and the shape of the molecule. But in the subconscious, rationality could not censor the connection, and so when he woke up he was no longer able to ignore its possibility. According to this perspective, truly irrelevant connections dissolve and disappear from memory, while the ones that are robust survive long enough to emerge into consciousness.[47]

Is it possible that elements of a memory are shuffled in relation to one another while a person is dreaming? Is this more or less random reassortment of ideas perhaps the genesis of creativity? Can sleep inspire specific insights?

The idea that sleep inspires insight is supported by an experiment described earlier.[48] Subjects were asked to solve a mathematical puzzle that was time-consuming but not difficult. A shortcut or trick was built into the puzzle so that someone with insight could solve the puzzle in much less time than would otherwise be required. Each subject completed a training period with the puzzle so that their speed could be assessed, then people were tested later. Some people slept between training and testing, whereas other people didn't have time to sleep. In the absence of sleep, people solved the puzzle about 6% faster upon retesting, whereas subjects given a chance to sleep solved the puzzle 17% faster, even without insight. This shows that sleep can enhance learning, but it doesn't prove that sleep enhances insight. However, among people who didn't have a chance to sleep, 23% gained insight into the workings of the puzzle, whereas 59% of the sleepers gained insight. This finding is particularly noteworthy because people were never told that there was a shortcut, so they didn't know there was a problem to solve. No one went to bed thinking that they could "sleep on the problem" because everyone thought that they already had a good understanding of the problem. Yet insight led to a mental restructuring of knowledge with a sudden increase in insight, and this insight was facilitated by sleep. In other words, sleep helped people to see that there was a hidden structure to the problem.

The sleeping brain and the dreaming mind retain secrets still, yet there is consensus that sleep is crucial to memory processing and may also be important in stimulating insight and creativity.[49] Problem-solving ability and fluid intelligence increase after even a short interval of REM sleep but not after an equivalent period of non-REM sleep. People are more able to solve word anagrams when awakened from REM sleep than when awakened from non-REM sleep. Thus, highly associative properties of the mind are enhanced in REM sleep, and REM sleep seems to facilitate cognitive flexibility.

Memory—and perhaps creativity—is enhanced by sleep. Yet it still seems that sleep should be an evolutionary burden, stringently selected against by hungry predators, unless the human brain had evolved some sort of protection, some way to assure that the sleeper was not simply slaughtered. Perhaps the evolution of sociality in humans can be thought of, in part, as a solution to the vexing problem posed by this loss of consciousness. Perhaps the human social habit is a form of safety in numbers, which compensates for our unrelenting need to sleep.

CHAPTER 11

EMOTION AND THE SOCIAL BRAIN

I t often seems that emotions are a disagreeable thing: the pain of separation and loss can last longer than the joy of being with a loved one. The elation and euphoria of new love is rather too quickly replaced by the emotional work of an ongoing relationship or the emptiness of a failed relationship. The mother's bond to an infant can be soured by the child's rebuff of the mother, from the growing child's terrible two's through the narcissism of the teenage years. Many people have a greater ability to feel anger than serenity, to feel resentment than altruism, to feel pain than pleasure, to feel sorrow than joy. It has even been proposed that social attachment can become an addictive disorder. What good are human emotions?

It seems likely that the human brain evolved under strong selective pressure to make a social mind. Affiliation is crucial to human survival, whether as an attachment to one's mother during infancy, as part of a "pack" during adolescence, as a partner in a pair bond during adulthood, or as a parent caring for a growing child. Although affiliation is not uniquely human, humans have developed unique strategies to induce, affirm, and strengthen social bonds. These strategies rely upon the capacity to experience and share emotion. It is surely not an exaggeration to claim that affiliation is necessary for human survival and that the social brain is what enables humans to bond. Yet, before we can discuss the evolution of the social brain, we must first understand what may be the most basic human emotion of all: stress.

STRESS AND HUMAN SURVIVAL

Does stress have a physical effect on the brain? Can stress even have a physical effect on the body? Or is stress just another emotion that can safely be ignored or suppressed? Why are some people able to cope with extreme stress while others are derailed by rather minor stress? What happens to the body when high-level stress cannot be avoided or ignored? Stress and emotion would seem to stand at the intersection of brain and mind; the brain's response to stress can have an impact on the mind's palette of emotions, just as emotions can feed back upon the stress response of the brain.

A key concept that relates to stress is homeostasis. This is the tendency of an organism—be it a bumblebee or a person—to maintain a constant internal environment in the face of an ever-changing external environment. In a person, blood pH, body temperature, plasma glucose, calcium levels at the synapse, and a host of other physiological parameters are maintained within narrow limits or "set points." For some of these parameters, even a relatively small change from normal can be fatal; if blood glucose levels fluctuate in a diabetic, such changes can lead to coma and death. Because set points are so critical to continued survival, homeostasis is a central concept in physiology and medicine.

Related to the concept of homeostasis is the concept of allostasis. This is the capacity of the body to achieve stability of critical parameters by varying the parameters that are less critical.[1] In other words, allostasis is how homeostasis is achieved. In allostasis the body responds to stress in an adaptive fashion; during a close encounter with danger, both blood pressure and heart rate will increase, delivering sufficient oxygen to muscles to power the fight-or-flight response.

Allostasis is under the control of the hypothalamic-pituitary-adrenal (HPA) axis, which is the master control for the brain-body. The HPA axis modulates the cardiovascular, metabolic, reproductive, and immune systems by secreting various hormones and peptides. "Allostatic load" is the wear and tear that can result from chronic over- or underactivity of the HPA axis.

There are perhaps as many as 11 different hormones, chemicals, or peptides involved in the human stress response,[2] but we won't review them all. Some of these molecules have not been proven to play a key role in the human stress response, and others are probably important only during extreme stress. The most vital molecule in the typical stress response is cortisol, which is secreted by the adrenal gland, near the kidney. In the short term, cortisol can activate certain parts of the brain (e.g., the hippocampus, amygdala, and prefrontal cortex). Cortisol can increase arousal and vigilance, focus attention, enhance memory formation, mobilize and replenish energy stores, and modulate the immune system. Yet excessive and sustained cortisol secretion can have serious side effects, including hypertension, insulin resistance, poor cognitive function, suppression of the immune system, and eventually atherosclerosis and heart disease.

THE TOXIC EFFECTS OF STRESS

Allostatic load is defined as the cumulative physiological burden placed on the body by the ongoing attempt to adapt to life's constant demands.[3] If the adaptive response to stress falls chronically outside normal set points, wear and tear of the body can occur. To determine if the accumulation of allostatic load has long-term health effects, allostatic load was characterized in a group of 1,189 healthy adults in their eighties. Ten different factors were measured, including systolic and diastolic blood pressure, total blood cholesterol, high-density lipoproteins, and so on. A composite measure of allostatic load was calculated for each person, then the cohort of people was followed for another 7 years. At the end of this time, there was a 23% higher risk of mortality for people with high allostatic load scores, even when the effects of age, sex, ethnicity, education, income, and baseline health were factored out. The higher the allostatic load, the greater the risk of death; people at the highest level of allostatic load were 6 times more likely to die than people at the lowest level of allostatic load. High levels of allostatic load were associated with an increased risk of heart attack and stroke, a decline in overall physical performance, and a decrease in cognitive function. Clearly, stress can have a major impact on people, and the toxic effects of stress are worse when stress is chronic. It is unlikely that stress would have as strong an effect on young people, since mortality may only occur after many years of chronic stress, but this is still a sobering indictment of our high-stress lifestyle.

There is also new evidence that sudden emotional stress can result in heart injury.[4] A small study of 19 patients, all of whom experienced a sudden onset of severe chest pain after an emotional stressor, found the most common stressor was the death of a loved one. There was clear and unmistakable evidence of dysfunction in all patients, suggesting that the heart injury was an abrupt and exaggerated response to stress. All patients were diagnosed with "myocardial stunning," a reversible but nevertheless serious form of injury that typically affects the ventricle of the heart.

Does stress also cause mental health problems? Since few things are as stressful as military service in a combat zone, where a momentary lapse of attention can lead to death, soldiers have been carefully studied to see if chronic stress has an impact on mental health.[5] A huge group of soldiers was screened for mental health issues immediately before deployment to Iraq and Afghanistan, and a second enormous group was screened 3 to 4 months after their return from the same combat areas (it is worth noting that the same people were not screened before and after deployment, which would have been a stronger study design). When the two groups were compared, clear-cut differences emerged that are likely to be the result of combat stress. For example, the rate of clinical depression among soldiers prior to deployment to Iraq was 5%, whereas the rate of depression increased to 8% after their return from deployment. Similarly, the rate of anxiety increased from 6% before to 8% after deployment. But the major difference was in posttraumatic stress disorder, which increased from 5% before deployment to 13% after deployment. Posttraumatic stress disorder (PTSD) can be very debilitating; common symptoms include intrusive and distressing dreams or memories of a traumatic event (including flashbacks), persistent avoidance of anything likely to trigger a flashback, and various symptoms that can be associated with depression, such as irritability, insomnia, hypervigilance, and poor concentration. PTSD usually occurs only after severe stress but, among the soldiers deployed to Iraq, stress was severe. Fully 95% of returned soldiers had seen human remains, 89% reported being ambushed or attacked, 86% knew someone who had been killed or injured, 48% had killed an enemy combatant, and 22% had lost a buddy to death or injury. The greater the number of stressful events that a soldier experienced, the higher the risk of PTSD, and 19% of soldiers who had experienced 5 or more firefights eventually had mental health problems. The rate of PTSD among soldiers returning from Iraq was at least 5-fold higher than the population at large, even though soldiers are typically among the toughest, best adapted, and most resilient members of society.

Another study, not of soldiers, but of 6,700 refugees who had resettled in Europe after exposure to war or terrorism, found that roughly 9% of refugees had PTSD.[6] Another study of survivors from war-torn Yugoslavia, where violence was savage and genocide was common, found that 33% of war survivors had PTSD symptoms.[7] The recent increase in terrorism worldwide will likely lead to more mental health problems, even in locations far away from a war zone.[8] In a national sample in the United States, 44% of adults reported having at least one stress symptom after the terrorist strike on New York City on September 11, 2001.[9] Severe and especially chronic stress can lead to serious mental health problems, although some people are better able to tolerate fear and regulate emotion. Research shows that resistance to the effects of chronic stress is generally better in people with high intelligence, a positive self-concept, and some degree of optimism.

Among people with PTSD, memories can be reactivated by cues associated with the original trauma. This may be due to a normal process of memory reconsolidation whereby old memories are recalled and rehearsed to make them more durable. There is a remarkable degree of lability of episodic memory, which can permit reorganization of an existing memory. Each time a traumatic memory is recalled, it is integrated into an ongoing perceptual and emotional experience, becoming part of a new memory. This may be the mechanism of "false memory" syndrome, in which an unskilled interviewer induces incorrect memories in a witness. This process of ongoing rehearsal may provide a mechanism both for the establishment of PTSD and for therapeutic interventions designed to moderate the force of persistent memory.

The Impact of Stress on the Brain

Brain imaging studies in patients using magnetic resonance imaging (MRI) have shown that the hippocampus may be reduced in volume among patients suffering severe stress. Women who were sexually abused as children have hippocampi that tend to be significantly smaller in volume than women who were not abused, especially if they have symptoms of PTSD.[10] Similarly, traumatized police officers with a diagnosis of PTSD have significantly smaller hippocampi than officers without a diagnosis of PTSD.[11] But the mechanisms that lead to hippocampal volume reduction are still controversial, since the studies done in humans are natural history studies, not experiments. It is impossible to intervene in a way that could damage a person's hippocampi, so such experimental studies must be done in animals.

One of the first indications that stress could have a direct impact on the brain came from a study of marmoset monkeys.[12] Captive monkeys were exposed to a social stress and then injected with bromodeoxyuridine (BrdU), which is incorporated into newly synthesized

DNA and which marks dividing cells. After only an hour in a high-stress environment, there was a significant reduction in the number of newly produced neurons in the hippocampus. In other words, stress interfered with neurogenesis in a part of the brain where neurogenesis is the rule rather than the exception. This is especially interesting in the context of other research that has shown that hippocampal cell division may be required for learning.

Many studies have now been done in rodents to determine if stress has an effect on the rate of hippocampal neurogenesis.[13] Stressful experiences cause an elevation in blood levels of cortisol, and cortisol can depress the rate of neuronal growth. Chronic stress results in chronic inhibition of neurogenesis, raising a possibility that stress could reduce cognitive ability. Chronic stress is known to depress learning in rats: when rats were stressed for 21 days by being forcibly restrained, their ability to learn a maze was impaired. Yet stress-induced impairment in maze performance was not permanent, as rats tested 18 days after the stress were normal again.

Significant stressors come in many forms. A recent study tested the effect on hippocampal development of separation of rat pups from their mothers.[14] The pups were separated from their mothers for a period of 3 hours, which is much longer than a rat pup would ever be left alone in the wild. To compensate for the effect of handling by the experimenters, the control rat pups were separated from their mothers for only 15 minutes, which is roughly equivalent to the length of time that a mother rat might leave the nest in search of food. When the pups attained adulthood, those rats that had been separated from their mothers for a long time were compared to the rats separated only briefly. When cortisol secretion by the two groups of rats was compared, there was no difference, so there was no lasting effect of stress on hormonal levels. However, those rats who suffered maternal separation had lower levels of hippocampal neurogenesis. When cortisol levels were experimentally reduced in the maternally separated rats, neurogenesis in the hippocampus increased and became more normal. What this suggests is that maternal separation had a lifelong effect by making maternally separated rats more sensitive to the effects of cortisol. Thus, adverse experiences resulted in abnormal stress reactivity. Stress reactivity could potentially account for some effects in human children who are neglected by their mothers.

Rodent studies suggest that stress can reduce the production of new neurons and can impair hippocampus-dependent learning. Interestingly, the effects of stress in rats can be counteracted by treatment with antidepressant medications.[15] Inescapable electrical shock has been used as an experimental stressor because it induces symptoms in rodents that are reminiscent of depression in humans, including decreased motor activity, loss of appetite and weight, decreased grooming and self-stimulation, and immune deficits. Yet inescapable shock is a flawed model in that many strong shocks are needed to induce the state of depression that has been called "learned helplessness" and, moreover, a great many variables affect the degree of depression. Nevertheless, when rats are subjected to inescapable shock, the rate of neuronal proliferation in the hippocampus is depressed. Fluoxetine, which is commonly used to treat depressed patients, stimulates neurogenesis in the face of stress. These findings suggest that inescapable shock, which is said to result in a "state of behavioral despair," can be prevented by treatment with antidepressants.

A failure of hippocampal neurogenesis has been implicated as a cause of clinical depression in people.[16] The evidence in favor of this conclusion is 3-fold: there is a slight reduction in hippocampal volume in clinically depressed patients; every treatment known to be effective in treating depression also increases the rate of hippocampal neurogenesis;[17] and treatments that block hippocampal neurogenesis also block the antidepressant effect. The idea that depression is a failure of neurogenesis is appealing because depressed patients often show memory impairment and poor cognitive flexibility. It has even been proposed that depression can be a "stem cell disorder," although this may overstate the case. Reduced hippocampal neurogenesis could be a cause, a consequence, or merely a correlate of depression. Plausibility alone is no substitute for a rigorous experiment and, since experiments cannot ethically be done in patients, a great deal of research will be needed in animals.

EVOLUTION AND THE SOCIAL BRAIN

It seems self-evident that the human brain evolved under strong selective pressure to foster social behavior. Social affiliation is a powerful antidote to stress as well as an effective means to stay alive in a world that is often hostile and full of predators. Knowing about cooperation, kin, and social status is essential for survival, so it seems likely that brain mechanisms evolved to help vertebrates—and eventually humans—to keep track of this vital information.[18] Social behavior is crucial for reproductive success, so the neural mechanisms of sociality are likely to be under strong selective pressure. Vertebrate sense organs evolved to provide information relevant to the social mind—the vomeronasal organ, which is sensitive to pheromones, is a good example of a sense organ whose sole purpose may be to provide social information. Vertebrate behavior patterns arose specifically to facilitate those dependencies critical to the social mind—altruism and reciprocity would have had great survival value. Vertebrate cognitive skills provided a clear benefit to the social mind—facial recognition and verbal communication clearly require keen cognitive ability and are of tremendous survival value. Vertebrate brain circuitry may even have developed specifically to reward social affiliation.[19]

Affiliation is clearly crucial to human survival, but it also carries a great risk of betrayal and exploitation. The majority of stimuli that are perceived as stressful in day-to-day life relate to social issues.[20] Social defeat is a potent stressor, but even relatively minor changes in a social hierarchy are intensely stressful. All social behaviors involve the same three basic stages: approaching a partner who is initially unknown and potentially treacherous; learning to recognize that partner; and investing time and effort in that partner to the exclusion of all others. If a person is to navigate the stages of affiliation successfully, it is essential to have a strategy to assess the goals and intentions of the prospective partner. In other words, if social attachment is to be successful, one person must be able to understand the emotions that drive another person.

Foremost among the strategies to assess partners and to foster cooperation is language. The range, nuance, subtlety, and complexity of human speech is unmatched in the animal kingdom. Even if we consider primates and dolphins, which have highly developed communication, our language stands apart in its sophistication. This may be a reflection of the fact that, all things considered, the human animal is a poor physical competitor; a lone human could easily be killed by a lion or hyena, so we must make alliances and cooperate in order to avoid extinction. The social habit, the drive to form social alliances, is far older than the human line, as many primates also have a complex social organization. The centrality of the social mind is of such paramount importance in humans that dysfunctions of it are likely to be at the root of psychiatric problems such as personality disorder, autism, and schizophrenia.

Emotion and Social Attachment

Brain mechanisms have developed that make social behavior intrinsically rewarding, which has led to a suggestion that some forms of social attachment can resemble an addictive disorder.[21] Addiction is a form of compulsive behavior that, though initially rewarding, eventually leads to a narrowing in the range of behaviors as the behavioral repertoire becomes more and more focused upon obtaining a specific reward. There can be a subjective sense of loss of control, with intense preoccupation, compulsive intoxication, and symptoms of withdrawal when the object of craving is withheld. This is a good description of addiction to heroin or cocaine, but it is also an accurate description of intense romantic love. It may be that drug addiction and romance both commandeer the same neural circuitry involved in forming social bonds. If this idea is true, then we would predict that emotional states such as maternal love will involve some of the same neural circuits as addiction. This prediction appears to be true, as we shall see.

If emotions evolved to facilitate social bonding, one would predict that people with certain emotional "types" would be less able to bond or less adept in social situations. This prediction also turns out to be true. Children were studied to determine if shyness can influence a child's ability to respond to social signals.[22] Children were shown pictures of human faces that were modeling joy, anger, or a neutral emotion and they were asked to describe which emotion was shown. While joy was rarely ever mistaken, shy children were less able to discern anger in the face of another person and less adept at reading the emotional content of a face. Yet reading a face is a key social skill: lack of insight into simple facial cues might mean that shy children do not know if people are angry with them. This might make shy children anxious in nonthreatening social situations, which could manifest as an avoidance of social contact.

If emotion evolved to facilitate social bonding, this could explain why people with different personalities respond to social stressors so differently. If people are split into two categories, those with high levels of aggression and those with low levels, the groups have very different stress responses.[23] Under conditions of chronic stress, aggressive people are prone to disorder from allostatic overresponse, with anger-management issues, hypertension, cardiac arrhythmia, chronic fatigue, inflammation, or sudden cardiac death being characteristic. Conversely, passive people are prone to disorder from allostatic underresponse, with anxiety, depression, psychosis, weight gain, and infection being common.[24] Evidence suggests that the way a man responds to a stressful situation will effect how his body ages.[25]

LOVE AND ADDICTION

The people of the Cook Islands in Polynesia have a word in their language that means "dying for love."[26] In the West, the Romeo and

Juliet story has taken many forms, but all of them are romantic and all involve heroes or heroines who are helpless in the face of a sometimes fatal attachment to another person. Notions of romantic love are common in at least 147 out of 166 (89%) contemporary human societies, while fewer than 5% of animals pair-bond.[27] Euphoria, obsessional thinking, psychological dependence, and physical craving—all of these are symptoms of romantic love, but they are also symptoms of drug addiction. Given that romantic love and addiction are both so powerful and so common, one can speculate that both tap into the same crucial part of the human brain.

It turns out that romantic love involves the reward circuitry of the brain, which has also been implicated in motivating addictive behavior.[28] This startling finding emerged from a study that used functional magnetic resonance imaging (fMRI) to evaluate 17 people who all claimed to be in the euphoric stage of romantic love. The fMRI method reveals those regions of the brain that are activated in a specific task, since activation induces an increase in the rate of oxygen consumption and oxygen delivery in the blood can be visualized by fMRI. To determine which regions of the brain are activated by romantic love, each volunteer was shown a picture of his or her beloved. These pictures alternated with neutral pictures of a stranger of the same age and gender as the beloved. When brain images acquired while the subject was looking at a stranger were subtracted from brain images acquired while looking at the beloved, the difference showed those parts of the brain that were activated by the object of adoration. Love lit up those parts of the brain that are involved in reward; in general, there was activation in brain regions that have long been implicated in drug-seeking behavior. What this suggests is that romantic love is actually a state of motivation rather than a specific emotion. Moreover, the reward associated with looking at the beloved had nothing whatever to do with whether the object of desire was truly lovely; both beautiful and not-so-beautiful partners activated the same brain regions. Love activates the anterior cingulate, a part of the brain that has been associated with obsessional thinking. Interestingly, romantic love lights up different areas of the brain than does lust. The degree to which "romantic" centers of the brain were activated by pictures was a function of how long the relationship had lasted: the longer the relationship, the smaller the activation, which suggests that romantic love is a state that sadly cannot be sustained.

If romantic love uses some of the same reward circuits that are co-opted by drug addiction, this could explain why falling in love can feel so overwhelming, so irresistible—it is literally like becoming addicted. It is worth noting that this brain circuitry did not develop to facilitate drug addiction. Quite the other way around, drug addiction is so compelling because it co-opts brain circuitry that may have evolved to reinforce social behavior. The fact that the human brain is so intensely motivated by social behavior confirms that we are indeed social animals.

"THEORY OF MIND" AND SOCIALITY

If one stares fixedly into the distance, other people will eventually turn to look in the same direction. Why are they interested in what you see? Perhaps what you stare at is a danger to you and, by extension, a danger to them. Thus, joint attention and shared interest can have survival value. Yet, in order to share interest, other people must have concluded that you are looking at something for the same reason that they would look at it. In other words, they must have concluded that your mind and their mind work in fundamentally the same way. The conclusion that another person's mental state bears some similarity to your own mental state has been called "theory of mind," to acknowledge that it is only a "theory" that we are of similar mind. Frankly, this terminology is problematic; to a scientist, a "theory" is a grand system of principles and rules that can be used to predict or explain a wide range of disparate observations. In contrast, the notion that I have a mind that is similar to yours is relatively trivial.

In formal terms, "theory of mind" is the attribution of a mental state to another person in order to understand and predict her behavior.[29] Yet any theory about the mind of another person is necessarily based upon an understanding of your own mind, with a further assumption that there is a commonality of cognition between your mind and another mind. This remarkable awareness of the content of another person's mind sometimes saves us from being boors. People with a poor theory of mind can be egocentric, insensitive to social cues, indifferent to the opinions of others, and socially withdrawn, or they may wrongfully project their own concerns onto others.

Theory of mind reasoning can be studied by asking people to attribute emotions to a person in a photograph or to tell a story about how a person might feel in a particular circumstance. It is becoming clear that theory of mind reasoning is a higher-level ability, which does not depend upon any single brain structure but, rather, depends upon a broad integration of abilities.[30] People who suffer damage to the amygdala—which is near the hippocampus—early in childhood may be unable to detect tactless comments, unable to interpret irony, and unable to understand the emotional state of another person. Yet people who suffer comparable damage to the amygdala in adulthood are essentially normal, as if the amygdala were no longer essential for this skill. Similarly, frontal lobe injury impairs the ability to infer the emotional state of other people, though damage to either the right or left lobe causes the same degree of impairment.[31] Yet extensive injury limited to the middle part of the frontal lobe causes no impairment of theory of mind reasoning.[32] This is reminiscent of the debate about where memory resides. The answer appears to be that neither memory nor theory of mind are locked into any one place but, rather, are properties of an intact mind.

The ability to understand the thoughts and motivations of another person is a vital skill, no matter whether the goal is to live in harmony with that person or to avoid being killed by him. Perhaps the theory of mind developed from something as simple as joint attention, but it is clear that it is much more than that. In a sense, actors depend for their living upon the fact that most people have a well-developed theory of mind. Were it not so, an actor would not be able to make you identify with a character, would not be able to make you feel the anger, joy, or anguish of another person. Drama may be a good way to give our theory of mind ability a workout, but it is also a good way to confirm that we are fully human.

CHAPTER 12

MOTIVATION AND THE
SOCIAL MIND

What motivates human social behavior? Did the human brain evolve specifically to facilitate sociality? Is it possible that human evolution was driven by the simple fact that a solitary human is a poor competitor in a world where many carnivores could kill humans quite easily and where even many herbivores are dangerous prey for a lone hunter? To overcome our inherent weakness, it may have been necessary for us to band together, to turn the paltry strength of an individual into the formidable strength of a tribe.

Thus, the pessimist could say that what motivates social behavior is fear and pain: the fear of predators and the pain of hunger have forced us to be social. Yet such negative emotions are far from the present experience of many people, who are nonetheless motivated to be social. This implies that there are positive emotions that also relate to social behavior and that continue to motivate us. Most people would agree that the emotional rewards that derive from affiliation and connection to others are a powerful motivator for human behavior. Thus, the optimist could say that what motivates social behavior is love and joy: love of kin and joy in their success has bound us into tribes. But the scientist could say that while both answers may be true, neither is satisfying, since the mechanism of motivation is never addressed.

THE MECHANISMS OF SOCIAL BEHAVIOR

Social behavior may be a function of "reward circuitry" in the brain—neurons that can act to reinforce behavior that is of benefit to an individual. This is a surprisingly contentious viewpoint.[1] These neurons can, under certain circumstances, perhaps be appropriated for other purposes. When the body is in pain, certain neurons—perhaps the neurons of the reward circuitry—can deaden that pain. And when the body is exposed to drugs that stimulate the reward circuitry—such as cocaine and heroin—these drugs can become powerful motivators of human behavior.

What goes wrong when the reward circuitry is subverted by a chemical that is not ordinarily found in the brain? How can cocaine craving induce a mother to abandon her child or a man to kill his own brother? Why do people choose cocaine over food, even if they haven't eaten for days? How can methamphetamines motivate someone to abandon a life that he worked so long and so hard to build? Why does alcohol induce a dependable man to ignore his responsibilities? What brain mechanisms account for the fact that an addict's life can become radically simplified to a consideration of where and how to score? How can an addictive substance be so intensely motivating that people seek the effect, even if they know it is killing them? Why are humans so vulnerable to addiction? And does this vulnerability tell us anything about the social mind?

The answers to these questions are remarkably complex. The reward mechanisms that motivate human social behavior share features in common with and may be identical to the mechanisms that make a person vulnerable to addiction. To understand the reward mechanisms that motivate human behavior and drug addiction, it may be necessary to understand those brain circuits involved in the response to pain. Only then can we explore how such circuits are subverted in addiction or implicated in motivating social behavior.

The conceptual wedge that will split apart these complex issues is the placebo effect. This effect—the brain's active response to an inactive drug—proves that the brain can reward itself. That such self-rewards usually occur in a social context is intriguing and may be very revealing.

REWARD CIRCUITRY AND THE PLACEBO EFFECT

There is a clear link between pain and placebo—often defined as an inactive medication—since it is well known that "sugar pills" can reduce the perception of pain. In fact, in one study of placebo for postoperative pain, placebo was 56% as effective as a dose of morphine.[2] This simple observation makes it clear that placebo should not be defined as an inactive medication: dulling pain was the desired action, so the placebo was active. Perhaps placebo can be defined as a desired medication, in that placebo can take the place of whatever medication is needed. It is thus true that the modern pharmacopoeia contains some drugs that are effective only because we expect them to be so and that the history of medicine was, for millennia, a history of the placebo.

There is also a link between pain and addiction, since chronic physical pain often results in addiction to pain-relieving medications. Furthermore, psychic pain, whether in the form of anxiety or depression, is often associated with addiction, especially to mood-altering substances. Depressed people are twice as likely as nondepressed people to become alcoholic,[3] and pain is a major risk factor for relapse in patients recovering from chemical dependency.[4]

Yet the idea that there may be a link between placebo and addiction is speculative, tentative, and likely to be controversial. Since placebo modulates pain perception, placebo may also have an effect on addictive behavior. Perhaps the strongest evidence for this link is the finding that a drug—naltrexone, an orally active form of naloxone—which blocks the reinforcement provided by the placebo effect, can help addicts to quit smoking,[5] reduce cocaine intake,[6] abstain from opiates such as heroin,[7] or moderate alcoholic binges.[8]

THE NATURE OF PAIN

In about 300 BCE, Aristotle proposed that pain is not a physical sense but an emotion. Today, we acknowledge that pain has both sensory and emotional features. The best-accepted current definition is that pain is an unpleasant sensory and emotional experience that is associated with actual or potential tissue damage.[9] In contrast, suffering is dependent upon an awareness of the future and the pain that time may bring. Not all pain causes suffering and not all suffering arises from pain, but suffering does tend to connote an enduring sense of pain. Suffering can be a perception of impending pain, as it depends upon consciousness and an abiding awareness of the future. The idea that suffering contains a kernel of consciousness is consistent with the fact that patients with schizophrenia, many of whom have an impaired level of consciousness, also have a reduced response to physical trauma.[10]

There is a major emotional component to pain: tribal initiation rituals, which might seem intensely painful to someone from another culture, may not be perceived as painful at all within a culture, since they indicate a "coming of age." In the not-too-distant past, the Meru tribe of Kenya practiced ritual circumcision of adolescent males as a way for a boy to become accepted as a man. As his foreskin was cut away, a boy was expected to remain dignified and stoic, showing neither fear nor pain, and his status within the tribe was determined by his ability to bear unflinchingly what others might think of as unbearable pain. Heroism is often a matter of ignoring pain: for example, many winners of the Congressional Medal of Honor have performed incredible feats of bravery while mortally wounded. Clearly, in terms of pain perception, the mind can override the brain.

It is worth noting that pain sensitivity is a trait, like height or weight, that seems to have a normal or "bell curve" distribution.[11] What this suggests is that pain sensitivity has a genetic component. This striking idea emerged from a study, which found that people who suffer from chronic pain may be more pain sensitive and more vulnerable to mild depression and anxiety. Variation in a specific human gene—known as the COMT gene—can explain up to 29% of the vulnerability to a chronic pain condition called TMJ or temperomandibular joint disorder. This finding is notable because, until recently, many people thought that pain was "all in your head."

Pain is a difficult thing to study, first and foremost because the mind can override the brain; the subjective nature of pain precludes a rigorous comparison of my pain to your pain. Second, situations can cause both pain and suffering, and suffering can make pain far worse. Third, pain is not a constant, even if the cause of pain is constant; pain waxes and wanes and is sensitive to a great many variables. Behavioral studies have shown that learning is involved in perception and that an injury can induce chronic pain through learning.[12] Injury is also associated with a stress response, since pain is intensely stressful and stress can modulate pain perception. Finally, the study of pain is made infinitely more complicated by the placebo effect.

THE PLACEBO EFFECT

The concept of the placebo effect is familiar to many because it is widely known that "sugar pills" can, under certain circumstances, have palliative powers. The mere fact that a patient is treated can make him feel better because there is an expectation of benefit. In a clinical trial, a placebo is generally an inactive treatment similar in appearance to the active treatment, which is given without any expectation of benefit.[13] Yet people will often report therapeutic benefit from receiving placebo treatment. Untreated patients can show improvement for any number of reasons, including the natural course of a disease; patients given inactive placebo for a cold will report that their symptoms abate after 10 days because that's how long a cold lasts. Placebos are also meant to control for the possibility of a beneficial effect to a patient from mere contact with a doctor: perhaps social support or wish fulfillment or the "laying on of hands" is responsible.

A recent systematic review of 114 clinical trials of patients treated for any of 40 different conditions found little evidence of a powerful placebo effect.[14] Yet it may be noteworthy that patients who received placebo probably knew that they were not getting active medication. A key element of the placebo effect is the patient's belief that he has received active therapy, so little or no placebo effect is expected if the patient does not anticipate benefit. Furthermore, few of these clinical trials evaluated quality of life, where placebo could potentially make a major difference. Nevertheless, this review found that there was a significant—albeit small—placebo effect, except in the case of pain, where placebo could be substantial. Thus, it is imperative that clinical trials include some patients who receive a placebo, since a placebo group is the only way to control for the possibility that a medication is ineffective. These results were corroborated in a recent study by the same authors,[15] though both studies were limited by the quality of the clinical trials that they reviewed.

Other reviews suggest that the placebo effect is far more crucial, that the mind is a participant in every clinical trial, whether acknowledged or not.[16] The physiological events induced by a placebo are often similar to the physiological events caused by the drugs that they replace. Thus, placebo opiates can cause respiratory depression, just as real morphine does; placebo L-dopa can moderate tremor in Parkinson's patients, just as real L-dopa does; placebo cyclophosphamide can cause immunosuppression, just as the real chemotherapeutic agent does; and placebo analgesia can reduce heart rate in a person experiencing pain, just as a real painkiller does. Studies with positron emission tomography (PET) show that the same brain regions are engaged by placebo painkiller as are involved in real analgesia, which demonstrates that placebo uses normal brain circuitry for analgesia.

Real therapies administered in a hidden way—without knowledge of the patient receiving the therapy—tend to be less effective than normal.[17] We know this because several well-known analgesics were used to treat pain from dental surgery, but the drugs were given in two different ways. One group of patients was told that they would receive a potent painkiller, then the drug was injected into their intravenous

(IV) line as they watched: because the patient could see the procedure, this was referred to as an open injection. Another group of patients got the same drug at the same time and at the same dose level, but without being told that they were receiving it. The drug was injected directly into the IV line of each person in a manner that was hidden from view so that they were unaware that they had received any drug. Within a few hours, both the open and the hidden injection of analgesic were equally effective, showing that the drugs were effective painkillers. Yet there was a big difference in the patient's response to a drug in the short term. When patients knew that they were getting painkiller, their pain subsided quickly, yet when patients received a hidden injection of the same drug, their pain subsided slowly. This difference, of course, can only be due to the placebo effect enhancing the speed of the analgesic effect. In the short term, an open injection of saline is as powerful as a hidden injection of morphine.[18]

THE POWER OF THE PLACEBO

One of the clearest examples of the power of the placebo comes from study of a drug known as proglumide.[19] Originally, this drug was thought to be an effective analgesic; when it was first tested in a clinical trial, it was better than a placebo. Fortunately, the scientists involved in this study did not stop probing at this point. To verify that proglumide had an analgesic effect, a double-blind trial was performed in 93 patients who had undergone lung surgery. One group of patients was told that they would receive a potent painkiller, then each of these patients witnessed an open injection of proglumide into his IV line. Another group of patients, who were also told that they would receive a potent painkiller, instead got an open injection of saline. Each member of a third group received a hidden injection of proglumide in his IV line without being told that he would receive it, while each patient in a fourth group received a hidden injection of saline in his IV line. With this convoluted but fascinating design, it was found that a hidden injection of proglumide was no better than saline. Thus, prog-lumide was shown to have no analgesic effect, though it did enhance the placebo effect when given as an open injection.

What this study suggests is that pain perception is a function both of pain pathways leading up from the body and of expectation pathways leading down from the brain. Proglumide had no effect on the pain pathways, as a normal analgesic should, but it had a very powerful effect on the expectation pathways, as a placebo does. In fact, proglumide enhances the placebo effect, and it does so in a dose-dependent manner: the greater the dose, the greater the effect when the drug is administered openly. If proglumide is given as a hidden injection, it is worthless. This thought-provoking study calls into question the way that clinical trials are usually done, since a hidden administration of drug may be needed to determine if a drug is truly effective. When we give a painkiller, we cannot know if it acts on the pain pathways, on the expectation pathways, or on both pathways. In the future, it may be necessary to use an open-hidden drug administration strategy, at least to test those drugs that are thought to act quickly.

Nowhere is the placebo effect more important or more controversial than in the treatment of clinical depression. This is because the course of the illness is quite variable, and depression usually will remit even without treatment.[20] Nevertheless, antidepressants are among the most frequently prescribed medications in the world. A systematic review of 75 recent placebo-controlled trials of antidepressants showed that about 30% of depressed patients who are given placebo report an improvement in their condition. In contrast, 50% of patients who receive an active medication report benefit. Thus, antidepressants are only somewhat more effective than placebo, and the incremental benefit from active medication is much smaller than we would like it to be. What is especially revealing is that the proportion of patients who report a benefit from placebo has been increasing steadily over time: in 1980, about 22% of placebo-treated patients reported a benefit, whereas in 2000, about 35% of placebo-treated patients reported a benefit.

How is it possible that more people now benefit from sugar pills than benefited in the past? It may be that people have come to expect more of their medication and it is this increase in expectation that has led to an increase in the efficacy of placebo. In short, what you really want is what you get. This finding has several significant implications, not least of which is that it would be very misleading to use historical placebo data in a clinical trial of a new antidepressant. Since the placebo effect was much smaller in the past, historical control data would enhance the apparent efficacy of a newer drug, even if it did not work.

It is a mistake to think of the placebo effect as being either trivial or somehow not "real." A study of patients who had wisdom tooth surgery showed that people who did not have any pain medication suffered gradually increasing pain as their general anesthesia wore off.[21] Patients who were told that they would receive analgesia but who were given placebo instead could be divided into two groups based on their placebo response. One group reported that their pain increased steadily, exactly as in untreated patients, while the other patients reported that their pain was stable or decreased. A patient was more likely to experience a placebo effect if his pain was severe before treatment; 39% of patients overall had a placebo response, but patients in severe pain had a 50% chance of getting a placebo benefit. In other words, the more patients needed a placebo effect, the more likely they were to get it. In many ways, a placebo is like an active drug; research has shown that placebos have a dose-response curve that is similar to an active drug and that patients can develop tolerance and even show side effects from a placebo.

The truly stunning thing is that the placebo effect can be blocked by treatment with a drug called naloxone.[22] Patients who are given naloxone after dental surgery report *more* pain than patients given placebo. The difference between open and hidden injections can be eliminated if a patient is given a dose of naloxone. We know now that pain induces the release of endorphins, which are naturally occurring molecules in the brain that reduce pain sensitivity. An opiate like morphine is effective precisely because it binds to the same brain receptors that bind endorphins, although morphine is more powerful and lasts longer than endorphins. Naloxone also binds to opiate receptors, thereby blocking the action of endorphins. Naloxone thus proves that the placebo effect is "real," a consequence of the fact that the brain attenuates its own pain.

PAIN AND ACUPUNCTURE

Acupuncture is clearly not the only treatment for which the placebo effect is important, but it is certainly a fascinating case study of just how subtle and magical the placebo effect can be. In 1998, a leading medical journal published the findings of a consensus development panel on acupuncture.[23] The panel was formed as part of an effort to stimulate scientists who work on acupuncture to come to a consensus on findings that could be trusted so that patients could be given clear advice on whether acupuncture works. Yet the panel was not a success, since it concluded that

> although there have been many studies of its potential usefulness, many of these studies provide equivocal results because of [problems with] design, sample size, and other factors. The issue is further complicated by inherent difficulties in the use of appropriate controls, such as placebos and sham acupuncture groups. However, promising results have emerged, for example, showing efficacy of acupuncture in adult postoperative pain and chemotherapy nausea and vomiting and in postoperative dental pain.

The failure to make clear recommendations was galling because, even in 1997, more than one million Americans received acupuncture treatment each year for a wide range of conditions.[24] While the occurrence of side effects of acupuncture was known to be very low, the evidence that it does any good was weak. Many factors can influence the efficacy of acupuncture, including the relationship between the clinician and the patient, the degree of trust, the expectations of the patient, and the compatibility between the backgrounds and beliefs of the clinician and the patient. Because so many variables are potentially crucial, few of the studies performed before 1998 were accepted by the panel. In the years since this consensus panel, there have been a number of high-quality clinical trials of acupuncture. One of the key advances has been that recent studies are placebo controlled so that patients don't know if they have received "real" acupuncture or "sham" acupuncture. In fact, the best studies are now "double-blind," meaning that neither the patients nor the clinicians know if the patient received real acupuncture. This is crucial, since the expectations of a physician might bias how a patient experiences her symptoms.

One of the first high-quality clinical trials of acupuncture tested the ability of acupuncture to relieve the nausea and vomiting caused by high-dose chemotherapy.[25] The patients were all women who received an aggressive regimen of chemotherapy (cyclophosphamide, cisplatin, and carmustine) in treatment for metastatic breast cancer, a treatment regime that is known to produce nausea and vomiting in most patients. A total of 104 women were randomly allocated to receive either electroacupuncture at classic antiemetic acupuncture points, or minimal needling with mock electrical stimulation, or no acupuncture. Women were also given standard antiemetic therapy, since to deny standard treatment would be unethical. Women who received traditional acupuncture had significantly fewer episodes of vomiting than women who got no acupuncture. Yet the women who received mock acupuncture got some benefit from placebo treatment: they suffered more vomiting than women who got acupuncture but less than women who received standard antiemetic therapy. In other words, sham treatment was not as good as acupuncture, but it was better than nothing. However, this study had weaknesses; though patients didn't know which treatment they received, the acupuncturists did know, so they could conceivably have communicated an expectation of success to the patients.

Another clinical trial, which suffered from the same problem of incomplete blinding, tested the effect of acupuncture on 177 patients with chronic neck pain.[26] Patients were randomized to receive either acupuncture, "sham" treatment, or neck massage for 3 weeks. One week after the last treatment, the acupuncture group showed improvement relative to the massage group but not compared to the "sham" treatment. Thus, massage had little or no benefit, even though it is one of the commonest forms of treatment for neck pain, while both acupuncture and "sham" treatment were safe and apparently effective. Positive benefits of acupuncture were stronger in patients with a long history of neck pain and in patients with a generalized form of pain known as myofascial pain syndrome. These results are hard to interpret, other than to say that acupuncture is no better than placebo, but placebo is better than nothing.

One of the first clinical trials that made an earnest effort to blind patients as to what treatment they received was a study of acupuncture for postoperative pain and nausea.[27] A total of 175 patients, all of whom were scheduled to receive abdominal surgery, were allocated to get either acupuncture or shallow needling that did not penetrate tissue as deeply as acupuncture. Needles were placed in the patients' backs so that patients could not see the needles as they were inserted. As a testament to the success of blinding, most patients could not tell which treatment they received. Furthermore, the evaluating physicians didn't know which treatment the patient received: only the acupuncturist knew for sure, and the acupuncturist did not assess treatment efficacy. Patients were allowed to self-administer intravenous morphine to control their own pain, and the amount of morphine they used was recorded as a measure of how well acupuncture worked. Morphine use was reduced 50% in patients who received acupuncture, and postoperative nausea was 25% lower compared to sham-treated patients. It may be worth noting that this study, which reported a far greater benefit from acupuncture than most other studies, was done in Japan, where patients probably have a stronger expectation of benefit from acupuncture than in the United States.

An innovative technical advance in 2004 made it easier to do real placebo-controlled studies of acupuncture.[28] Previously, it had not been possible to blind both the patient *and* the therapist as to whether a patient got acupuncture: though needle insertion is rather painless, the patient can still watch the needle going in and the therapist, of necessity, watches needle insertion. This all changed when a novel method was introduced to blind both patient and therapist. With this new method, acupuncture sites are covered by a small plastic ring that supports a thin piece of plaster above the skin. As the needle penetrates the plaster, the tip of the needle cannot be seen. A "placebo" acupuncture needle was made by modifying a typical needle so that the needle retracts into the handle, much the way a carnival sword retracts into the grip when a "victim" is stabbed. The patient still feels some pain as the needle penetrates the cutis of the skin, but the "placebo" needle never penetrates deeper into tissue. The sensation of cuticular penetration cannot be distinguished from deeper penetration, so neither the patient nor the acupuncturist knows if "sham" acupuncture was done.

Using this improved placebo, a total of 220 surgical patients received either acupuncture or sham treatment for postoperative nausea.[29] The incidence of nausea and vomiting was the same in both groups if the patient was unaware of which treatment was received. Even with the large sample size in this study, there was no significant difference in the number of patients who required additional pain medication. Thus "sham" acupuncture is as good as "real" acupuncture in reducing postoperative pain.

There are now a number of large, well-designed, and carefully controlled studies that have failed to find a benefit from acupuncture. A trial of acupuncture in 100 fibromyalgia patients found that the subjective pain rating among patients who received acupuncture was no different from patients who received sham therapy.[30] In 302 migraine patients, there was no difference in the number of days with headache when comparing sham treatment with acupuncture, but both groups enjoyed a significant benefit compared to an untreated group.[31] In this study, the proportion of responders was 51% with acupuncture, 53% with sham acupuncture, and only 15% in the untreated group. Placebo treatment was thus just as effective as several medications that are widely accepted for migraine prophylaxis.

These findings suggest that the context of treatment matters a great deal. Patients receiving placebo could benefit from the healing touch of the acupuncturist, the therapeutic environment of the clinic, or even the fact that they were enrolled in a clinical trial. In a study of acupuncture for the pain of dental surgery, those patients who believed they had received "real" acupuncture had less pain than patients who believed they had received placebo.[32] Belief had a stronger effect on pain perception than anything else, since the *perceived* assignment to a treatment group had a greater effect than did the actual treatment. And yet there is uncertainty: a recent clinical trial of acupuncture for chronic headache found that acupuncture was effective while placebo was not. The study, which followed 401 patients over a period of 12 months, found that patients who received acupuncture used 15% less medication, made 25% fewer trips to a doctor, and took 15% fewer sick days.[33] Nevertheless, on balance, the evidence to date suggests that acupuncture is largely a placebo effect.

ALCOHOLISM AS A MODEL OF ADDICTION

Why are human beings prone to addiction? Addictions come in many forms, some innocuous and some deadly: coffee, chocolate, sex, food, alcohol, cannabis, and cocaine. We will focus on alcoholism because it is the most common, best understood, and most costly to society, but many truths about alcoholism are true of other addictions as well. The *New York Times* reported that, according to the National Survey on Drug Use and Health, alcohol is the addictive substance that is most likely to cause problems for Americans.[34] Using the government's definition that an addict is someone who has used a drug in the last month—which is a commonly used but overly broad definition—then 60% of Americans are addicted to alcohol. Furthermore, 27% of Americans went on an alcohol binge—defined as 5 drinks on one occasion—in the last month. By comparison, the addiction rate for other drugs is lower: cigarettes (37%), marijuana (15%), painkillers (10%), crack cocaine (8%), methamphetamine (5%), and heroin (3%).[35] We note that, by this definition, everyone is addicted to food and water.

The critical issue becomes, how damaging is the addiction in question? Alcohol abuse is causally related to more than 60 different potentially fatal conditions, including cancer, stroke, heart disease, diabetes, and cirrhosis.[36] The antialcohol campaign in Russia showed that a 25% drop in consumption of alcohol was associated with a 40% drop in the homicide rate. Across the globe, the burden of disease from alcohol (4.0% of all deaths) is exceeded only by malnutrition (9.5%), unsafe sex (6.3%), high blood pressure (4.4%), and tobacco (4.1%).[37]

Depression and alcoholism frequently co-occur. It was once standard wisdom that depression responds poorly to treatment in alcoholic patients, but this is apparently not true.[38] Antidepressant medication is about as effective in alcoholic and drug-dependent patients as it is in nonaddicted people, with treatment resulting in a 25% reduction in substance abuse. If medication is effective in treating depression, it also tends to diminish the severity of alcohol abuse. This argues strongly that alcoholism is an effort to self-medicate the symptoms of depression.

Chronic alcoholic patients show a decline in cognitive ability that may reflect a direct effect of alcohol on learning.[39] Given that learning can cause neurogenesis, it may be that alcohol depresses the rate of neurogenesis in the brain. This hypothesis was tested in rats that were made alcohol dependent by force feeding with alcohol for 4 days. As the rats were withdrawn from alcohol, the rate of neurogenesis in the hippocampus was measured. Right after weaning, the number of new cells in the hippocampus was reduced. Seven days later, there was a burst of neurogenesis as the rate of formation of new neurons doubled. These results suggest that alcohol intoxication is associated with a depression of hippocampal neurogenesis, whereas abstinence can return neurogenesis to a normal level. Whether a similar effect would be seen in humans is not known, since such experiments are not possible.

If alcohol use induces a change in the plasticity of the brain, then addiction shares features in common with learning. Perhaps this is why alcoholics find it so hard to learn adaptive behaviors as they attempt to abstain from alcohol. Recent research suggests that exposure to amphetamine, cocaine, nicotine, and morphine results in persistent changes in neuronal structure in the reward circuits of the brain.[40] The magnitude of structural change is related to drug dose, the number of drug treatments, and the pattern of drug abuse, and it is not known whether neurons can return to normal after periods of abstinence. Different drugs act in different ways, but all of the drugs have an effect on neurons that would likely impact the way those neurons work. Thus, some of the cognitive deficits in addicts may be due to limits on synaptic plasticity imposed by earlier drug use. The effects of drug use may therefore last far longer than the use itself.

Dopamine pathways are thought to mediate the rewarding effects of alcohol and other drugs of abuse.[41] Alcohol intake indirectly stimulates dopamine release in the brain, suggesting that if dopamine release could be blocked, this might lessen the reward obtained from alcohol. This idea was tested using a drug called topiramate, which blocks dopamine release in the brain. Topiramate has few side effects and has been in clinical use for years because it reduces the incidence and severity of epileptic seizures in some patients. Epileptic patients given topiramate reported that it helped them to fight food cravings and lose weight. A 12-week clinical trial was done using a placebo-controlled double-blind study design to test the effect of topiramate on alcohol consumption in 150 alcoholic men. Results were encouraging: treated men had 2.9 fewer drinks each day, 28% fewer days of heavy drinking, and 26% more abstinent days. Topiramate may reduce alcohol craving by inhibiting the alcohol-induced release of dopamine so that alcohol is no longer as rewarded as it once was. In short, alcoholics have fewer reasons to abuse alcohol when its reinforcing effect is blocked by topiramate. This finding could herald a new direction in addiction treatment. Traditional therapies have focused either on substituting a less-addictive substance in place of a more-addictive substance (e.g., methadone for heroin) or have sought to treat the symptoms of withdrawal without reducing the cravings. In contrast, topiramate may directly block the mechanism that reinforces alcohol addiction.

REWARD CIRCUITRY IN THE BRAIN

A person primed to show a placebo effect is probably in a state of willing expectancy. An expectation of benefit from treatment is much like an expectation of "reward" from any other behavior, so the placebo effect may engage the "reward circuitry" of the brain. This circuitry is activated in subjects seeking or experiencing a reward, and it probably evolved to motivate such survival-enhancing behaviors as eating and sexual activity. Thus, the placebo effect may depend upon ancient circuitry that has been crucial for human survival.

How does addiction co-opt the reward circuitry of the brain? Originally it was thought that reward is a function of dopamine release in the brain, and that dopamine is a "hedonic signal" that is associated with a sense of pleasure.[42] This idea is not as well accepted now, however, because it fails to account for a core feature of addiction—addiction is a stubbornly chronic condition and addicts are at risk of relapse for years after they stop the addictive behavior. It is now thought that dopamine release in the brain promotes reward-related learning and that this learning becomes compulsive and resistant to extinction. It is perhaps noteworthy that addiction is not just a matter of physiological dependence; certain drugs that do not cause addictive behavior can nevertheless induce withdrawal symptoms when a dose is missed (e.g., tricyclic antidepressants and serotonin reuptake inhibitors), whereas other drugs are strongly addictive but do not cause withdrawal symptoms (e.g., amphetamines).

Drug-seeking behavior is now thought to be due to two linked processes: an increased motivation to use the drug of choice based on the rewarding properties of that drug, and a decreased ability to inhibit drug use, even if the addict clearly understands that drug use can be lethal.[43] Addicts consistently choose small immediate rewards over large delayed rewards, so they may have a diminished capacity to evaluate the consequences of their actions. This inability to delay gratification can occur even if a patient knows that his choice is inappropriate. Addicts are thus similar to patients with obsessive-compulsive disorder. Drug relapse is rarely related to craving (7%), but is far more often a function of compulsive drug use (41%).

The brain reward circuitry may also be abnormal in people who are addicted to nicotine.[44] Approximately 25% of American adults smoke regularly. Among smokers, the rate of clinical depression ranges up to 60%, whereas depression affects about 25% of nonsmokers. Smokers with depression are far more likely to become dependent upon nicotine, to progress to serious dependence on nicotine, to experience severe withdrawal symptoms upon quitting, and to have some relief of depressive symptoms while smoking. When the brain reward system of clinically depressed people was probed by injection of radioactive amphetamine, it was found that people with depression, whether smokers or not, had an abnormally quiescent brain reward system. Since nicotine can induce dopamine release in the brain, these results suggest that smokers use tobacco to self-medicate for depression.

Reward circuitry is involved in the placebo effect, and reward circuitry is also important in addiction. Thus, the placebo effect is implicated in the reinforcement of addictive behavior. This idea is attractive for several reasons: addicts are likely to have experienced a good deal of suffering before their addiction, so addiction could be an effort to self-medicate; addiction commonly co-occurs with other conditions that respond to placebo treatment (e.g., depression); naltrexone, which blocks the placebo effect, also reduces addictive behavior; there is a learned component both to addiction and to the placebo response; both addiction and placebo activate some of the same brain regions; and both addiction and placebo response are more prevalent in people with a high level of anxiety and impulsivity.

We know something about how reward circuitry in the brain controls addictive behavior, but we know little about the relationship between reward circuitry and social behavior. Perhaps the same endorphins that account for the placebo effect also regulate mood and even social behavior.[45] This idea could offer insight into some difficult and critical questions: What makes people so vulnerable to addiction? Does early exposure to drugs of abuse make a person forever more vulnerable to addiction? Is the risk of addiction greater among people with deficient social skills? What treatments will be most effective for addiction? Can chronic pain be treated without increasing the risk of addiction? Is there a way to help people who suffer from the effects of self-induced social isolation? Do certain genes or environmental factors increase the risk of social isolation?

CHAPTER 13

GENES, ENVIRONMENT, AND HUMAN BEHAVIOR

F ossil evidence suggests that the distant ancestors of humans had rather small brains and that the brain became progressively larger as it became increasingly more human. Such incremental change in form could have been the result of a gradual process of evolutionary change.

Yet, if the human brain has evolved, this implies that many—perhaps even most—properties of the brain are under the control of specific genes. If genes could not control what the brain is—and perhaps what the mind can become—then it would be impossible for the brain to change as a result of evolution. In short, to understand the brain, we must acknowledge that it can evolve, that what it is and what it does is controlled by the genes in the human genome.

Such considerations resurrect an old debate about whether genes or the environment have a dominant influence over human behavior. This argument last erupted in 1994, when there was virulent controversy surrounding publication of *The Bell Curve*.[1] That book argued that genes limit intelligence and tend to structure a class system in the United States. It thereby endorsed a defeatist public policy that has little basis in truth.[2]

Since it is easy to argue from an ideologue's viewpoint and hard to capture the nuances of truth, most scientists have avoided getting embroiled in an argument about the relative role of genes and the environment in the formation of the mind. Yet science has much to say about this subject. Genes simply cannot explain all of human behavior: if they could, life would be simpler and more predictable than it is. But on the other hand, environment cannot explain all that we become. The most recent information suggests that roughly half of what makes us who we are is inherited from our parents. This, of course, means that roughly half of what we are is a result of the environment acting upon us.

PATTERNS OF BRAIN GROWTH AND MATURATION

There is a complex pattern of change in brain volume over time as a healthy young brain matures. Intracranial volume increases nearly 3-fold from birth to age 5, but there is only about a 10% increase in brain volume thereafter, and total brain volume is stable by the age of 12.[3] On average, boys' brains are about 12% larger than girls', even after factoring in height and weight. Yet brain size is a poor correlate for intelligence, so it is likely that the volumetric difference between boys and girls is not important.

Gray matter volume of the brain is at a maximum by roughly 10 years of age, but decreases thereafter.[4] At the same age, white matter volume is only about 90% of the adult value. There are rapid changes in the relative proportion of gray matter to white matter through adolescence, and the rate of change does not slow until age 25.[5] Because total volume of the brain is stable by age 12 at the latest, any decrease in gray matter volume after age 12 must be offset by an increase in volume of either white matter or the cerebrospinal fluid (CSF) that surrounds the brain. Although CSF volume apparently does increase somewhat during adolescence, the increase is too small to account for the decrease in gray matter volume.

The most reasonable hypothesis to explain gray matter volume loss in adolescence is that it corresponds with white matter volume increase, in an ongoing process of myelination.[6] Because a myelinated neuron transmits action potentials as much as 100 times faster than an unmyelinated neuron, myelination may be related to the cognitive maturation of the brain. White matter myelin does not fully mature until perhaps as late as 48 years old.[7] It remains unknown whether the decrease in gray matter volume between age 20 and age 40 is entirely explained by white matter myelination or whether there is also a component of gray matter atrophy. However, over the age range from 20 to 77 years, the annual volume loss rate in gray matter averages less than 1%.[8] Careful work has shown that frontal and prefrontal gray matter are both more prone to age-related loss in volume than is gray matter elsewhere in the brain, which may explain why frontal functions of the brain (e.g., planning and organization) are more likely to be impaired in old age.

There is a general pattern for brain regions to mature from the back of the head to the front.[9] Cortical regions that perform such basic functions as sensing and moving, which tend to be at the back of the brain, mature sooner, and brain regions that integrate basic functions—by performing associative functions—mature later. Those parts of the brain that are "ancient," in the sense that they resemble parts of more primitive organisms (e.g., the "dinosaur brain"), mature early in development. Yet certain parts of the brain do not mature until adulthood, and maturation of these parts may, in fact, be key to defining adulthood. For example, the dorsolateral prefrontal cortex (above and outside the eyes) does not mature in structure until the third decade. This brain region is thought to inhibit impulses, weigh consequences, prioritize, strategize, and make decisions— the relative lack of maturity of this tissue in adolescence will not surprise anyone who has tried to raise a teenager.

Direct evidence linking brain development to behavior may be at hand in a study that related white matter maturation to reading ability.[10] Although most children learn to read rather quickly, perhaps 10% of children have problems reading, which cannot be explained by poor schooling, lack of intelligence, or inadequate opportunity. A group of 32 children of varying reading skill levels was evaluated by magnetic resonance imaging (MRI), using a new method called diffusion tensor imaging (DTI), which visualizes white matter. As white matter tracts become more adult in structure, the DTI signal changes in a predictable way. Immature white matter shows less uniform structure, whereas mature white matter shows tightly packed, myelin-covered nerves. Any disturbance in myelination is obvious, provided that all subjects are of the same age. Children who were imaged were also tested for reading skill using a word identification test. A strong correlation was found between reading ability and the degree of maturity in a specific white matter tract in the brain. Structure in this specific tract did not correlate either with age or nonverbal intelligence but did explain 29% of the variation in reading ability. While this may not seem like very much, it was statistically significant and could well be clinically significant. Thus, maturation of white matter appears to play a role in the acquisition of reading ability.

Cognitive skills can apparently only be learned when the brain structures needed to support that skill have matured.[11] Thus, development of an adult capacity for working memory is contingent upon the maturation of white matter in the frontal lobe. The DTI method suggests that the degree of myelination in frontal white matter correlates with working memory. This is interesting because variation in working memory is not well explained by age alone, confirming a perception that maturation is only partially explained by the chronological age of a child.

There is emerging evidence of critical periods of brain growth, such that children who grow well during a critical period will eventually be cognitively superior to children who grow poorly during that period.[12] This potentially controversial finding is based on a longitudinal study of 221 children studied over a period of 9 years. Cognitive ability was related to head size at age 9 and to head size at 9 months of age but not to head size at birth or in utero. Head size is a surrogate measure for brain volume, so this study suggests that children who enjoy a rapid rate of growth during early childhood are more likely to have good cognitive ability. Thus, it seems that brain growth during infancy may be more important than brain growth during fetal life.

These findings are consistent with a great deal of evidence showing that children who have delayed growth during infancy are more likely to be cognitively impaired or to need remedial help in school. Why are such findings controversial? In the recent past, politicians have argued that Head Start and other such programs that intervene in the lives of disadvantaged children are a waste of money.[13] This argument, which was always weak, can no longer be sustained in the light of new evidence. Children who participate in a preschool intervention for as little as one year have a significantly higher rate of school graduation, better success in education, and lower rates of juvenile arrest, especially for violent crime.[14]

HOW DO GENES BUILD BRAINS?

We know very little about how genes interact to control brain growth, although there have been some tantalizing clues lately. The most distinctive traits of the human brain are the enormous size, relative to the size of the body, and the enormous complexity, relative to even genetically similar animals such as the chimpanzee. Since mutation of any one of 6 different genes is associated with a medical condition known as microcephaly, it appears that these 6 genes regulate brain size in humans.[15] In microcephaly, there is a striking reduction in the volume of the brain combined with severe mental retardation, even though brain structure is normal and there are few if any effects elsewhere in the body. If a patient has a mutation of one microcephalin (or MCPH1) gene, brain volume may be reduced to 400 cubic centimeters (cc), as compared to a normal volume of 1,400 cc. Thus, mutation of one gene leads to a 70% reduction in total brain volume. The MCPH1 gene may control the rate of cell division of neuronal stem cells during neurogenesis, but this is not known for certain.

One particular form of the MCPH1 gene is more common than any other form, at least according to one study that sampled DNA from 86 different people from around the world.[16] The abundance of this particular form was far higher than would be predicted by chance alone, suggesting that this mutation is under natural selection. This particular variant is a new mutation that differs from the ancient form. Natural selection has apparently acted quickly to increase the abundance of what may be an adaptive gene. Analysis of the degree of divergence between genes—based on the assumption that mutations occur at a relatively constant rate—has revealed the approximate date of this mutation's earliest appearance. This analysis suggests that the MCPH1 mutation first appeared in humans only about 37,000 years ago, even though the human genome is about 1.7 million years old. Interestingly, the date of this mutation roughly coincides with the time at which Europe was first colonized by people emigrating from Africa. Although we cannot be sure that MCPH1 increased in frequency due to selection for brain size, it appears that there is ongoing natural selection operating on the human brain.

Another study confirms and extends the previous study, reporting that a second gene has also undergone recent and dramatic selection in humans.[17] This gene, known as ASPM or MCPH5, is also one of the 6 genes that—in some forms—can produce human microcephaly. The ASPM gene was sequenced in the same group of people, and again it was found that one variant was far more common than any other. Analysis of this variant suggests that it appeared a mere 5,800 years ago, after which it underwent a very rapid increase, perhaps because the new variant of the gene acted to increase brain size. If the date of origin of the new variant is correct, this implies that the human brain is undergoing extremely rapid evolution, perhaps even now, due to selection for a trait associated with increasing brain volume.

Researchers are continually adding to the list of identified genes that contribute to the normal development of the human brain.[18] Certain genes influence development of the cortex, with effects on the growth of neural stem cells, the maintenance of neural integrity, or the migration of neurons from their site of birth to their site of incorporation into the neural matrix. A mutation in any of these genes could lead to characteristic abnormalities in brain growth and development. As our understanding increases, it may become possible to predict which parents are at risk of having an infant with an abnormality, although it may never be possible to "cure" an infant born with a syndrome of abnormal cortical development. The process of wiring the cortex is so exceedingly complex that it seems unlikely that physicians will ever be able to direct this process in the absence of an intact genetic program. Once aberrant connections are made, it is unlikely that they can be unmade and a normal cortex constructed—although it may become possible at least to ameliorate some of the clinical problems associated with cortical miswiring.

Clearly, a great many genes are involved in building the human brain. A recent comparison of the human to the primate brain showed that at least 91 genes are expressed differently in humans and, of these, at least 90% are more expressed in humans.[19] By contrast, in human liver and human heart, there were essentially no differences in gene expression compared to primates. The fact that so many genes in the human brain are increased in expression suggests that neuronal activity is greater in humans than in primates. It seems likely that very few genes are unique to humans,[20] so the differences between humans and primates may arise mostly from different levels of expression of similar genes.

Gene expression controls how the brain wires itself. We know very little about how the rat or mouse brain makes specific cell-to-cell connections during growth and development, and we know essentially nothing about this process in humans. Yet the pervasive differences between the human brain and other primate brains could be explained by a relatively small number of genes.[21] Genes must be responsible for every feature of the human brain, since they: specify whether and where neurons will form circuits of specialized function; encode synaptic architecture and control the degree of neuronal plasticity; regulate the metabolic rate of neurons; dictate the balance between neurons and supportive cells; influence the growth of axons and dendrites; control how distant neurons can interact with one another; confer pattern on the developing brain; and influence how neurons attach to one another and how sensory organs wire themselves into the growing neural network.

Though we are in the dark about how genes build minds, we have gained some understanding of how genes build components of the brain. There have been elegant studies of how sensory organs wire themselves into the brain. For example, 50,000 ganglion cells in the rat retina send out axons that find their way to the visual center of the rat brain.[22] The pathfinding ability of these axons is remarkable: during fetal life, neurons send out cell processes that grow all the way up the optic nerve before they can even approach the right part of the brain. Then these growing sensory neurons must connect to specific higher-order neurons in order for visual signals to be properly processed. There are more than a dozen molecules that guide this process, and the process remains poorly understood, even in rats. Yet the stakes are very high; if it became possible to control the events that enable a ganglion cell to make the right connections within the brain, it might then become possible to induce axons to regenerate after injury. Prospects for this are far in the future: in humans, over a million retinal ganglion cells must navigate to the visual cortex during fetal life. The distance they must cover is far longer than in rats, and the process occurs quite early in embryonic development, so the mechanisms are inaccessible to study.

For many years the accepted model of how the brain wired itself assumed that there is an "initial exuberance" of synaptic connections and that these connections are progressively pared back as the brain continues to develop.[23] According to this model, many more synapses are present at birth than are needed at maturity, and experience acts to strengthen those synapses that are needed and weaken those that will eventually be discarded. Yet there has been little evidence to support this model, and scientists now think that even the first connections are made with a high degree of specificity. But plasticity is still required of neurons, or else the synaptic remodeling that enables us to learn would not be possible. The immature visual cortex is capable of a great deal of remodeling, as shown by the fact that sensory deprivation early in development remodels the neural connections of the visual cortex in a permanent way. Monocular deprivation studies in cats show that the wiring pattern established by visual neurons during the first few days of life has an irreversible effect on the adult visual cortex. These results suggest that gene expression in growing neurons is evoked by—and perhaps controlled by—the environment.

DO GENES SHAPE WHO WE ARE?

The first and most robust evidence for the importance of genes in human behavior came from the study of twins separated at birth and reared in different homes.[24] Because these children were reared apart, they experienced different environments, even though they shared the same genes. "Split-twin" studies characterize the importance of both nature and nurture, since the effects of genes and the environment can be separated. These experiments are best if they involve both identical and fraternal twins, since this makes it easier to pull apart the effect of genes. Identical twins are derived from a single fertilized egg, which splits into two separate embryos that share identical genes. In the biological literature, identical twins are usually referred to as monozygotic (or MZ) twins, since they derive from a single egg. Fraternal twins, often called dizygotic (or DZ) twins, are instead derived from two different fertilized eggs. Since two eggs must be fertilized by two different sperm cells, the genetic differences between fraternal twins can be extensive. By having MZ and DZ twins with different degrees of genetic similarity in the same study, it is much easier to determine the relative importance of genes to development.

Any time a split-twin study is done, the study should include as many twins as possible. This is because there is always some degree of randomness and uncertainty in the expression (or in the measurement) of human traits, and this randomness can only be factored out if a study includes many subjects. It is simply inadequate to describe similarities between individual twin pairs, no matter how striking the coincidences. From a scientific standpoint, such anecdotal evidence is simply not worth reporting. It is very difficult to do split-twin research because it is usually hard to find an adequate number of twins who were separated at birth. Only by having access to information on millions of births through a central registry of birth data is it possible to accrue the many subjects necessary to make a strong split-twin study.[25]

Comparing twins split at birth with twins reared together provides insight into the role of genes and the environment in determining human traits. If genes are central in determining human traits, then identical twins will be more alike than fraternal twins reared under the same circumstances. If identical twins reared together have similar intelligence but the intelligence of identical twins reared apart differs, this would mean that the environment is crucial in determining intelligence. If fraternal twins reared together are similar in intelligence, even though they are not genetically that similar, this would also be convincing evidence that intelligence is molded by the shared environment of the home. Alternatively, if identical twins reared apart are as similar to each other as identical twins reared together, this would argue that the environment is far less important than genes in determining intelligence.

Split-twin studies have confirmed that genes are essential in shaping human intelligence.[26] The famous "Minnesota Study of Twins Reared Apart" found that identical twins reared apart were nearly as similar to each other as were identical twins reared together. Variations in intelligence were calculated to be about 70% due to heredity. This implies that education, social interaction, nutrition, and a generally supportive environment together were only half as important as genes in the genesis of intelligence. Traits such as personality, temperament, occupation, leisure activities, and even social attitudes were also similar between identical twins. Split-twin experiments may thus have led to the pervasive modern bias that all human traits are inherited.

Yet the very latest results imply that split-twin studies can overestimate the importance of genes. This is because there are many problems and difficulties associated with doing a proper split-twin study. Even with a well-executed study, it is still hard to understand the influence of the environment. Any study of twins, whether identical or fraternal, must confront the fact that similarities in the environment of two different people can arise in numerous ways. Two strangers, born on the same day in the same place, will likely share a certain portion of their environment. And while young children do not experience much outside the home, their parents do, and parents bring external influences into the home. Any child who was born in the United States after 1950 experienced the corrosive paranoia of the cold war. To a certain extent, the vaguely sensed dread of our parents, the bomb shelters and air-raid drills, the press coverage of the red menace, and the general frenzy following the Russian launch of *Sputnik* weave all baby boomers together in a shared experience. Similarly, the Great Depression was an experience that had an impact on virtually every household in the early 1930s, leaving a mark on nearly every child born in that era. Today, many children share an experience of poverty and deprivation, and this experience is likely also to leave a profound mark. People of different backgrounds but similar ages share many experiences and may develop similar personality traits in response to those shared experiences.

THE HERITABILITY OF TRAITS

When a large number of identical and fraternal twins are studied, the data can be analyzed to determine the "heritability" of a trait, or the extent to which heredity controls expression of that trait. However, a number of problems with this approach remain. Perhaps the biggest problem is that the "heritability" calculated for a trait is only accurate for the people from whom data were drawn in the first place. For example, if we find that weight is 70% heritable among a group of people in New York, we might then assume that weight is also 70%

heritable among subsistence farmers in Somalia. Yet we cannot assume that the weight difference between an overfed New Yorker and an underfed Somali is caused by genes. Clearly, genetic differences exist between these groups, but the average difference in weight is far more likely to be due to shortage of food in Somalia. Every time a scientist calculates heritability in a group of twins then projects that heritability to another population, the results should be somewhat suspect. If the heritability calculated from a group of white male twins is projected to black male and female nontwins, the results are likely to be substantially wrong. Yet this kind of thing is done all the time, even by highly reputable scientists.

These caveats are not meant to imply that split-twin studies are invariably weak or that some other type of genetic study is inherently stronger. Split-twin studies, when properly done, are an astonishing window into the degree to which human traits are controlled by genes and how genes are influenced by the environment. It is quite likely that split-twin studies will remain one of the key ways to attack the problem of nature versus nurture for years to come. The caveats are simply meant to illustrate the difficulty of doing science properly.

There are many other issues that can undermine a study of behavioral genetics. For example, the mathematical model used to calculate the heritability of a trait can fail if certain general assumptions are violated. Most heritability models assume that all genes act additively, that individual genes simply sum up to some larger effect. The models fail to allow for more complex interaction between genes, particularly the likelihood that some genes negate the effect of other genes. In real life, genes can interact in subtle ways that the models do not acknowledge. Let us assume that there are two genes that can produce obesity; perhaps one gene dulls the sense of satiety while another gene affects metabolism so that fat deposition is more likely. A person inheriting either of these genes might have a tendency to become somewhat obese, but a person inheriting both genes might have a strong tendency to morbid obesity.

Even more problematic, the mathematical model used to calculate heritability assumes that "assortative mating" does not happen.[27] This means that the model assumes that people mate at random and that children therefore show a random mixture of all possible genes. Yet assortative mating happens all the time. If a woman is very tall, she may only be attracted to tall men. Intelligent women are more likely to bear the children of a man who is intelligent, while dull women may be unable to attract an intelligent man. Yet for the heritability model to work well, mating must be random, since the model assumes that all genetic combinations are possible. If selective mating rules out some possible gene combinations, then assumptions inherent to the model are violated. If mating is not random—as it surely is not—then the model will fail and the genetic component of behavior will be underestimated. Thus, even with perfect data—and data are never perfect—the estimated heritability of a trait would be incorrect.

In addition to split-twin experiments, there are several other methods to determine the link between a gene and a particular human trait. These other methods are built upon the foundation of molecular genetics. The workhorse of molecular genetics is linkage analysis, which assumes that related individuals who share a trait in common also share a set of genes for that trait.[28] The trait and the linked genes are passed from generation to generation and will virtually always appear together. Linked genes may or may not actually code for the trait in question, but they are at least physically close to the relevant gene. Thus, whenever a trait is present, so are the linked genes. To explain this concept by analogy, consider the association between tattoos and antisocial behavior. All tattooed men are not antisocial, nor are all antisocial men tattooed, but antisocial men often have very elaborate tattoos. Thus, the presence of very elaborate tattoos may tell us something about whether an individual man is antisocial. The key insight is that elaborate tattoos serve as a marker—admittedly fallible—for antisocial behavior. An obvious marker may be less obvious and just as fallible as a tattoo, but it can reveal the presence of a hidden gene.

Linkage analysis proposes an explanation for the inheritance of a particular trait, and then tests whether the proposed genetic model can explain the inheritance of that trait. To increase the chance of success when doing linkage analysis, it is important to collect as many pedigrees as possible from large families. Computer programs are available to sort through a mountain of this type of data quickly to help researchers find the most plausible model of inheritance. Hundreds of human traits have been analyzed in this way, but the greatest success has been achieved for traits that are coded for by a single gene. Several other methods exist to determine the relationship between traits and genes, but the lesson should be obvious: this type of research is difficult. Molecular genetics methods are hard to apply, require very large sample sizes, and are somewhat vulnerable to finding a false association between a trait and a gene, so genetic explanations of human behavior are at least somewhat suspect.

THE GENETICS OF MENTAL DISORDER

Human beings are not tabula rasa, upon which the environment writes. We all carry genetic baggage, and there is a hereditary component to every human behavior that has been studied. For example, a recent study of split-twins tested whether there is a genetic risk for a constellation of related traits.[29] The traits of interest were alcohol and drug dependence, conduct disorder, and antisocial personality disorder. Past studies have shown that these traits run in families, although it has not been clear whether this is due to shared genes or to shared environment. To determine whether "nature or nurture" is more crucial, researchers studied a sample of 542 twins from the Minnesota Twin Family Study, including 357 MZ or identical twins, and 185 DZ or fraternal twins. Each twin and each parent answered a long list of questions designed to assess the degree to which they displayed the traits of interest. Genes were parsed apart from environment and both shared and nonshared environments were characterized using a sophisticated mathematical approach to analysis. Overall, it was found that the heritability of traits was 60–80%, which means that there is a substantial genetic component to all of the traits. For each of the different traits, the estimate of heritability was lower than the aggregate heritability, which may mean that children inherit a general proneness to these traits. The most likely interpretation of this finding is that siblings, friends, neighbors, and neighborhoods all interact to determine the expression of traits, independent of either the parental genes or the environment of the parental home. In short, children can inherit a vulnerability to addiction, but the environment helps to determine whether alcoholism or drug dependence will develop.

The heritability of a broad range of brain disorders is shown in the table below.[30] For most of these diseases, the heritability estimate is based on well-designed, well-replicated split-twin studies, and usually the heritability estimate is quite high. For example, the overall heritability of Alzheimer's disease is 53%, meaning that about half the risk of disease is transmitted through the genes:

	Heritability	Annual cost to US healthcare system ($ billion)
Addiction	40%	544.1
Alzheimer's disease	53%	170.9
Anxiety disorder	30%	82.6
Schizophrenia	70%	57.1
Depression	40%	53.1
Stroke	10%	27.0
Parkinson's disease	10%	16.0
Multiple sclerosis	40%	7.6
Seizure	60%	1.0
Huntington's disease	100%	0.2

Brain and nervous system disorders may cost the United States economy as much as $1.2 trillion annually, and millions of Americans are afflicted each year. Twin studies suggest that, overall, about 40%

of the societal burden of brain disorder is dependent on genes.[31] Much of this burden arises from the complex effects of multiple genes, rather than from any one gene.

How the Environment Elicits Behavior

"Environment" is often taken to mean just the social environment that envelops and protects a child from birth to independence. This includes early interactions with parents and siblings, as well as the more sporadic interactions with whatever members of the extended family happen to be present. Somewhat later the environment expands to include teachers and friends, and these parts of the social environment assume greater and greater significance with the passing years. Eventually, people with whom an adolescent or young adult forms lasting friendships will come to play an increasingly important role.

But considering environmental influences to be synonymous with social influences is very narrow and restrictive. Instead, environmental influences should be broadly defined as any and all influences not explicitly genetic.[32] This opens a door to many factors that might otherwise be overlooked or undervalued, including factors in the physical environment. It also makes it easier to examine the complex interactions between genes and the environment.

The environment can have a profound effect on a child before the child is even born. Toxins or drugs can pass through a mother's body and have a profound and direct effect upon the unborn child. An example of this is the drug thalidomide, which was taken by pregnant women 50 years ago to control morning sickness, but which had a devastating effect on fetal development. The hormonal environment, which the child encounters in utero, is also potentially important, and maternal hormones may be able to modulate the fetal genome. In addition, the nutritional status of the mother can have a profound impact on the development of a fetus. Developmental patterns established early in fetal life may take an entire lifetime to work out.

After birth, the newborn infant is extremely vulnerable to the vagaries of the environment. The nutritional history of the mother can be important if she is nursing, and the nutritional history of the baby is obviously critical. But feeding and freedom from physical harm are not enough; the infant must be loved, touched, cleaned, warmed, kept free of disease, and given adequate stimulation and rest. Only when all of these physical conditions are satisfactorily met can the child begin to fulfill its genetic potential. Since mere survival can occur under conditions that are far from optimal, this implies that the environment can hinder a child from fulfilling her genetic potential.

Direct interactions between the infant and the mother immediately after birth are crucial, and the relationship with the mother remains crucial for many years. The personality of the child and the personality of the mother can interact in a crucial way to give this relationship a structure that may last a lifetime. Neonatal behaviors naturally elicit warmth and nurturance from the mother. But if a child is somehow deficient in giving these signals, or if a mother is deficient in responding to them, this can have a major effect on a child. In fact, the ability to elicit or give nurturance may be hereditary, so that the same thing that interferes with a child's ability to elicit nurturance may interfere with a mother's ability to give it. It is likely that hereditary components of a social behavior—such as warmth or empathy—are first experienced in interactions with the maternal "environment."

While it is clear that maternal acceptance and nurturance of the infant is critical, the optimal development of a child may require a "father" or an extended family. It is reasonable to suppose that the infant is sensitive to the social supports that surround her mother. Rearing an infant is extremely stressful, involving many sleepless nights, many disruptions of prior routine, and many new anxieties, leaving a new mother with little time or energy to deal with her own emotions. The importance of an effective social network around the mother cannot be overestimated. It is possible that infants are sensitive to this support network in some way or another, since maternal anxiety can be transmitted to the infant by many different cues. In fact, maternal stress can even reduce milk output, so that a nursing mother who is badly stressed may not be able to meet the nutritional needs of her child.

The family unit forms a sort of collective environment that can be rich and supportive or full of tension and dissension. Little is known about the sensitivity of the newborn to family cues, but it is clear that even very young children are exceedingly sensitive to the nuances of family life. Family discord may lead directly to parental neglect, or it may cause the child to suffer indirectly, since an emotionally drained mother may be unable to provide adequate care for her child. If there is family discord, the child may be drawn into it as part of the power struggle between parents, or the child may be excluded or marginalized, since neither parent may have the energy to meet her needs. Thus, family discord can potentially have damaging consequences for a child.

THE ENVIRONMENT OUTSIDE THE FAMILY

The larger social context around the family is a moderating influence that can be benign or malignant but is only rarely beneficial to a child. The social context can seldom compensate for an inadequate family life, but it can negate the best efforts of a family to protect a child. A child born to the terror of the Warsaw ghetto or the civil war in Darfur may never reach her full potential because of awful childhood circumstances. On the other hand, a child in a larger social context that is highly supportive, such as a kibbutz in Israel or a Head Start Program in Harlem, may derive little additional benefit if her family is already supportive. But as young children grow into young adults, the larger social context grows increasingly important. Teachers and friends assume an ever-larger role for the child, and parents may begin to feel extraneous as a child reaches adolescence. The parental feeling of irrelevancy is almost certainly not accurate, since parents may remain important in the lives of their children even as the children become adults. Yet we know almost nothing about how the family environment and the social context interact in fostering the maturation of a child, nor do we know how this interaction changes as the child ages.

In addition to these direct effects of the environment on the child, there can be a complex interplay between genes and the environment. How a parent chooses to bring up a child forms the environment of that child, but it arises in part from the genome of the parent. Most parents provide their children with both genes and a home environment, and the two reinforce each other in subtle ways. The interaction can be passive, such as when children receive both "smart genes" and an enriched environment from their parents. But there can also be an active interaction between genes and the environment; a child with a particular behavior can evoke something from the environment that actually reinforces that behavior. For example, children with high verbal ability are more likely to spend time reading, thereby further enriching their verbal abilities. A child who is more intelligent may be more likely to test her environment continually and to elicit more interesting responses from parents and others around them. Alternatively, a child who is inherently somewhat hostile is more likely to evoke hostility from the people around her, which may in turn lead to reinforcement of her hostile behavior. And the complex interplay between genes and environment can take a lifetime to play out. Antisocial boys are more likely to experience social rejection as adults, and they have a higher-than-normal rate of divorce, unemployment, and criminal behavior.

How a child structures his environment is highly significant and may radically affect how different children experience the same environment. We are not passive participants in our own lives. We actively structure the environment around us, and children are certainly capable of this as well. The structure a child chooses to impose on his environment is the structure with which that child feels most comfortable, and this is likely to have something of a genetic component.

At every step during brain development, the environment can conceivably have an impact on gene regulation through subtle mechanisms that remain very poorly understood. Clues about the molecular mechanisms that might regulate brain growth can be derived from the study of patients in whom an environmental pathogen has produced abnormal brain development. For example, fetal alcohol syndrome results from in utero exposure to alcohol, which causes characteristic facial deformity, learning disability, mental retardation, and brain malformation.[33] Fetal alcohol syndrome (FAS) has also been linked to attention deficit hyperactivity disorder, to depression, and even to psychosis in adults. A single episode of moderate to heavy alcohol consumption by a pregnant mother can raise her blood ethanol level to a point that neural death might be induced in the fetal brain. Since different neuronal populations may be at risk at different times during pregnancy, this could explain why FAS causes a broad spectrum of clinical problems. Alcohol interferes with the binding of several neurotransmitters so that even if interference is brief, it could induce neuronal death. Interestingly, certain drugs of abuse (e.g., ketamine and PCP) and anesthetics (e.g., nitrous oxide) block the same receptors as alcohol and may cause fetal brain injury by a similar mechanism. Children with FAS have problems that can last a lifetime: FAS is associated with a 122-fold increase in risk of developmental delay, a 66-fold increase in behavior problems, a 37-fold increase in language delay, a 12-fold increase in mental retardation, and an 11-fold increase in school failure.[34] A child with FAS is also much more likely to be removed from his home environment and to grow up in foster care.

DNA and Destiny

Behavioral genetics is a brand-new science, which has so far provided us with more questions than answers.[35] Virtually never is a scientific study so well designed that it is the last word. Sincere scientists acting in good faith report data that later proves to be incorrect, while other scientists may attempt ambitious studies that fail ultimately to obtain a clear-cut answer to any question. Overcautious scientists may be reluctant to report their findings, while reckless scientists may report preliminary findings from a small study as if these findings were established fact. Some reporters accept preliminary findings at face value, failing to see the problems that ultimately undermine a badly done study. Many newspapers and television news shows are guilty of presenting a simplistic summary of a complicated issue in order to fit the story into a particular newspaper column or bit of airtime. And some science writers willfully distort or misrepresent the truth in service to a political ideology. The layperson is bombarded with utterly fallacious pseudoscience while standing in the grocery checkout line. Many stories presented even in respectable newspapers are incomplete, distorted, or incorrect. Most people simply lack the time, energy, or inclination to separate grains of scientific truth from the chaff of misinformation. But this doesn't stop the average person from forming an opinion.

Some people see infants as blank slates to be written upon by experience: if we simply give a child the right set of experiences, that child can grow to be an athlete, an artist, or an astronomer. Other people seem to see nature as constantly at war with nurture for control of the individual, giving the phrase "nature *versus* nurture" an entirely new meaning. Still others seem to see human beings as automatons whose every action is controlled by genes, irrespective of what choices the environment presents. The pendulum of public opinion has swung back and forth, often ignoring and sometimes villifying scientists who seek to study the origins of human behavior. Yet people are still fascinated with rumors of genetic proclivity. There has come to be a subtle but pervasive bias that nature—in the form of hereditary forces at work in the individual—dominates nurture, that human traits result from the unfurling of an immutable program borne in the genes. Yet the truth is far more interesting and far more subtle than that.

We are an amalgam of ability and opportunity, a result of chance and choice, necessity and opportunity.[36] We are not extraordinarily precise machines, functioning in a way decreed by the genes—if we were, then continual change in the human condition would not be as inevitable as it seems to be. Yet neither are we entirely free agents, devoid of genetic baggage and free to make whatever choices we want—if this were true, we would not see families beset by genetic ills and ravaged by the same conditions generation after generation.

To understand the human condition better we must develop a keener appreciation for the subtle interactions possible between nature and nurture. As the late psychologist Donald O. Hebb of McGill University said, "Heredity determines the range through which environment can modify the individual." In part, this simply concedes that we cannot make every child into an astrophysicist; but there is more to it than that. This quote also acknowledges that there is an ongoing interplay between the genes and the environment, such that we are each a product of both forces. There is a continual tension between the possible and the actual, with the possible determined by genes and the actual determined by environment.

This tension between genes and environment is nothing new. The environment has been allowing adaptive genes to survive and selecting against maladaptive genes for millennia. And genes have responded over time in many ways: some genes are increased in abundance, some are decreased, and some have probably become extinct, since they were first expressed in our distant ancestors. The human being is a work in progress, evolving still, and the brain may be the part of the body that is evolving most rapidly. Perhaps the enormous range of ills that can affect the structure of the brain—and the function of the mind—is a result of the newness of it all.

CHAPTER 14

NEUROLOGY AND ILLNESSES OF THE BRAIN

The human brain may be evolving still, as there is evidence that several key genes that set us apart from primates arose by mutation within the last 40,000 years. Thus, the brain is changing rapidly in form—as we have seen—and may be changing in function. Rapid evolution could be to blame for the fact that the human brain is subject to an enormous range of ills. Perhaps the brain is simply under-engineered, less robust, and less resilient than it would be were it not such a rapidly changing thing.

While the brain has an incredibly broad range of things that can go wrong with it, it has a rather limited number of ways to show distress. The restricted palette of symptoms that a distressed brain can show has led to two competing and often incompatible explanations for the genesis of brain illness. Just as the paucity of facts has permitted Marxists and capitalists to interpret the same economic observations in very different ways, the symptoms of brain illness can be seen through very different lenses.

291

NEUROLOGY AND PSYCHIATRY

Historically, neurology grew out of psychiatry, which was already well established as a medical specialty by the early nineteenth century. The split between neurology and psychiatry may have originated in the effort of neurologists to relate brain malfunction to anatomic pathology in the belief that every symptom has an anatomic locus. Psychiatrists, on the other hand, were more impressed by the integrated or holistic nature of the patient, and their beliefs culminated in the development of psychoanalysis. The general acceptance by psychiatrists of psychoanalysis, which lacked a firm scientific foundation, drove a wedge between neurology and psychiatry that persists to this day, even though many of the ideas of psychoanalysis have been repudiated.

After World War II the division between neurology and psychiatry became explicit and sometimes even acrimonious. An academic journal known as the *Archives of Neurology and Psychiatry* split into two journals. The separation widened as the American Academy of Neurology was founded in 1948 and departments of neurology sprang up in medical schools across the United States. Yet in recent years, there has been rapprochement between the fields, as the biological underpinnings of mental illness have become better understood. Now, there is a broad acceptance among psychiatrists that clinical psychiatry will inevitably move closer in practice to clinical neurology.[1]

In discussing illnesses of the mind and the brain, we will maintain the traditional dichotomy of neurology and psychiatry, not because it is inherently correct but, rather, because it is more familiar to most people. By tradition, illnesses of the brain are treated by neurologists and illnesses of the mind are treated by psychiatrists; yet illnesses of the brain and mind impact the same organ. The perception that illnesses of the brain are "organic" or somehow "real," whereas illnesses of mind are "functional" or "all in your head," is simply wrong. This formulation is often taken to mean that mental illness is moral weakness or a character flaw, which has perpetuated the stigmatization and economic victimization of people with mental illness. In fact, any separation of brain from mind is potentially a problem if it perpetuates the "arbitrary and baseless cleavage of brain-based disorders into two disparate medical specialties."[2]

Nevertheless, physicians who treat the ills of the brain/mind can be separated on the basis of which particular symptoms are most common among their patients. Neurologists tend to focus specifically on motor and sensory disorders; these include disorders of muscular strength or coordination, decrements in overall muscle function, and deficits in tactile perception. Thus, neurologists tend to treat symptoms associated with a loss of physical function. Conversely, psychiatrists tend to focus specifically on mood, emotion, thought, and behavioral disorders; these include disorders of emotional regulation, reality perception, and behavioral control. Thus, psychiatrists tend to treat symptoms associated with a loss of mental function. There is also a broad range of symptoms treated by both neurologists and psychiatrists, which includes disorders of attention, awareness, perception, memory, comprehension, speech, and motivation.

We must explicitly note that what follows in this and the next chapter is not prescriptive. It would be foolish to diagnose problems without consulting a physician, dangerous to think that medical conditions can be treated without the help of an experienced clinician, and potentially life threatening to minister to problems on the basis of what is learned from a book. Medicine evolves extraordinarily rapidly, and there could be new diagnostic methods, new insights that affect the risk-benefit analysis, and new treatments available by the time this book is published. Therefore, we will deliberately avoid discussing medications unless the efficacy of a medication amounts to a test of an explicit hypothesis that relates to disease etiology.

THE PROCESS OF CELL DEATH

There are many potential ways to approach the subject of neurologic illness, but one possible organizing principle is the process of cell death. There are basically two ways that a brain cell can die: by necrosis or by apoptosis. Each process is associated with a range of illnesses that fall under the rubric of neurology.

Necrosis is the more familiar process, as this is what typically happens after a stroke when large portions of the brain may be deprived of blood supply, so that many cells die in a short period of time. Necrosis has been known for many years, since it is easy to see, either by magnetic resonance imaging (MRI) or at autopsy. Necrosis involves cell swelling, membrane breakdown, and the leakage of cell contents into the surrounding tissues. It is associated with rupture of those cell organelles that lock away enzymes that degrade proteins—in necrosis, cells undergo a process that is essentially one of self-digestion. This process typically results in liquefaction of brain tissue, with a complete loss of structure, although occasionally there can be dry, hardened bits of coagulated necrosis. Necrosis is not coordinated in any way; it is simply the response of brain cells to an overwhelmingly toxic event, usually the loss of oxygen and glucose as the blood supply is interrupted.

Apoptosis is a less-familiar (and rather recently discovered) process that scientists now think is of tremendous importance in explaining how the brain succumbs to illness.[3] Apoptosis is very easy to overlook on a microscope slide, since apoptotic cells tend to be widely scattered rather than clumped together, and they look like a single shriveled cell among many cells that can look entirely normal. Apoptosis involves cell shrinkage without membrane breakdown; those organelles that contain the cell's degradative enzymes do not break open, so the apoptotic cell does not self-digest and liquefy. Instead, an ordered process occurs that has been called "programmed cell death" or "cell suicide," during which an apoptotic cell shrivels up and simply disappears. Apoptosis is a carefully controlled process that occurs rather rapidly, so there may be only a few apoptotic cells visible, even if the overall rate of apoptosis is high. In short, apoptosis is easy to overlook, which may account for why such a crucial process went unnoticed for so long. Apoptosis is probably involved in the response of brain tissue to many illnesses, including Parkinson's disease, Alzheimer's disease, Hunt-

Table 5. Prion diseases include Creutzfeldt-Jakob disease, kuru, Gerstmann-Straussler-Scheinker disease, and fatal familial insomnia. Tauopathies include Picks disease, corticobasal degeneration, and progressive supranuclear palsy. Data from Taylor (2002) Multiple sclerosis is not associated with the accumulation of mutant protein. SN = substantia nigra.

Illness	Mood effects	Cognitive effects	Specific pathology	Protein involved
Alzheimer's disease (AD)	Apathy, depression, anxiety	Agitation, disinhibition, delusions	Plaques, tangles, synaptic loss	Tau, Ab
Parkinson's disease (PD)	Depression, anxiety	Dementia, psychosis	Neuronal loss in SN, Lewy bodies	a-Synuclein, parkin 1-10
Huntington's disease (HD)	Apathy, depression, anxiety	Dementia, personality change	Neuronal loss in neostriatum	Huntingtin
Amyotrophic lateral sclerosis	Depression	Subtle loss of verbal fluency	Loss of motor neurons	Ubiquitin, SOD1, neurofilamin
Prion diseases	Apathy, depression	Dementia, personality change	Variable tissue loss	PrPsc
Tauopathy	Apathy, depression, anxiety	Agitation, disinhibition, dementia	Frontotemporal atrophy	Tau, ubiquitin

ington's disease, and amyotrophic lateral sclerosis. Apoptosis may even play a role in normal maturational processes in the brain.

It should be noted that there are other possible organizing principles relevant to how the brain succumbs to illness, with an obvious choice being protein accumulation (table 5). A number of common neurologic illnesses including Parkinson's and Alzheimer's are characterized by the accumulation of toxic proteins in brain tissue.[4] As it turns out, several genes linked to neurodegenerative disease code for enzymes that, if mutated, cause abnormal accumulation of misfolded proteins. Such proteins evade the normal quality control systems that are designed to eliminate faulty protein. When these proteins accumulate in sufficient quality, the misfolded proteins are prone to aggregate; these aggregates may then become toxic. Because protein aggregates are visible under the microscope, neurologists claim most of the related illnesses as neurological, even though many of these illnesses are associated with psychiatric symptoms.

NECROSIS AND STROKE

Stroke is the third most common cause of death worldwide, following ischemic heart disease ("heart attack") and cancer. Stroke is defined as the abrupt onset of symptoms of brain injury unexplained by trauma and that lasts longer than 24 hours or leads to death.[5] There are three types of stroke: ischemic stroke results from obstruction of a blood vessel (about 80% of patients); intracerebral hemorrhage results from the rupture of a blood vessel within the brain (about 15% of patients); and subarachnoid hemorrhage results from the rupture of a blood vessel coursing over the brain surface (about 5% of patients). Tiny undiagnosed strokes called lacunar infarctions can occur in elderly people or in people with various illnesses; for example, lacunar infarction is seen in about 24% of children with sickle-cell anemia.[6] All stroke types cause the death of brain cells if the supply of oxygen and glucose is interrupted for a sufficient period of time. Typically, death of brain cells is rapid, beginning within as little as 4 minutes after blood flow is interrupted. Because cell death occurs without any metabolic warning, there is no time for apoptosis to occur, and cells simply die as they run out of energy. Blood flow to the brain is ordinarily about 100 milliliters per 100 grams of brain tissue per minute. Brain cells may begin to die if the blood-flow rate falls to around 25 mL/100 g/min., since neurons will lack sufficient energy to maintain electrical activity. Brain perfusion relies upon only three major arteries (fig. 14), so constricted blood flow in any of these vessels, as happens in many people with high levels of blood cholesterol, can put the brain at increased risk of stroke.

In one of the most startling successes in modern medicine, there has been a rapid and dramatic reduction in incidence, severity, and mortality of stroke, at least among patients who receive state of the art

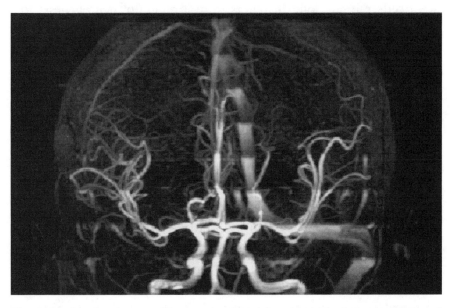

Figure 14. Another image of the arteries that deliver blood to the brain. The head is seen from the back, with the 3 main arteries that supply blood to the brain seen arising from the bottom center of the image. This image suggests that the arterial blood supply to the brain can be somewhat tenuous if there is damage to any of the 3 main arteries. This is the same subject shown in figure 5.

medical care.[7] In England, among people less than age 85 years old, the incidence of first stroke fell by 32% and the incidence of major stroke fell by 40% between 1981 and 2004. This is a result of the increased effectiveness of preventive measures and the reduced exposure to risk factors. At least half of all strokes are due to atherosclerosis, which is the accumulation of fatty deposits on the inner surface of blood vessels, many of which perfuse the brain. Medicine for hypertension and blood cholesterol has proven to be remarkably effective in reducing stroke risk, especially in combination with a 50% reduction in the smoking rate. There has also been a vigorous public education campaign about stroke, which has emphasized the need for rapid treatment—stroke has been called a "brain attack" in analogy to a heart attack.

It is becoming clear that the brain can recover from stroke to at least some extent by the action of neural stem cells. There is strong evidence of neuronal regeneration in animal models of stroke; neural stem cells migrate long distances from a region of neurogenesis to a region of stroke damage in rats.[8] When rats are subjected to an experimental stroke, there is a marked increase in the rate of proliferation of neural stem cells and an increase in the rate of migration of these cells into damaged regions of brain. Injection of bromodeoxyuridine (BrdU) shows that 5 weeks after an experimental stroke, roughly 42% of neurons in the stroke region are newly produced. Thus, ischemia induces neurogenesis, cell migration, and differentiation into mature neurons. Roughly 80% of the newly arrived neurons die within 6 weeks of the stroke, and only about 0.2% of dead neurons are replaced by new neurons. This suggests that the stroked region of a brain is actually toxic to newly produced cells, which would make sense; many end products of cell death may remain in that region, and the stroked region may still be poorly perfused.

It may become possible to use the lessons learned in rats to enhance stroke recovery in humans.[9] One approach that might be worthwhile is to clear a stroked region of cell debris so as to make the environment generally more hospitable to newly produced cells. This might enhance the survival of cells that migrate into the area. If more newly produced cells could survive and take on the duties of dead neurons, then recovery from stroke would improve. Certain molecules that enhance the rate of neurogenesis have already been identified, although it is not known whether these molecules are effective in humans, especially in the context of a stroke. The creation of a hospitable milieu for migrating stem cells in the ischemic brain will be a formidable clinical challenge. Yet we already know that certain strategies are fairly effective in enhancing neurogenesis. Exercise and environmental richness both increase neurogenesis, and depression suppresses it; efforts to enhance stroke recovery could potentially exploit these insights.

APOPTOSIS AND PARKINSON'S DISEASE

Early in the 1980s, 4 young patients arrived at the Stanford University Hospital, all of them with a previous history of heroin abuse, and all with a very recent onset of severe Parkinson's disease.[10] Symptoms included near-total paralysis, a complete inability to speak, striking muscular rigidity, tremor, a fixed stare, and constant drooling. A careful clinical history revealed that all 4 patients had been well but developed symptoms within a week after using a new form of synthetic heroin. The first symptoms included rapid onset of visual hallucinations, limb spasticity and stiffness, and muscular slowing that developed within a few days of drug use.

This striking case, as tragic as it was for the patients, was an enormous breakthrough in our understanding of Parkinson's disease (PD). All patients responded well to therapy with L-dopa, which remains standard therapy for PD, and their rapid clinical response proved that they had a condition that was indistinguishable from PD. Yet even 5 months later, none of the patients showed signs of remission, and when one of the patients was weaned off L-dopa, he reverted to his original state of immobility and rigidity, able only to move his eyes. Analysis of residue of the drug that the patients had taken revealed that it was contaminated with a substance called MPTP, which is usually removed from heroin during the process of synthesis. However, this particular sample of heroin had been made by an unscrupulous dealer experimenting with a synthetic shortcut, which wound up producing heroin badly contaminated with MPTP. When MPTP is injected into rats, it produces an excellent experimental model of the human disease. It is now known that MPTP selectively damages neurons in the substantia nigra, beneath the thalamus. The rapid action of MPTP proves that PD can result from environmental toxins.

PD is the most common serious movement disorder in the world, affecting about 1% of adults aged 60 years and older. A classic "split-twin" study was done to investigate the extent to which PD is a heritable illness.[11] A total of nearly 20,000 white male twins enrolled in the World War II Veteran Twins Registry were evaluated in this study. Medical records of the twins were examined to identify twins who had succumbed to PD and to determine whether the co-twin also had PD. Scientists found 193 men in the Twin Registry with PD, then determined the concordance rate between identical and fraternal twins. Among men older than age 50 at diagnosis, the concordance rate between MZ and DZ twins was the same, meaning that there is little evidence of a genetic cause of PD. However, among men younger than age 50 at diagnosis, a different story emerged. If one member of an identical twin-pair got PD, the risk that the other twin would also get PD was elevated 6-fold compared to normal. This means that in the early onset form of the disease, there is a substantial genetic component of risk, although the impact of genes in late-onset illness is minimal.

Overall, these findings are consistent with findings in many other diseases: the earlier the onset of disease, the greater the role of genes in disease causation, whereas late-onset illness is more likely to be caused by the environment. Yet it is not at all clear which environmental factors contribute to it; some things that might be expected to increase the risk of PD are actually protective. For example, cigarette smoking, coffee drinking, and moderate alcohol consumption all *decrease* the risk of PD. Similarly, severe malnutrition, intense psychological stress, frequent infection, and extreme exhaustion also do not increase the risk of PD, as proven in a study of men imprisoned by the Japanese during World War II.[12] Although dietary factors play a role in whether the brain ages successfully, there are few causes of PD that are due to diet.[13] Both high calorie and folate-deficient diets are risk factors for PD, and there is evidence that exposure to pesticides and trace metals contribute to it, especially in context with the wear and tear and metabolic stress of aging. Overall, a brain-healthy diet appears to be similar to a heart-healthy diet, but no dietary factors have yet been identified that strongly increase the risk of PD.

Typical symptoms of PD are rigidity, tremor, and bradykinesia or extreme slowness of movement.[14] PD is usually asymmetric, progresses rapidly, and responds well to treatment with L-dopa, making it rather easy to diagnose clinically. Mean age at onset is the early 60s, which is consistent with long-term exposure to environmental toxins. It is worth noting that PD frequently has psychiatric symptoms, particularly in advanced cases, with depression in half of all patients and dementia, hallucinations, and psychotic behavior in a smaller fraction of patients. A typical sign of chronic PD at autopsy is the accumulation of protein in small masses called Lewy bodies, but Lewy bodies were also found in up to 16% of healthy elderly people at autopsy. Studies of newly diagnosed patients with positron emission tomography (PET) suggest that the clinical signs of PD only become evident when roughly 70% of all dopamine-producing neurons in the substantia nigra have already been lost.

What causes PD? Clearly, age is a factor, as PD is so much more common in older than in younger people. About 15% of PD patients have a first-degree relative with the disease, so genes are also a factor, although most patients do not have a family history of disease. Though at least 10 different genes have been identified, each of which increases the risk of PD somewhat, the known mutations cause very few cases. It is possible that PD is the result of a slow process of apoptosis that proceeds for years before any symptoms occur. Autopsy studies of PD patients show that neuronal loss is restricted to the same brain regions impacted by MPTP.[15] It is noteworthy that MPTP has a direct effect on mitochondria—the cell organelles that generate energy to power the cell—and mitochondria can initiate apoptosis. Split-twin studies implicate an unknown environmental toxin in PD. The toxin must be common because PD is so common, but the toxin cannot be ubiquitous, otherwise genes would be *more* important in determining who becomes ill. A reasonable hypothesis is that exposure to a moderately common childhood illness is a factor; perhaps the illness causes flulike symptoms that are in no way notable but that initiate a slow process of apoptosis in some vulnerable people.

Because cell loss in PD is limited only to dopamine-producing cells, transplantation of such cells is an attractive therapy. However, a recent placebo-controlled clinical trial, which involved the transplantation of embryonic dopamine neurons to treat severe PD, found little benefit, except in patients younger than 60 years.[16] A total of 40 patients were randomly assigned to receive either a transplant of nerve cells or sham surgery, with the patient not knowing which treatment was actually done. One year later, transplanted neurons were alive in 17 of 20 patients. However, patients older than 60 years showed no clinical benefit from treatment, and younger patients showed very modest treatment benefits.

Another possible treatment approach is to prevent dopamine-producing cells in the substantia nigra from dying. To this end, some patients received an infusion of a growth factor called glial cell–derived neurotrophic factor (GDNF) directly into the brain.[17] GDNF is a small protein that somehow promotes the survival of dopamine neurons. When GDNF was infused into the ventricle (fluid-filled space), close to the substantia nigra, in 50 patients with PD, there was no improvement in symptoms, although there were substantial side effects. However, it is possible that GDNF did not penetrate to where it was needed. To address the issue of delivery, another study was done in which GDNF was infused directly into the brain in 5 PD patients.[18] After two years of continuous infusion of GDNF, there were no major side effects and patients showed a striking improvement in clinical symptoms. Improvement was noted especially in off-medication motor skills, which is to say motor skills during the period when daily L-dopa treatments are starting to wear off. Among control patients who were not treated with GDNF, there was a decline in virtually every measure of disease severity, including memory, attention, and executive function, whereas GDNF-treated patients had stable function in every domain and improved function in memory. It was concluded that GDNF infusion is safe, can be tolerated for at least two years, and leads to significant symptomatic improvement. A logical source of GDNF for PD patients might eventually be transplanted neural stem cells but, because stem cell research has been stifled by crass political considerations, this research has languished.

Death of dopamine neurons in PD has an impact on the mind as well as on the brain, which underscores that neurology can merge into psychiatry. Dopamine neurons are rather rare in the normal brain, but they are crucial for movement, motivation, and working memory.[19] Dopamine does not simply excite or inhibit other neurons; rather, it modifies the response of neurons to their milieu. Dopamine deficits account for the symptoms of PD, and L-dopa is a metabolic precursor that effectively replaces missing dopamine. Interestingly, dopamine deficit has been implicated in a number of other conditions, including drug addiction, attention deficit hyperactivity disorder (ADHD), and even schizophrenia.

HEALTHY AGING AND DEMENTIA

The brain is prey to all sorts of ills as we age, some of which are rather benign and some of which can be devastating. An abiding fear for many elderly people is that they will become demented, losing the ability to live independently or make decisions for themselves or relate to the people around them in a meaningful way. One of the most detailed and interesting studies of healthy aging and dementia is the Nun Study, a longitudinal study of 678 nuns, all of whom were in a single congregation.[20] The sisters ranged in age from 75 to 101 years old at study entry in 1991, and the oldest surviving sister was 107 years old in 2002.

For each sister in the Nun Study, there were 3 sources of information available: convent archives provided details about their early lives, including such things as a writing sample; annual examinations documented changes in physical or cognitive status during the study; and finally, all study participants agreed to have an autopsy after death. Sister Matthias, who lived to be 104 years old, was a paradigm of healthy aging: she was happy, physically active, vivacious, and cognitively intact. At autopsy, her brain showed mild atherosclerosis, a tiny stroke or infarct, and moderate evidence of the type of pathology that is associated with Alzheimer's disease. By contrast, Sister Agnes, who lived to be 92 years old, was so severely impaired in the last years of her life that she could communicate only with an occasional smile or an isolated word. At autopsy, her brain showed several small infarcts, a large hemorrhagic stroke, moderate atherosclerosis, and abundant evidence of the pathology associated with Alzheimer's disease. These cases suggest that cognitive impairment can be associated with a range of pathologies rather than any one type. Dementia is most severe when both vascular lesions and the lesions characteristic of Alzheimer's disease are present. Yet healthy aging is possible even when there is brain pathology, as several sisters had extensive pathology but were still cognitively intact. This implies that some people may be able to resist the effects of disease through mechanisms that remain unknown.

Dementia, which is defined as a general deterioration of mental abilities, affects about 1.5% of people by age 65 but up to 30% of people by age 80.[21] Among the different types of dementia recognized by neurologists are vascular dementia, Alzheimer's disease, Lewy body dementia, and AIDS dementia, but Alzheimer's by itself accounts for roughly half of all dementia cases. Psychiatric symptoms are present in up to 90% of dementia patients, which further blurs any distinction between neurology and psychiatry. The various dementias share certain risk factors, including hypertension and depression, with vascular injury of the brain present in up to half of all dementia patients.

Treating the risk factors for vascular injury should reduce stroke prevalence and could potentially reduce the prevalence of dementia as well. Additional risk factors for dementia are old age, male gender, heart attack, coronary artery disease, diabetes, atherosclerosis, smoking, high serum lipids, and a history of stroke. In short, most risk factors for dementia relate either directly or indirectly to vascular injury.

Vascular injury of the brain is present in most elderly people at autopsy.[22] Among 209 people who died at a mean age of 86 years, vascular pathology was present in 78% but dementia was diagnosed in only 48%. A brain imaging study of 1,105 people living in Rotterdam showed that dementia is also more common among people who suffer brain infarction, even if that infarction is "silent," or never clinically diagnosed.[23] A silent infarction can still be seen by magnetic resonance imaging (MRI)—often as a small area of injury in white matter—even if it produces symptoms so mild that a stroke is never diagnosed. Patients with a silent infarction at the start of the study were twice as likely to develop dementia during the 4-year follow-up; moreover, decline in cognitive function was restricted to patients who had an infarction during the follow-up. Thus, evidence is strong that dementia is related to vascular injury.

ALZHEIMER'S DISEASE

Alzheimer's disease (AD) is one of the most common and most dreaded outcomes of aging. It affects roughly as many people as stroke, and even among Caucasians older than age 65 who live independently, up to 10% have AD.[24] Alzheimer's is associated with death of neurons in the hippocampus and the cortical gray matter, which results in impaired cognition, brain atrophy, and eventual death. Possible risk factors for AD have been explored for many years, with disappointing results: age, dementia in a close family relative, and the E4 allele of the APOE gene are the only confirmed risk factors. The APOE gene codes for apolipoprotein E, a blood protein that is involved in lipid metabolism.

Several additional risk factors interact with the APOE gene, including female gender, low serum lipids, herpes infection, or a history of head injury. In general, the risk of developing AD before age 90 depends upon the number of disease-susceptibility genes present, but the known risk factors account for only a small percentage of all AD cases.

In a large study of elderly people who died at an average age of 86 years, there was clear pathological evidence of early-stage AD in 70%.[25] Among elderly people who were not demented, AD features were present in 33%, so some of these people probably had preclinical AD. Among demented people, 61% had neurofibrillary tangles, which are often taken to be diagnostic of AD.

As is true of most other dementias, AD is more common in the context of vascular or brain injury. This was clearly shown in the Nun Study, where stroke-free sisters were able to tolerate more AD lesions before exhibiting symptoms of dementia.[26] Even a small infarction may trigger expression of AD symptoms, especially if infarction occurs deep in the white matter. In general, greater AD pathology translates into worse memory impairment, yet 8% of sisters with severe AD pathology still had no signs of memory loss. It is unknown if a moderate burden of fibrillary tangles can cause cognitive impairment in the absence of any vascular pathology.

There is some evidence that high intelligence in early life protects from the effects of AD.[27] An analysis of writing samples in the Nun Study, collected when the sisters were only 22 years of age, showed that writing with a low density of ideas or low grammatical complexity was associated with lower cognitive test scores at about 80 years of age. Impoverished imagination in early life thus had a strong and consistent relationship with impaired cognitive function in later life. AD was present in all of the sisters with low idea density in early life but in none of the sisters with rich ideation. Evidence shows that participation in cognitively challenging activities can reduce the risk of AD.[28] A relatively small increase in mental activity is associated with improved cognitive function, especially in working memory and perceptual speed. This finding recalls studies in rats that showed that environmental enrichment enhances neurogenesis, but it has not yet been proven that cognitive stimulation in humans is also protective. It could be that people in the earliest stages of AD, when there would be no clinical symptoms, are still impaired enough that they are less likely to read a newspaper or seek other mental challenges.

Brain imaging by MRI has provided a new window into the AD process.[29] By imaging a large number of people, it has become possible to describe general trends in healthy aging and to identify how these trends are altered by disease. In patients with AD, there is a spreading wave of gray matter loss as the disease progresses. Tissue loss begins in the temporal lobe and spreads into the frontal and occipital lobes, avoiding the sensorimotor cortex. The frontal lobes are spared early in the disease course but later show more than 15% loss of gray matter volume. The pattern of volume loss in the AD brain is distinct from the pattern seen in healthy aging, so it may eventually become possible to diagnose AD by MRI.

Currently, a definitive diagnosis of AD can be made only at autopsy by finding the plaques and tangles diagnostic of AD.[30] Both plaques and tangles are comprised of proteins that are abnormally deposited in brain tissue. Plaques are large extracellular clumps of a protein called amyloid beta (Aβ), which are not unique to AD (table 5). Tangles are also protein, but tangles are actually inside neurons and are composed of a protein called tau. Tau is a normal and essential component of neurons, but in AD patients, tau forms abnormal clumps. Tangles are also not unique to AD, but AD is unique in causing plaques and tangles at the same time.

One of the most exciting recent developments in AD is that there are new methods that may make it possible to image plaques and tangles in the living human brain. This is encouraging since it could help us to understand and control the early stages of disease, and it could also provide a useful endpoint in clinical trials. A radioactive molecule called Pittsburgh Compound B is injected into the bloodstream of patients with possible AD, then positron emission tomography (PET) is used to image the brain.[31] Since the compound binds to Aβ plaques, radioactivity is retained in the brain of AD patients, while it quickly washes out in patients free of plaques. So far, there is only a 2-fold increase in brain activity in AD patients, which is a small signal to detect by PET, but it may be possible to increase the affinity of Compound B for Aβ protein.

Because AD is associated with risk factors for atherosclerosis, vascular injury, and vascular inflammation, it may be possible to mitigate the impact of AD by treating the underlying risk factors.[32] Therapies for atherosclerosis have also shown promise against AD. For example, statins are used to treat high cholesterol, but statin therapy may also be associated with a 40–70% reduction in risk of AD. In addition, nonsteroidal anti-inflammatory drugs (including aspirin) may reduce the impact of AD.[33] In a meta-analysis that pooled results from 13,211 patients, nonsteroidal anti-inflammatory agents reduced AD risk by 28%. Short-term use reduced AD risk by only 5%, but use for longer than 2 years was associated with a 50% reduction in AD risk. Even aspirin alone was able to reduce AD risk by 13%.

However, treatment of AD is not yet very effective.[34] Three drugs are currently in wide use for AD (donepezil, rivastigmine, and galantamine), but the evidence supporting their use is rather weak. Though all of the trials reviewed were placebo-controlled and double-blinded, they were too small or too flawed to be truly informative. Even more recently, a large clinical trial of donepezil found little long-term benefit from treatment.[35] Donepezil therapy was associated with a lower rate of progression to AD over the first year of treatment, but the rate of progression over three years was no different than in people given placebo.

How strong is the evidence that AD is associated with apoptosis? Perhaps the best evidence is that the tau protein is cleaved by an enzyme—called an "executioner caspase"—that is also involved in apoptosis. This caspase is induced by Aβ protein. Previously, it was not known if caspases cause tangle formation or if they are induced by tangles, so the new work suggests that apoptosis plays a part in the early stages of AD. In short, Aβ accumulation triggers apoptosis, which then results in tau cleavage and tangle deposition. Apoptosis is an intriguing explanation for why in AD specific neurons are deleted in an otherwise normal brain, without inflammation and without necrosis.

It may be tangles, not plaques, which determine the degree of dementia in AD patients.[36] In one study, 97 people agreed to an autopsy after death, and each was given a battery of cognitive tests about 8 months prior to death. At autopsy, the volume of brain plaques and tangles was measured. It was found that virtually all cognitive impairment could be explained by the tangles alone, not the plaques.

Yet puzzling experimental evidence from animals has raised questions about the importance of tangle formation in AD.[37] Since tangles are clumps of tau protein, transgenic mice bred to make large quantities of mutant tau are an accepted model of human AD. Expression of mutant tau in mice causes age-related neuronal death, forebrain atrophy, and tangle accumulation. Yet if these mice are treated to suppress tau expression, memory function recovers despite continuing accumulation of tangles in the brain. This could mean that memory loss in AD is a function of reversible neural dysfunction rather than irreversible structural degeneration, or it could mean that the toxic element is the tau protein itself rather than the neurofibrillary tangles. It is even possible that the tangles are an effort by the brain to protect itself by causing tau to aggregate. Alternatively, it could mean that the mouse model of AD is not a good model for human disease.

AD is a disease widely acknowledged to be neurological, yet psychiatric symptoms are quite common in the earliest stages. Among 100 cases proven by autopsy to be AD, behavioral symptoms were present at diagnosis in 74% of patients.[38] The most common problem was apathy or depression in 58% of patients, but hallucinations were present in 25% and delusions in 20%. Although AD can be seen under a microscope—which would put it in the traditional realm of neurology—it is clear that patients often show symptoms consistent with an illness of the mind.

Interactions between Brain and Mind

We have not discussed many neurologic illnesses that illustrate an interaction between the brain and the mind. Huntington's disease—a degenerative disease associated with spasmodic and involuntary movements of the limbs and face—is a good example. It is a neurological illness, in that it affects cells in a limited portion of the brain, but it often has psychiatric symptoms as well.[39] As another example, multiple sclerosis is a neurological illness with psychiatric symptoms that may result from an abnormal immune response.[40] Brain tumors often cause neurologic symptoms, and recent evidence suggests that certain brain tumors arise from the neural stem cells involved in stroke recovery.[41] Traumatic brain injury is associated with a delayed but striking increase in risk of such psychiatric illnesses as major depression, alcohol abuse, panic disorder, and psychosis.[42] The spongiform encephalopathies—prion diseases such as Creutzfeldt-Jakob disease and kuru—are neurological illnesses with psychiatric symptoms that may be transmitted by consumption of food.[43]

The mind and the brain are inseparable, so it seems likely that the futures of neurology and psychiatry will merge in the discipline of neuropsychiatry. There is a broad overlap of causes, symptoms, and treatments of illnesses once thought to be either neurological or psychiatric, so it seems inevitable that there will be more cooperation between fields in the future.[44] Yet such cooperation will be difficult in practice, since many physicians already feel overwhelmed by the knowledge base required to treat patients with neuropsychiatric illness. Both psychiatry and neurology rank at the top of the list in terms of the difficulty that clinicians report in learning their specialty.[45] Yet patients may benefit from new perspectives that could arise from such synergy. Although neurologic and psychiatric disorders together account for only 1.4% of patient deaths, they account for a remarkable 28% of all patient years of life lived with disability.[46]

CHAPTER 15

PSYCHIATRY AND ILLNESSES OF THE MIND

The idea that every brain illness can be linked to a part of the brain, that every behavioral anomaly must have a corresponding brain abnormality, is the basic assumption that created the field of neurology and that still sets it apart from psychiatry. Because of the axiom that every behavioral anomaly must have a physical locus, those illnesses that form the core of neurological practice are "focal" in nature—they have a point of origin. In contrast, those illnesses that form the core of psychiatric practice are diffuse in nature, since they typically lack a (known) focus.

The culmination of this thinking is the idea that neurologic illnesses are "organic," whereas psychiatric illnesses are "functional." This dichotomy has caused endless problems through the years, not least because of its impact on patients. There is often a tacit assumption that a disease without a locus is somehow less "real" than a focal disease—if you can't point to the injury, then it doesn't really exist. Psychiatric illness is often assumed to be a character flaw, an expression of weakness, or even a divine judgment. It is an axiom that psychiatric illness is just as real, just as damaging, just as free of moral overtones, and perhaps just as much locked into a locus—though not as yet pinpointed—as is any neurologic illness.

The human brain is central to all neurological and all psychiatric illnesses. It occupies a place in the body that is no more privileged, no more sacrosanct, than the heart or the lungs, even though the brain seems to have evolved more rapidly and more recently than all other human organs.

The idea that the brain is somehow distinct from the body can be traced back to the French philosopher René Descartes, who wrote in the *Discourse on Method* in 1637 that "I am a substance the whole nature and essence of which is to think, and which for its existence does not need any place or depend on any material thing." From this arose the Cartesian mind-body dualism, the idea that the human body has a corporeal solidity, has length and breadth and depth, but the mind has no need of such rude physicality.[1] Rather, Descartes conceived of the mind as being an essence somehow free of a body: he believed his thought was irreducible and therefore more real than his brain. Descartes could doubt the existence of physical objects, but he could not doubt the existence of his own skeptical mind. From this reasoning arose the aphorism, "I think, therefore I am." Were Descartes to see a psychiatrist today, this belief system might well get him a prescription for antipsychotic medication.

ARE MENTAL ILLNESSES MYTHICAL?

Mental illness is the leading cause for hospital admissions in the United States. According to the National Institute of Mental Health, mental illness (including suicide) accounts for roughly 15% of the global burden of disease, more than that caused by all cancers combined and only slightly less than that caused by all cardiovascular conditions combined. Clinical depression is the second-leading factor associated with disability worldwide, trailing only heart attack, and well ahead of alcoholism, traffic accidents, lung cancer, arthritis, and diabetes. About 90% of people who die by suicide had a severe mental illness at the time of death, and more teenagers and young adults die by suicide than from cancer, heart disease, AIDS, birth defects, stroke, influenza, and pneumonia *combined*. As the population of the world ages, the burden of psychiatric illness is likely to become heavier still, since mental illness and suicide are both more prevalent in the aged.

A recent analysis shows that up to 5% of all patients with schizophrenia will die within 5 years, usually from suicide.[2] Other studies estimate that the lifetime risk of suicide among schizophrenics may be as high as 10%, but this is probably due to some confusion between the percentage of the dead who died by suicide versus the percentage of a group of patients who died by suicide. Patients with schizophrenia have a suicide rate that is 10-fold higher than normal, and virtually all suicide mortality occurs within 5 years after diagnosis.[3] In comparison, roughly 5% of all patients at high risk of heart disease will die within 5 years, and no one would argue that heart disease is mythical.[4]

Yet in 1960, Thomas Szasz published a paper whose inflammatory title was "The Myth of Mental Illness." In this paper, later a book of the same title, Szasz made the claim that

> not so long ago it was devils and witches who were held responsible for men's problems in social living. The belief in mental illness, as something other than man's trouble in getting along with his fellow man, is the proper heir to the belief in demonology and witchcraft. Mental illness exists or is "real" in exactly the same sense in which witches existed or were "real."[5]

Later, Szasz noted that

> our adversaries are not demons, witches, fate, or mental illness. We have no enemy whom we can fight, exorcise, or dispel by "cure." What we do have are problems in living—whether these be biologic, economic, political, or sociopsychological . . . mental illness is a myth, whose function it is to disguise and thus render more palatable the bitter pill of moral conflicts in human relations.[6]

It is hard to believe that Szasz ever met a patient with schizophrenia if he thinks that it is simply a problem in living. And yet his writings are still quoted today by those skeptical of psychiatric medicine. We cannot know how influential Szasz has been; clearly, his writing preceded and may have fueled the movement to deinstitutionalize the mentally ill, many of whom had been warehoused in grim facilities that were more concerned with controlling and coercing than with curing. Yet, deinstitutionalization may have been an inevitable movement, based on two perceptions: endless confinement of the mentally ill, often against their will, is a violation of their civil rights; and the cost to society of endless institutionalization had become an intolerable economic burden. Nevertheless, the night streets are now full of homeless people, many of whom are mentally ill and desperately in need of medical care and a safe place to sleep.

Part of the problem is that physicians are uncomfortable with the realization that to diagnose a psychiatric illness, you must rely upon a description of symptoms made by someone who is not completely competent. Reliance on such an inherently fallible witness would seem to open the door to malingering or hypochondria, on the one hand, or unscientific subjectivity on the other. Most illnesses that physicians treat have signs, or observable and objective indications of disease, often including a valid and repeatable measurement (e.g., body temperature, serum cholesterol, blood pressure). A "heart attack" is associated with an abnormal electrocardiogram (EKG), signs of cardiac ischemia, and elevated blood serum levels of proteins such as troponin or creatine kinase. Yet psychiatric illnesses typically have only symptoms, or nonobservable and subjective indications of disease, often including a patient's own description of his misery, but rarely involving a valid and repeatable measurement. True, psychiatrists can make measurements, but these are usually based upon a patient's answers to a structured series of questions. For the most part, it is not yet possible to make an objective measurement *free of input from the patient* that characterizes the severity or even the nature of psychiatric illness.

Yet it is possible to observe the pain of psychiatric patients: if they are homeless, friendless, jobless, inarticulate, miserable, and scared, this can be observed. A physician cannot know why this is so; ultimately, one must rely upon the patient's own account, and this makes many physicians uncomfortable. Yet this does not make the illnesses any less real.

How Prevalent Is Mental Illness?

As simple as this question seems, the answer is highly controversial. The boundary between mental illness and normal mental struggle has become a battle line, dividing psychiatrists and even the lay public into two warring camps.[7] At the core of the debate is a simple question that lacks a simple answer: How severe must a mental struggle become before it should be called a mental illness? By current guidelines, severity is assessed by determining if the struggle interferes with normal life. Thus, a college student who drinks too much beer on Sunday night and misses his chemistry test at 9 AM the next morning, thereby lowering his grade, or a middle-aged professional who smokes a joint before driving to a restaurant, thereby risking arrest for driving while impaired, would have a diagnosable substance abuse disorder.

Many people assume that the best measure of whether or not a person has a mental illness is whether the person seeks treatment, but most psychiatrists would argue that mental illness is dramatically undertreated. Yet a satisfying overall definition of "mental illness" is probably not possible, since so many different conditions must be included. And changes in the definition of any one condition cause major changes in the estimate of overall prevalence of mental illness. For example, in a 1984 survey, the prevalence of social phobia was measured by asking a large sample of people whether they had an excessive fear of public speaking. By this narrow definition, roughly 1.7% of people have social phobia.[8] Yet researchers soon realized that fear of public speaking is not the only manifestation of social phobia; many people have other fears, including a fear of eating in public or of using public washrooms. When in 1994 these other fears were also assessed, the estimate of social phobia increased to 7.4%.

The most recent large-scale assessment of the prevalence of mental illness used definitions of mental illness established by the *Diagnostic and Statistical Manual of Mental Disorders (DSM-IV)*, which has become the agreed-upon standard, the bible of psychiatry.[9] This manual is a masterpiece of clear writing and clear thinking, yet it is not the last word on mental illness. Nevertheless, the *DSM-IV* is easy to use, consistent when used by different clinicians, well validated in clinical practice, and a major improvement over what was available before it.

The lifetime prevalence of mental illness in the United States was estimated to be about 30% in both 1990 and 2003.[10] This has been a contentious conclusion because it implies that a significant proportion of all people are affected by mental illness at some point during their lives, and no one wants to think that he or she could become mentally ill. Part of the problem is that if people think about mental illness, they think of schizophrenia or bipolar illness or some other debilitating illness. However, the estimate of 30% prevalence also included people with milder illnesses, such as substance abuse, social anxiety, or an extreme phobia. The lifetime prevalence of serious mental illness (e.g., schizophrenia, bipolar illness, or a suicide attempt) is roughly 5%, while the lifetime prevalence of moderate mental disorder (e.g., clinical depression) is 14%, and that of mild mental illness is 11%. There was no significant change in prevalence of mental illness over the time between 1990 and 2003, but there was a dramatic increase in the percentage of people who sought treatment. Overall, 12% of the US population aged 18 to 54 years old received treatment for mental illness between 1990 and 1992, whereas 20% of the population received treatment between 2001 and 2003. Among patients with mental illness, 20% received treatment in 1990, while 33% received treatment in 2001. What this means is that the majority of patients with mental illness still do not receive any treatment. These conclusions are based on a nationally representative face-to-face household survey of 9,282 people, so the results are likely to be reliable and robust.

Several other interesting findings emerged from this national survey.[11] The best predictors for which patients would receive treatment were age greater than 24 years, female sex, non-Hispanic white race, and married. The increased rate of treatment in recent years may be due to aggressive direct-to-consumer marketing by drug companies, public education programs that clarify which conditions can be treated, expansion of community screening and other outreach services, greater availability of mental health services, and/or new legislation to promote access to mental health services, including changes in how mental health services are reimbursed by insurance.

WHAT IS DEPRESSION?

Depression is the most common of all psychiatric disorders. Major depression is clinically defined by the *DSM-IV* as involving 5 or more of the following 9 symptoms, with symptoms present nearly every day over a 2-week period:[12]

- depressed mood most of the day, by subjective report or by observation;
- anhedonia, or markedly diminished interest or pleasure in most activities;
- significant weight loss or weight gain (>5% body weight), without dieting;
- insomnia or, less frequently, hypersomnia;
- observable restlessness (i.e., not just feelings of restlessness) or slowing;
- sustained fatigue or loss of energy, often without exertion;
- feelings of unrealistic worthlessness or inappropriate guilt;
- diminished ability to think or concentrate or be decisive;
- recurrent thoughts of death or suicide (with or without a specific plan).

At least one of the symptoms must be depressed mood or anhedonia; the symptoms must cause significant distress or impairment (in any domain of function); the symptoms cannot be directly due to the effects of substance abuse or another medical condition; and the symptoms must not be reactive (i.e., in response to a depressing life event). In children and adolescents, irritability can take the place of depressed mood. A key issue is whether symptoms are sufficiently severe to cause a subjective sense of distress, even if treatment is never sought. Another problematic issue is in determining when reactive depression (e.g., bereavement), which is considered normal, crosses a threshold to become something worse: the threshold is generally considered to be 2 months, but this seems inadequate if, for example, a parent has unexpectedly lost a child to death.

Major depression is more prevalent in women: the lifetime risk is 10% to 25% for women and 5% to 12% for men.[13] At any given time, 5% to 9% of women and 2% to 3% of men are depressed. The typical age of onset is in the mid-20s, with some indication that people are becoming depressed at a younger age. Some people have an isolated episode of depression, whereas other people have to battle recurrent depression, often with episodes occurring more frequently with the passing years. At least 50% to 60% of people who have one episode of depression will have another at some point later in life. Episodes of depression tend to resolve spontaneously, even in the absence of treatment, which is why it is so important to have a placebo arm in any clinical trial of antidepressant medication. The natural history of depression is such that about 40% of untreated patients will still meet criteria for depression one year after their first episode, 20% of untreated patients will continue to have some symptoms but will no longer meet full criteria, and 40% will have fully resolved all symptoms. The first episode of major depression often follows a severe stressor, such as the death of a loved one or a divorce, but later episodes are less likely to be reactive in nature. Depression is up to 3-fold more common among people with a first-degree relative who has suffered depression.[14]

Depression in adolescence has been controversial recently, but there is good evidence that about 5% of adolescents are depressed at any given time.[15] Adolescent depression, though it is likely to self-resolve, is a significant risk factor for suicide in adolescence, and suicide is the third-leading cause of death among adolescents. Therefore, it is critically important to diagnose and treat adolescent depression, since the suicide rate is compelling evidence that the illness is a terrible burden to bear.

CAN MERELY TALKING ABOUT DEPRESSION CURE IT?

This question is liable to be as inflammatory as Szasz's question about whether mental illness is mythical. Yet there is surprisingly little hard evidence available about the therapeutic efficacy of psychotherapy's "talking cure." The placebo effect is very powerful, as we have seen, and it is very difficult, perhaps to the point of impossibility, to design a clinical study that would be able to detect a placebo effect from the "talking cure." There can be no question that a therapist offers hope, and that hope itself is therapeutic; in fact, it could be argued that hope *is* the placebo effect. But what could possibly constitute a placebo control for the act of discussing problems with a willing and sympathetic listener? Would it perhaps be sitting in the presence of such a listener but not actually discussing your problem? Or perhaps talking to someone who is unsympathetic or doesn't understand your words? Yet a placebo control is badly needed, since untreated depression tends to wax and wane on its own; the fact that depression resolves does not prove that psychotherapy is effective.

In short, we cannot assume that the "talking cure" is effective for depression without doing placebo-controlled clinical trials. To prove that psychotherapy works, "talking" must be objectively compared to a placebo-treated group in a randomized clinical trial, yet no one has thought of a way to have a placebo control for psychotherapy. However, there may be a way to parse apart the effects of "talking." Many double-blind, placebo-controlled clinical trials have shown that about 50% of patients who receive antidepressant medication report improvement in their depression over the short term, whereas only

30% of depressed patients given placebo pills report improvement.[16] Thus, we know that antidepressant medication is at least moderately effective.

Because antidepressant medication is effective, reasonably cheap, and fairly free of side effects, we can reformulate the original question about the efficacy of psychotherapy. It seems legitimate to compare the talking cure directly to medication, provided that there is a placebo control for the medication. If psychotherapy is not more effective or more enduring than pharmacotherapy, how can we justify the greater time and greater expense of psychotherapy? A collation of clinical trials—each of which was completed in the past 20 years (when newer antidepressants first became available) and all of which included a drug-treatment arm, a "talking cure" arm, and a placebo arm—shows that *all treatments are better than placebo* (table 6).

In most cases, there is not a significant difference in efficacy between drug treatment and the talking cure. However, in every case where there was a significant difference, drug treatment alone was superior to the talking cure alone. Even in the earliest studies tabulated, which used drugs that are no longer considered state of the art (tricyclic antidepressants like imipramine,[17] amitriptyline,[18] and nortriptyline,[19] or monoamine oxidase inhibitors like phenelzine[20]), drug treatment was at least as good as the talking cure. More recent antidepressants—selective serotonin reuptake inhibitors (SSRIs) such as paroxetine[21] and fluoxetine[22]—tend to be significantly better than psychotherapy, relative to placebo (table 6). Overall, the remission rate for drug therapy alone is about 46%, for the talking cure alone about 46%, and for placebo about 24%.[23] Thus, while the "talking cure" works, there is no compelling evidence that psychotherapy is any better than drug therapy for major or minor depression, whether in adults or adolescents.

It is noteworthy that psychotherapy generally takes longer and is more expensive than drug therapy.[24] In a recent clinical trial of paroxetine, drug therapy reduced symptoms of depression within 3 months, whereas cognitive behavioral therapy (CBT) took 5 months to alleviate

Table 6. Summary of clinical trials of depression in the last 20 years that had a drug treatment arm, a therapy arm, and a placebo arm. The "Effect of psychotherapy" is the ratio of the improvement on psychotherapy, divided by the improvement on drug treatment. In no case was therapy significantly better than drug treatment, and therapy was significantly worse in those ratios marked with an asterisk (CBT = cognitive behavioral therapy; PSRs = percentage of patients with symptom resolution; DFDs = depression-free days).

Study author (reference)	Patient diagnosis	Patients enrolled	Treatment arms	Relative improvement	Effect of psychotherapy
Elkin (1989) (Note 17)	Mixed depression	57	Drug treatment (imipramine)	42 (PSRs)	
		59	CBT	36	86% (=36/42)
		61	Interpersonal therapy	43	102%
		62	Placebo	21	
Scott (1992) (Note 18)	Major depression	26	Drug treatment (amitriptyline)	58 (PSRs)	
		29	Cognitive therapy	41	71%
		29	Treatment as usual	48	
Mynors-Wallis (1995) (Note 18)	Major depression	31	Drug treatment (amitriptyline)	52 (PSRs)	
		30	Problem solving	60	115%
		30	Placebo	27	
Schulberg (1996) (Note 19)	Major depression	91	Drug treatment (nortriptyline)	48 (PSRs)	
		91	Interpersonal therapy	46	96%
		92	Treatment as usual	19	
Reynolds (1999) (Note 19)	Major depression	25	Nortriptyline + psychotherapy	80 (PSRs)	
		28	Nortriptyline + clinic care	57	
		25	Placebo + psychotherapy	36	*45%
		29	Placebo + clinic care	10	
Jarrett (1999) (Note 20)	Major depression	36	Drug treatment (phenelzine)	42 (PSRs)	
		36	Cognitive therapy	44	105%
		36	Placebo	19	
Williams (2000) (Note 21)	Minor depression	137	Drug treatment (paroxetine)	61 (PSRs)	
		138	Primary care counseling	52	*85%
		140	Placebo	40	
March (2004) (Note 22)	Adolescent depression	109	Drug treatment (fluoxetine)	61 (PSRs)	
		107	Fluoxetine + CBT	71	
		111	CBT alone	43	*70%
		112	Placebo	35	
DeRubeis (2005) (Note 21)	Mixed depression	120	Drug treatment (paroxetine)	50 (PSRs)	
		60	Cognitive therapy	43	*86%
		60	Placebo	25	
Revicki (2005) (Note 21)	Major depression	88	Drug treatment (paroxetine)	258 (DFDs)	
		90	CBT	251	97%
		89	Community care	225	

symptoms to the same degree. Furthermore, the cost per depression-free day was 9% less for drug therapy than for CBT, which is a compelling consideration for someone who lacks medical insurance. Other clinical trials have verified that symptom resolution tends to be faster with drugs than with therapy.[25] Any clinically depressed patient would sincerely hope for rapid recovery, since the depths of depression can be the pit of hell.[26] Finally, there is evidence that some CBT therapists are more effective than others, since therapy given at one major university was more successful than CBT given at another major university.[27] It seems likely that therapists at either university are probably superior to most community therapists, yet, because the patient has no way to decide which therapist is good and which is not, a patient who prefers CBT must choose a therapist in ignorance. This, too, is a compelling consideration for people who have a marginal ability to pay for medical care; drug therapy can be administered effectively by the family physician, whereas counseling may be ineffective if it is administered by an inept practitioner, but it is usually costly anyway.

Such findings may explain why there has been a national trend away from psychotherapy.[28] From 1987 to 1997, there was a 3-fold increase in the proportion of people who received treatment for depression. The proportion of treated patients who received antidepressant medication rose from 37% to 75%, while the proportion of treated patients who received psychotherapy fell from 71% to 60%. The family physician is now much more likely to be involved in depression treatment than in the past.

Yet the talking cure is not without value. When drug therapy is combined with CBT for depressed adolescents, the combination had a response rate of 71%, which was superior to the response rate with either drug therapy alone (61%) or CBT alone (43%).[29] Similarly, in adults the combination of drug therapy with CBT had a response rate of 73%, whereas the response rate with both drug therapy alone and CBT alone was only 48%.[30] Combining drug therapy with psychotherapy improves clinical response—and it may help patients to adhere to their prescribed medication.[31] There is also evidence that when drug treatment is combined with psychotherapy the long-term recurrence of depression is reduced. The 3-year recurrence rate of depression with combined therapy was half what it was with either drug treatment alone or psychotherapy alone.[32] Psychotherapy may also help to stabilize a patient who has tapered off antidepressant medication.[33]

ANTIDEPRESSANT USE IN ADOLESCENTS

There has recently been a storm of controversy surrounding the prescription of antidepressant medication to depressed adolescents. There is concern that antidepressants paradoxically *increase* the risk of suicide among newly treated adolescents.[34] Due to this concern, the Food and Drug Administration (FDA) required a "black box warning" to be added to all antidepressants. This action, which was strenuously opposed by some clinicians, must be kept in perspective. Suicide is the third-leading cause of death among unmedicated American teenagers, and many of these suicides could probably have been prevented if these teens had been treated. According to the Youth Risk Behavior Surveillance, up to 8.5% of high school students attempted suicide in 2003, so suicide is a compelling concern in the absence of treatment.[35] Fluoxetine in adolescents is more likely than placebo to result in clinical improvement, but it is also somewhat more likely to result in a "suicidal event," though it should be noted that there were *no completed suicides* in the data evaluated by the FDA. Yet, among teenagers treated with antidepressants, depression improved 4 times more often than suicidal behavior developed, so this seems to represent an acceptable risk-benefit ratio.[36]

Fewer physicians are prescribing antidepressants to depressed adolescents since the black box warning was required by the FDA.[37] This could be a terrible mistake. Nationwide, suicide rates tend to be lower in counties where the prescription rate for newer antidepressants (e.g., SSRIs) is high; high rates of suicide occur in counties with fewer such prescriptions or with more prescriptions for the old-style tricyclic antidepressants.[38] SSRIs are associated with 43% fewer suicides than tricyclics overall, although there was a suggestion in the data that the rate of "non-fatal self-harm" may be higher in depressed adolescents treated with SSRIs.[39]

In summary, the risks of antidepressant medication in adolescents are real, albeit small, but there is probably a greater risk in doing nothing.[40] This is especially true given that depression is a risk factor for substance abuse, early pregnancy, poor academic performance, and social isolation. But caution is required; although psychotherapy has been strongly urged for any adolescent patient receiving medication, patients must make up their own minds based on the available evidence.[41]

Is Depression a Focal Illness?

One of the key distinctions between neurologic and psychiatric illnesses has been that neurologic illnesses are focal, whereas psychiatric illnesses are diffuse. This dichotomy is probably responsible for the notion that neurologic illnesses are "organic," whereas psychiatric illnesses are "functional." Depression is generally seen as a systems-level disorder that affects broad regions of the brain, including cortical, subcortical, and limbic sites. Yet recent results suggest that treatment-resistant depression may actually have a locus in the brain, calling into question the whole dichotomy between neurologic and psychiatric illness.

Up to 20% of depressed patients fail to respond to current treatment.[42] Treatment-resistant depression is defined as depression that fails to respond to at least 4 different treatments, including antidepressant medication, psychotherapy, or electroconvulsive therapy. Positron emission tomography (PET) shows that these patients often have a small area of the brain (below the corpus callosum and between the eyes) that is metabolically overactive. In one study, a group of 6 depressed patients who met strict criteria for treatment resistance were treated in an entirely new way by having an electrode implanted in their brain, which allowed "deep brain stimulation." Chronic electrical stimulation of the overactive area was associated with a striking and sustained remission of depression in 4 of the 6 depressed patients. Deep brain stimulation was associated with a *decrease* in blood flow specifically to the area that had been metabolically overactive. Not only were there behavioral and physiological changes in treated patients, but the patients also reported subjective changes in their emotional state. After the surgery was done but before patients were able to leave, researchers had an opportunity to do a placebo-controlled trial. The patients could not sense the stimulation directly, so it was possible to turn the current on and off and ask the patients how they felt under each condition. In the placebo-current condition, patients reported no benefit, but when electrical stimulation was used, the patients reported benefit:

> All patients spontaneously reported acute effects including "sudden calmness or lightness," "disappearance of the void," sense of height-ened awareness, increased interest, "connectedness," and sudden brightening of the room, including a description of the sharpening of visual details and intensification of colors in response to electrical stimulation.[43]

Many of the patients also enjoyed a cognitive benefit from treatment and no side effects were reported other than a risk of infection at the site of implantation. Electrical stimulation of this brain locus thus caused a sudden change in mood for the better, and the mood changes endured for months, long after the electrical current was switched off. This exciting research may herald a new treatment for depression, but it also suggests that depression really is "organic," and not merely "functional." As of yet, this therapy is still experimental and still controversial.

WHAT IS SCHIZOPHRENIA?

Schizophrenia is a disorder of perception or consciousness that has been rigorously defined in the *DSM-IV*.[44] Schizophrenia involves 2 or more of the following 5 symptoms, each present for a significant portion of time during a 1-month period:

- delusions, or false beliefs;
- hallucinations, or false perceptions;
- disorganized speech, including derailment or incoherence;
- grossly disorganized or catatonic behavior;
- negative symptoms, including anhedonia or cognitive impairment.

In addition, there must be social or occupational dysfunction for a significant portion of the time since symptom onset; continuous signs of disturbance for at least 6 months (with at least 1 month of symptoms severe enough to meet the criteria above); no evidence of significant mood swings that might suggest a mood disorder with psychotic features; no evidence of drug abuse or another medical condition that might explain symptoms; and no evidence of a developmental disorder.

Delusions are erroneous beliefs that usually involve the misinterpretation of a perception or experience. The content of a delusion often entails a sense of persecution, and the patient may believe that she is being followed, spied upon, harassed, tormented, or ridiculed. There can instead be grandiose delusions, such as a patient's firm belief that he is actually Napoleon or Christ. Alternatively, there can be somatic delusions, including a conviction that one's internal organs are rotting or have been replaced by the organs of another person, or that one's thoughts are being controlled or "inserted" by the FBI. Referential delusions are also common: the patient may believe that a character in a television show is talking to her, or that a book or newspaper article refers specifically to him. Sometimes, it can be difficult to discern between a delusion and a fixed belief because some fixed beliefs are firmly held in the absence of any evidence (e.g., All tax cuts are good!). A fixed belief system can assume the dimensions of a delusion, although belief systems are usually less resistant to contrary evidence.[45]

Hallucinations are erroneous perceptions that seem as real as any other perception; they can take any perceptual form, including auditory, visual, olfactory, gustatory, or tactile. Patients are often completely unable to tell the difference between hallucination and reality. The most common hallucinations in schizophrenia are auditory, and can involve hearing critical voices, often several at a time, that offer a running commentary on thoughts or behaviors. A striking feature of many schizophrenic patients is that they have essentially no insight into the nature of their illness. For example, one patient wrote in a diary about the fantasy men who trouble her:

> I have written this to expose what someone has done. These are the things that I have been experiencing for the past 10 years. My perception is still off and some things may be hard to understand. . . . Whoever reads this may experience minor perception changes and other minor things they would not normally experience. . . . These things I have written are the true and accurate events that took place—they are not delusions or hallucinations. I may still have trouble discerning between true and untrue in some situations, but I can discern that these things are *true*.

This passage was written after the patient had begun to relapse, following an initially successful treatment course. This quote, and a good deal of other evidence, suggests that schizophrenia is, at least in part, a disorder of consciousness or self-awareness.

Schizophrenia affects somewhat less than 1% of people, with some variation in prevalence from place to place; it is more prevalent in cities than in urban settings, and there are pockets of high prevalence, which may be related to shared genetic or environmental risk factors. The best guess is that the prevalence of schizophrenia is about 0.7% of the population, which means that over 2 million Americans may have the disease. The age at onset of this illness is typically about 24 years old, with onset slightly younger for men than for women. However, patients have been diagnosed as young as 5 years old. It is intriguing that symptoms often begin to emerge at about the time of puberty, but whether or not puberty actually "causes" schizophrenia remains unknown. Men tend to be schizophrenic more often than women, although this has also been contentious. The course of illness is typically chronic, meaning that patients may wax and wane or have severe exacerbations, but tend not to recover fully, even with medication. It has been estimated that the annual economic cost of schizophrenia to the US economy may be as high as $62.7 billion.[46]

Schizophrenia tends to run in families, and split twin experiments show unequivocally that there is a substantial genetic component to the illness, even though there is also a role for certain (largely unknown) risk factors in the environment.[47] The estimated heritability of schizophrenia is 81%, while the estimate for environmental effects is only 11%. Known environmental risk factors include maternal starvation during pregnancy, prenatal stress, perinatal obstetric complications, pre- or postnatal exposure to infection, and exposure to environmental toxins; each of these known factors increases the risk of schizophrenia slightly. But having a first-degree relative with schizophrenia increases the risk of disease roughly 10-fold.[48]

There has been a great deal of argument about whether schizophrenia is a developmental or a degenerative disorder. Virtually all pathological studies of schizophrenia have involved patients who had the disease for many years, so if degenerative brain changes are present they could result from drug abuse, malnourishment, homelessness, or even medication. Yet several recent studies show that brain changes are present at diagnosis, including brain atrophy and changes in brain chemistry.[49] It remains unknown whether these changes are developmental, arising at birth, or degenerative, arising somewhat before diagnosis.

In a sense, schizophrenia is the most human of illnesses. The underlying nature of the disorder remains unknown to a degree that is alarming, in part because there are no good animal models of this disease. Since it is not possible to tell whether a rat is hallucinating or delusional, it is impossible to tell if it has schizophrenia. It has been proposed that schizophrenia is a disorder of information processing and attention.[50] This could explain why patients feel as if they are bombarded with stimuli and why they often misinterpret information (leading to delusions), confuse internal and external stimuli (leading to hallucinations), and withdraw from the world (leading to anhedonia and avolition). Alternatively, schizophrenia can be interpreted as a disorder of working memory, in that patients are unable to plan their speech (leading to disorganized speech and thought disorder), maintain a clear idea of what to do (leading to behavioral disorganization), or compare internal and external stimuli to past recollections (leading to altered consciousness). Finally, schizophrenia has been proposed as a disorder primarily of cognition and consciousness, in that patients are unable to find a boundary between internal and external experiences (leading to hallucinations and delusions).

Perhaps because our understanding of the disease is so poor, clinicians have had rather little success in treating schizophrenia.[51] Most patients improve following initial treatment with the newer generation atypical antipsychotics. There can be substantial improvement within a month, but the duration of successful treatment is typically quite short. Depending upon the medication used, successful treatment averages from 5 to 9 months and is rarely more than 12 months. Overall, 74% of patients discontinue antipsychotic treatment entirely in less than 18 months. Most of the time, the patient himself discontinues medication, presumably because of a sense that the medication is ineffective. However, it should be noted that schizophrenic patients have an astonishing lack of insight, so it could be that patients elect to discontinue medication even though their medication may actually be working.

There are many reasons why antipsychotic medication might be discontinued. Typically, medication is better at alleviating some symptoms than others. Hallucinations and delusions may resolve, but the patient may still be troubled by symptoms such as loss of a sense of pleasure (anhedonia), loss of the ability to persist in goal-directed activity (avolition), loss of the range of emotional expression (affective flattening), a pervasive sense of emptiness (apathy), or poverty of speech (alogia). In addition, antipsychotic medication can be associated with distressing side effects in 66% of patients, including hypersomnia or sedation, insomnia, constipation or urinary hesitancy, decreased libido, dramatic weight gain, and metabolic changes that produce diabetes, tremor, and dizziness.[52] Finally, it may be crucial that comorbid (co-occurring) mental illness is commonplace among schizophrenics. In one study, 29% of all patients were drug dependent or drug abusing, 28% were depressed, 25% were alcohol dependent or alcohol abusing, and 14% had an anxiety disorder. Typically, antipsychotic medication does not help much with co-occurring complaints.

IS SCHIZOPHRENIA A FOCAL ILLNESS?

The currently dominant model is that schizophrenia is a neurodevelopmental disorder that is present from birth but evolves over the life span. The neurodevelopmental model would suggest that those brain changes associated with schizophrenia—because they may occur during development—are broadly distributed over the entire brain. Yet the neurodevelopmental model may be incorrect, as the model does not readily account for some key features of schizophrenia, such as the protracted period during childhood when symptoms are dormant and the fact that clinical deterioration after diagnosis can be progressive and, in some cases, rapid. Nevertheless, there is a consensus that schizophrenia is a diffuse disease affecting broad areas of the human brain. It has even been proposed that schizophrenia is a disorder of consciousness or integration, characterized by a pattern of defective interaction among widely scattered brain areas.[53] All of these models are inconsistent with the idea that schizophrenia is a focal illness.

Yet there is also new evidence suggesting that schizophrenia may involve apoptosis.[54] Apoptosis is the process of programmed cell death, which we discussed in relation to neurologic illnesses such as Parkinson's and Alzheimer's disease. If apoptosis plays a role in schizophrenia, this would represent a convergence of neurology with psychiatry. Apoptosis could explain why it has been impossible to find pathological evidence of schizophrenia, since apoptosis could affect a relatively few cells broadly scattered throughout the brain. One might not see clear-cut pathological evidence of apoptosis in schizophrenia for two reasons: relevant changes might all occur in the time period surrounding diagnosis, whereas the brain would usually not be autopsied until many years later, and apoptosis could occur quite rapidly (apoptotic cells are typically cleared in less than a day in a rat), so pathological examination of a brain (perhaps even at the time of diagnosis) might still not show unmistakable evidence of apoptosis.

How good is the evidence that apoptosis contributes to schizophrenia? As of yet, there is nothing definitive, but some intriguing observations have been made.[55] First, there is strong data that the brain undergoes a process of volume loss around the time of diagnosis. No one knows which tissues are shrinking, but volume loss would be consistent with apoptosis. Second, there has never been any evidence of a necrotic process in the schizophrenic brain to account for volume loss, so apoptosis would seem to be the only way to account for such loss. Third, several environmental insults that increase the risk of neuronal apoptosis also increase schizophrenia risk, such as obstetric complications or ischemia. Fourth, there is evidence that certain proteins may be altered in the schizophrenic brain in a way consistent with apoptosis. For example, the Bcl-2 protein, which is a potent inhibitor of apoptosis, is reduced by about 25% in the temporal lobe of schizophrenic patients. Finally, patients with schizophrenia may simply be more vulnerable to oxidative stress—the stress caused by poor oxygen delivery to the brain—as oxidative stress is known to induce apoptosis in experimental animals. Some scientists have concluded that apoptotic mechanisms could provide a plausible explanation for several key features of schizophrenia.[56] On balance, the evidence is still weak, but the hypothesis is intellectually appealing and worthy of further study. It would be a breakthrough to find that neurological and psychiatric illnesses are not that different. At the very least, this would remove some of the stigma of an illness that many people mistakenly believe to be "functional" rather than "organic."

WHY DO MENTAL ILLNESSES EXIST?

It seems that evolution should act against mental illness, since people with mental illness are less likely to survive to reproductive age, less likely to have consensual sex, and less likely to have children.[57] In fact, schizophrenic men are half as likely as healthy men to have children. Yet mental illness is present in every human society with more or less equal frequency. This would seem to be an evolutionary enigma.

It has been proposed that people who carry genes that predispose to mental illness may enjoy some competitive advantage over people who do not carry such genes. This is an analogy to the argument as to why sickle-cell disease has not been eradicated from the human population. Sickle-cell disease is a serious blood disorder that increases the risk of pediatric stroke 250-fold in comparison to healthy children, but people with sickle-cell trait (i.e., gene carriers for sickle-cell disease) have an increased resistance to the malarial parasite. Where malaria is epidemic, gene carriers enjoy a significant competitive advantage over people who lack the sickle gene, which means that the gene is maintained in the population even though sickle-cell disease itself can be fatal. Several other genes that should also be strongly selected against may be maintained in the human gene pool in the same way; a gene for Tay-Sachs disease is thought to confer resistance to tuberculosis and a gene for cystic fibrosis may be associated with resistance to cholera.[58]

It is an appealing idea that genes associated with mental illness are somehow linked to enhanced creativity or greater environmental sensitivity, so that gene carriers might be at an evolutionary advantage. There are no data to support this hypothesis strongly, but there are at least a few facts. It is well known that certain highly creative people have had a psychotic break (e.g., Isaac Newton, Ludwig von Beethoven, Vincent van Gogh, and Nobelist John Nash) and that other highly creative people fought lifelong battles with depression (e.g., William Blake, Edgar Allan Poe, Lord Byron, Charles Darwin, Virginia Woolf, and Ernest Hemingway). In a sample of 291 men—all selected for world-class ability in science, philosophy, politics, or art—depression and alcoholism were more prevalent than normal in visual artists and in creative writers.[59] Depressive disorders afflicted writers almost twice as often as men with any other creative achievement. A study in Iceland, where the inheritance of psychosis has been studied in great detail, found that healthy relatives of psychotic patients are far more likely than normal to excel in creative writing, scholarly achievement, and mathematical ability.[60] One could imagine that low-level mania—which might make someone more likely to fill a blank page or an empty canvas—would also be attractive to members of the opposite sex, at least during the yearning years of early adulthood.

An early study of the link between creativity and psychosis compared creative writers to patients diagnosed with schizophrenia or bipolar disorder.[61] On a standardized test of object sorting, writers and bipolar patients tended to show conceptual overinclusion, lumping things together that did not necessarily belong together. But the writers differed from the bipolar patients in showing more richness of ideas, while the patients showed more idiosyncratic thinking. Schizophrenic patients tended to be underinclusive rather than overinclusive, and showed less richness of ideas than either writers or bipolars. The conceptual style of creative writers therefore resembles mania more than schizophrenia. Interestingly, if overinclusiveness is truly an index of thought disorder, then bipolar patients have a more florid thought disorder than schizophrenics.

Another possible way to explain why mental illness has not been eliminated from the human population is that the genes that are mutated in mental illness could somehow be more fragile, or more prone to mutation. This hypothesis suggests that mental illness would often result from a fresh mutation. This idea is inconsistent with the known heritability of depression and schizophrenia, but it could explain why mental illness can occur in people with no known genetic risk factors. Many of the capacities that set us apart from other animals, even the higher primates, relate to advanced social abilities such as the ability to infer what others think and feel ("theory of mind"), to read subtle facial signals that reveal emotions, and to discern when others are lying. These traits may have developed rather late in human evolution, so it is possible that the genes for these traits are "under-engineered," or somehow less robust than normal.

A final possibility for why mental illness has not been more strongly selected against is that the genes that are abnormal in mental illness may be closely linked to other genes that are strongly benefi-cial. In other words, it may be impossible to select against mental ill-ness without also selecting against some crucial feature of human physiology. The upshot is that every one of us may have some degree of vulnerability to mental illness; whether the vulnerability is con-verted to frank illness may have something to do with the environ-ment that we experience.

WHAT IS MENTAL HEALTH?

Since we have spent so much time discussing ills of the brain and the mind, it is a fair question to ask, what is mental health? This is a far more difficult question to answer than it might seem, as there are sev-eral obvious pitfalls.[62] There is a tendency to assume that mental health is the opposite of mental illness, although this greatly underes-timates human potential. There is an inclination to equate average with healthy, even though average may be far from optimal. There can be a failure to distinguish "trait" from "state," so that there is confu-sion between how someone usually is and how he is right now. And there can be a reluctance to incorporate different cultural norms and differing societal expectations into a single definition of mental health. Even Freud, who had an opinion on almost everything, dismissed mental health as "an ideal fiction."

It would be a mistake to assume that mental health is the opposite of mental illness, as if health were somehow a dichotomous variable—that a person is either completely well or completely ill. Mental health is a spectrum, and the variables that define mental health form a broad distribution. When someone is mentally ill, she has simply crossed a threshold, beyond which symptoms are expressed in a more florid fashion than is considered "normal." If a patient is successfully treated, she may cross the threshold back to normalcy, but she may never achieve "wellness." Yet wellness may be required for the achievement of full potential.

It is also potentially problematic to equate average with healthy. For example, the "average man" has borderline hypertension, yet it is known that even small elevations in blood pressure increase the risk of heart disease and stroke. Should we accept borderline hypertension, or should we work to control hypertension, even if it is rather mild? Our society has decided that, from the standpoint of individual benefit and from the standpoint of societal economic burden, it makes more sense to treat mild hypertension. By the same token, doesn't it make more sense to strive for more optimal mental health? It seems inadequate to accept the average level of function as sufficient, because the average includes many people who are not mentally ill but who are not functioning at their best. We should not consign people who endure mild depression to a lifetime of suboptimal functioning. Perhaps mental health should be understood as being *above normal*, as achieving a mental state that is objectively desirable.[63]

It can be very difficult to discern trait and state, to differentiate between how someone usually is versus how he is now. Every person goes through periods of depression, when their function is suboptimal, when they feel that the world has closed ranks against them; this is not mental illness unless it is persistent; that is, when it is a trait rather than a state. But how can we define trait and state? It is entirely arbitrary to require that symptoms of mental illness be present for a certain number of months before a diagnosis can be made, but that is the practice nonetheless.

Finally, it will be hard to incorporate different cultural norms into a definition of mental health.[64] Colonial America did not recognize alcoholism as an illness, but that society had many problems related to alcoholism. The values of a culture can be parochial: most Americans have a degree of optimism that is seen as hopelessly naïve by Europeans, while many Americans think of Europeans as being stultified in the past. But there are more significant and insidious differences between cultures that can actually lead to inappropriate diagnoses of mental illness. Certain cultures accept the idea that God can speak to individual people, whereas Western culture tends to view this as a sign of mental illness. A person imbued with the alternate culture could receive an inappropriate diagnosis from a healthcare provider who is insensitive to cultural differences. Nevertheless, there are shared values across cultures, and it is important to incorporate these into a definition of mental health that will work in any culture.

Ultimately, one must also question the motivation for defining mental health.[65] If mental health is "good," what is it good for? For the self or for society? For fitting in or for creativity? For happiness or for survival? Who should be the judge of normalcy and how should this determination be used? Should treatment ever be considered for someone who is already functioning at an adequate level? There are well-accepted psychological interventions that can improve already-adequate intelligence and social skills; most people could benefit from a tutorial on learning skills or stress management. But there are no well-accepted medical interventions that can improve already adequate physiological functions. If heart rate is normal, if lung function is average, if perception is accurate, if mood is typical, it could be disastrous to intervene. In the healthy rested patient, virtually any psychopharmacological intervention will, over time, make brain function worse.

Although a definition of mental health is ultimately somewhat arbitrary in placing higher value on certain traits than on others, one well-accepted set of criteria for emotional maturity was first proposed by William Menninger.[66] These criteria include:

- the ability to deal constructively with reality
- the capacity to adapt to change
- the gift to love and be loved in return
- the ability to relate to other people in a consistent and satisfying way

- the capacity to find more satisfaction in giving than receiving
- the gift of being relatively free from anxiety and tension
- the ability to redirect hostile feelings into constructive actions

If pessimism is the dominant cognition of the depressed, optimism appears to be the dominant cognition of the mentally well.[67] And mental health has an impact on physical health. In the Nun Study, among those nuns who expressed the most optimism, only 24% died by age 80. In contrast, among nuns who expressed the least optimism, 54% were dead by age 80. Mind and brain are clearly linked, but new evidence suggests that mind and body are also linked; therefore, one of the strongest motivations for mental health may be that it contributes to physical health as much as to psychic well-being.

Chapter 16

Intelligence
and Sociality

It should be clear that human intelligence did not evolve so that we could build cities and towns, write books and symphonies, or design planes and cars. These things are instead fruits of emergent complexity, the unintended outcome of a brain that developed in order to do something else entirely. This is proven by the fact that there are people in the world who live as they did 100,000 years ago, without cities, books, or planes, unable to count past 5,[1] unable to build anything more advanced than a rude hut. Yet in every respect they are fully human, sharing with us the structure of their brain and the function of their genes. And they share another key human feature with us as well: they are deeply social, committed to life in communities, which may be based on families but which can still show cooperation across family lines. It is arguably true that humans are more social than any other organism, and it seems likely that our brain evolved in order to foster this sociality.

Our ability to cooperate with other human beings across genetic lines is quite unique; humans have built societies that are largely (albeit not completely) blind to genetic relationships. We congregate in groups to gain protection from predators, just as wildebeest, zebra, and antelope do. Yet unlike the herding animals, which tend to be genetically linked through extended family relationships, we congregate in far larger numbers to form nations and states, and we collaborate with people we've never seen before and with whom we have no definable genetic relationship. We combine forces to build complex structures, just as ants, bees, and termites do. Yet unlike the social insects, all of whom are genetically identical to others in the nest or hive, we work with people from divergent genetic backgrounds. We collaborate in groups to hunt, just as lions, wolves, and dolphins do. Yet unlike pack-hunting animals, which are usually closely enough related to other members of the pack to be considered extended family, we can collaborate with people from far beyond our families.

How is this level of cooperation among humans possible? We will argue that what makes collaboration across genetic lines possible is our intelligence. Our intellect makes it possible to see the benefit of collaboration, to communicate with exquisite detail about how to cooperate, to work out implicit and explicit rules for alliances, to externalize those rules in governments and legal systems, to internalize those rules as a moral sense, and to deal with those people who transgress the rules in a way that fosters stability. Only because of the unparalleled nature of human intelligence is this level of sociality possible. How else to explain the success of a species that is, on the basis of physical gifts alone, ill-equipped for success in competition with chimpanzees and ill-suited to avoid becoming prey for lions? The fact that our intelligence also makes it possible to create and to maintain a rich multitude of cultures and to develop a high degree of technical expertise makes us unique, but this is not what makes us human.

Do Humans Actually Think Differently Than Primates?

The traditional idea that humans differ from all other animals in being able to use tools has fallen on hard times since it has become clear that a great many animals, not just primates, are capable of tool use. Birds poke sticks into crevices to get at insects, and some birds have even been seen to modify a stick before using it; such modification of a found object to make a tool would seem to define tool-using capacity. Among primates, tool use can be fairly sophisticated, although somewhat unpredictable. Wild chimpanzees in West Africa are known to use a pair of stones as hammer and anvil to break open nuts from the oil palm tree. This behavior has never been reported in Central or East Africa, even though the same species of monkey and the same species of oil palm are present.[2] Scientists see this as evidence of culture—a behavior pattern transmitted by social interaction—in chimpanzees. What is interesting is that some chimpanzees, even in West Africa, never use stone tools. West African chimpanzees have acquired the basic behavioral repertoire to crack nuts by age 2.5 years, and most such chimpanzees combine the behaviors to make nut cracking possible by age 3.5 years. But a small number of chimpanzees never learn how to crack nuts. This proves that nut-cracking behavior is not somehow "hardwired." Although some chimpanzees fail the intelligence test of nut cracking, most learn to use tools quickly, proving that chimpanzees both use tools and share culture.

In comparing humans to chimpanzees, researchers point to a number of similarities in the way we think.[3] Humans can (and regularly do) interpret the thoughts, wishes, and beliefs of other people in a type of projective thinking that has been called "theory of mind." Experiments show that chimpanzees can also interpret the thoughts of other chimpanzees in a "theory of mind" task, although their ability is rather poorly developed as compared to humans. In fact, chimpanzees are less adept than dogs in using the direction of eye gaze to infer the location of hidden food. Nevertheless, one chimpanzee can infer what another is thinking based solely on knowledge of what the second chimpanzee can see. In short, chimpanzees recognize the act of seeing as a proxy for the mental state of knowing, just as humans do.

Chimpanzees even seem to have a primitive sense of numeracy. Chimpanzees can perform two numerical operations in common with humans; they can understand small numbers (less than 5) in an exact way, and they can understand larger numbers in an approximate way.[4] For example, when 3 free-ranging males from one colony find a lone male from another colony, the 3 males will kill the lone chimpanzee. If a recording of a strange male chimpanzee is played in the forest, groups of fewer than 3 males will remain silent and still, whereas groups of 3 or more males will call back and move toward the speaker, preparing for an attack. Thus, chimpanzees understand that it takes 3 males to kill a lone male safely. Chimpanzees also have the capacity to understand larger numbers in an approximate way: forced to choose between a large number and a larger number of bananas, the chimpanzees will always select the largest number, even if the ratio of numbers is close. Chimpanzees can remember a number sequence up to 5 digits long, which is not all that different from humans, who can typically remember a number sequence of 5 digits as preschool children, and up to 7 digits as adults.[5] Yet teaching chimpanzees to count is a grueling process. It took 20 years and thousands of sessions to train a chimpanzee to count to 9, and each number in the series took the same amount of time to learn. In contrast, a 4-year-old child who has learned the numbers 1, 2, and 3 can usually acquire all the other numbers without further training simply by insight into the nature of numbers. Furthermore, chimpanzees are unable to represent large numbers in an exact way, even though children usually become quite adept at this.

Yet there are humans that also have a poorly developed sense of numeracy. One Amazonian tribe lacks words for numbers greater than 5, even though tribe members can perform simple mathematical operations with larger numbers.[6] The Munduruku of Brazil fail in any operation requiring exact arithmetic for numbers greater than 5 but they can still determine relative magnitude by approximation. The Munduruku have a primitive "number sense" that enables them to pick the larger of 2 groups, even if there are as many as 80 objects in a group, and they can mentally add numbers to an approximation. Similarly, the Piraha of Brazil have a "one-two-many" system of counting, but they are able to deal with numbers larger than 2.[7] In one experiment, the Piraha were allowed to inspect a pile of nuts briefly, then the nuts were placed in a can and withdrawn one at a time. Subjects were asked, after each withdrawal, if there were any nuts remaining in the can or if it was empty. Although performance was predictably worse for large numbers, the subjects were still able to perform at a rate better than chance for numbers larger than 2. These findings suggest that numerical competence is a basic skill independent of language, and that elementary calculation is possible, even without words to represent the numbers.[8] This conclusion seems to disprove what is called the strong Whorfian hypothesis—that thought is impossible without language. Although our thoughts may seem inseparable from the words we use to express them, it is apparently true that thought can occur in the absence of language. This is an important conclusion because we therefore cannot exclude a possibility that animals can think, even if they lack a language.

THE CENTRALITY OF LANGUAGE

The greatest single difference between humans and nonhumans may be in language use. But how can we define language to focus this discussion? We will assume that language is a system of communication that enables one to understand, predict, and *influence* the action of others. Inherent to this definition is a concept of theory of mind: if communication is instinctual rather than having a purpose, then it should probably not be considered a language. If communication has a purpose, this assumes an awareness of other independent actors, whose actions can potentially be influenced. Interestingly, this definition of language could include such things as painting and music, since art self-consciously manipulates emotion. Finally, for communication to serve the needs of the listener as well as the needs of the speaker, the listener must be able to understand what the speaker is "really" saying. It is not enough to understand the literal meaning of speech. The listener must also be able to understand the emotional subtext—what is in the speaker's mind—so that the listener is not too often deceived. This requires a theory of mind.

Although bees dance, birds sing, and chimpanzees grunt, these systems of communication differ fundamentally from human language. Bees dance in a way that may be entirely instinctual, and birds sing as a result of hormonal fluxes. Furthermore, these types of communication lack the richly expressive, infinitely diverse capacity for expression that even a young child has.[9] An important issue is whether human communication is strictly similar to animal communication or whether human communication is so different that it should be considered a uniquely human adaptation. Clearly, human language is quantitatively different from animal language: an average high school graduate knows 60,000 words, and this vocabulary is achieved with relatively little effort. In contrast, chimpanzees that receive intense instruction in sign language are able to achieve a vocabulary of no more than a few hundred words by adolescence. Thus, the human vocabulary is more than 100 times larger than even the most extravagant claim for vocabulary in a trained primate. Moreover, the issue of whether a chimpanzee can put together a grammatical sentence has fanned controversy. Whether human language is also qualitatively different from animal language remains unknown. Certainly, animals have some of the prerequisites of language: many primates (and even some parrots) can discriminate words of human speech; some animals have a finely tuned ability to imitate human speech (although, strangely, not primates); primates can discriminate sentences from two different languages based on rhythmic differences; and various species (again, strangely, not primates) possess the laryngeal structure necessary to finely modulate sound production.

It could be that humans have no truly novel traits that equip us for speech, although we do have an enormously expanded capacity relative to other species. Nevertheless, it has been hypothesized that humans have an innate ability to learn language. It is noteworthy that a child is exposed to only a small proportion of all possible sentences in a language, yet most children can eventually produce sentences that are grammatically correct, though they have never heard them before. This can only happen because the child has learned rules of grammar, even though mathematical analysis suggests that such rules cannot, in principle, be entirely deduced from heard speech. For this reason, Noam Chomsky has proposed that humans have a "language organ"— that language is somehow innate to the human condition:

> To come to know a human language would be an extraordinary intellectual achievement for a creature not specifically designed to accomplish this task. A normal child acquires this knowledge on relatively slight exposure and without specific training. He can then quite effortlessly make use of an intricate structure of specific rules and guiding principles to convey his thoughts and feelings to others, arousing in them novel ideas and subtle perceptions and judgments.[10]

This proposal has been elaborated in a claim that language is "hardwired," but this is somewhat hard to accept. First, it seems to invoke "intelligent design," an intellectually unappealing sham theory and, second, it ignores the fact that some apparently normal children do not learn to speak if they are not exposed to language at the appropriate age. Perhaps there exists a window of opportunity that closes as we age, so that children unexposed to language while the window is open are unable to learn language later. We note that Chomsky's "language organ" is just a metaphor for the structures within the human brain that facilitate language.[11]

Although Chomsky has argued that language "is 'beautiful' but in general unusable," we do not concur.[12] Language is a complex adaptation for communication that was probably cobbled together from other skills, but which has conferred a powerful selective advantage during human evolution.[13] A capacity to vocalize is necessary for language—since sounds can be used to convey meaning—but it is not specific to language, since sound can also be used to startle a lion about to prey upon us. A capacity to perceive fine gradations of sound is necessary for language—since perception enables us to discern the meaning of similar words and to hear words against a noisy acoustic background—but it is not specific to language, since perception is also used in learning about the world. A capacity to form concepts is necessary for language—since concepts provide the system of meaning that language expresses—but it is not specific to language, since concepts are also used in reasoning about the world.

One of the few adaptations that is necessary for speech, but that may not be useful for any other purpose, is the enlarged brainstem region that enables us to have voluntary control over breathing. Breath control is required for speech production, but it is hard to imagine why humans need to control their breathing for any purpose other than language. Similarly, the nerves and muscles that enable humans to control facial and throat movements to the degree necessary to form words precisely may be unique to speech. Nevertheless, it is possible that both breath control and fine muscle control evolved after humans were already vocalizing and merely made the process of speech more precise.

Perhaps human language is an example of emergent complexity, a property of the human brain made possible by the huge expansion of neurons that occurred in our evolutionary history. If this is true, then the differences in mode of thought between humans and other primates must arise from differences in brain structure or function. But is there any concrete evidence of such differences between humans and primates?

DOES THE STRUCTURE OF THE HUMAN BRAIN DIFFER FROM THE PRIMATE BRAIN?

Magnetic resonance imaging (MRI) has opened a door to understanding how the human brain differs from the primate brain by making it possible to image the living brain with incredible detail and precision. In the past, the human brain was studied only after death, which typically occurs in old age or after the ravages of disease or injury. Our understanding of the adolescent brain was based entirely on autopsy data, so the enterprise was open to a criticism that the study material was from an inherently biased sample, since most young people do not die. Now, MRI enables us to study a truly representative sample of living brains so that we can understand the ways in which humans differ from each other and from primates.

Recently, an MRI study was done on 44 different subjects from 11 different primate species, all healthy and alive (i.e., squirrel, capuchin, and rhesus monkeys, baboons, gibbons, orangutans, chimpanzees, bonobos, and gorillas, as well as humans).[14] This study showed that the human brain is much larger than would be expected based on the trends among primates. When brain volume is plotted as a function of body weight, the primates form a neat line. However, the human brain is at least 3 times bigger than expected. For example, orangutans are slightly larger than humans in body size, but their brains are 69% smaller than human brains. The disproportionate human brain could explain why we are so vulnerable to whiplash injury, since our large head is at the end of a slender neck that perhaps evolved for a smaller head. The overall increase in brain volume is driven, in part, by an increase in volume specifically of the forebrain, suggesting that there has been a selective pressure operating on that part of the brain that controls language and social intelligence. Furthermore, relative to other primates, the human brain is more deeply folded or gyrified. According to one explanation of the reason for cortical folding, gyrification should enhance the speed of cortical processing. Finally, there is evidence that white matter volume in particular is increased in humans, which suggests that the number of connections between neurons has increased faster than the number of neurons. Thus, the human brain may have changed in ways that would specifically foster connectivity, which might also enhance what we know as consciousness. Interestingly, there is evidence that a reduction in the size of the chewing muscles in humans may have enabled brain volume to increase substantially.[15] The gene mutation associated with muscle reduction in the jaw probably occurred about 2.4 million years ago, which is about when the human line diverged from the line that led to modern primates.

There are several types of neurons that may be unique to the brain of humans and certain higher primates known as "great apes." A recent study identified neurons called "spindle cells" in the anterior cingulate of humans and great apes; these neurons are not present in the brains of lower primates.[16] In fact, spindle cells are not observed in any other mammalian species, and it may be that these cells only arose after the humans and great apes split away from other primates about 20 million years ago. The anterior cingulate, where these neurons are found, is a unique region of the brain. It serves several functions that are properly regarded as "primitive," including control of heart rate, blood pressure, and digestive function. But the anterior cingulate does far more than that. It has been proposed that spindle cells may be able to integrate emotion with motor function to control facial expression and vocalization. In support of this idea, injury to the human anterior cingulate is associated with a form of mutism, as well as with stupor and stereotyped behavior, and such injury can cause symptoms reminiscent of schizophrenia. It may also be noteworthy that spindle cells in the anterior cingulate are vulnerable to degeneration in Alzheimer's disease.

Another unique type of neuron, called a "mirror neuron," has been found in the monkey premotor cortex.[17] Mirror neurons are activated when a monkey performs a specific action, sees someone else perform that action, or even hears sounds associated with that action. In other words, mirror neurons are activated by specific actions, whether those actions are seen, heard, or performed. Mirror neurons therefore encode not only the performance of an action but also the idea of an action; such neurons could by used to plan actions or to recognize when others are performing those actions. Interestingly, mirror neurons are located in a part of the monkey brain that is homologous to that part of the human brain involved in speech production (Broca's region). This suggests that mirror neurons could potentially encode abstract ideas and also have access to the auditory equipment necessary to describe those ideas.

There is exciting evidence that although the genes expressed in the human and the primate brain are quite similar, gene expression levels are remarkably different. Humans and chimpanzees are 98.7% identical in terms of gene sequences, but there are clear differences in the patterns of gene expression.[18] Scientists compared gene expression levels in human and chimpanzee blood, liver, and brain and found that expression differed more in the human brain than in the liver or the blood. Mathematical analysis shows that genes in the human brain are typically expressed at a much higher level than in the chimpanzee brain, whereas genes elsewhere are expressed at roughly comparable levels in chimpanzees and humans.[19] Thus, there are extensive changes in brain proteins but smaller changes in the proteins expressed in other tissues. These findings suggest that the human brain is evolving quite dramatically relative to other tissues in the human body. Moreover, there is a higher rate of protein evolution in primates than in rodents and a higher rate in humans than in primates.[20] The rapid changes in human brain structure, which are clearly seen in fossil skulls, apparently correlate with rapid changes in gene expression.

Between humans and chimpanzees, roughly 10% of genes differ in expression in at least one brain region.[21] Gene expression differs in a systematic way between individuals and between species. The genes involved in protein synthesis and turnover tend not to differ much, whereas the genes involved in neurogenesis and neurotransmitter function differ quite a bit. There is only a 3% to 5% increase in gene expression levels in the human Broca's region, despite the fact that this is the brain region involved in speech production. There is greater variability in gene expression in the human cortex than in the chimpanzee cortex, which may mean that the human cortex responds more to differences in the environment. Out of 2,014 genes that are expressed at different levels in humans and chimpanzees, a total of 1,270 genes are more highly expressed in humans, while 744 are more highly expressed in chimpanzees.

Gene expression differences between human and chimpanzee are not distributed randomly over the genome. Instead, there are "hot spots" of change, and these "hot spots" coincide with chromosomal rearrangements. When the sequence of genes on a chromosome is rearranged, a small number of genes may become duplicated so that these genes are present in two locations on the chromosome. Most increases in gene copy number are thought to arise from duplication of large segments of chromosomes, so segmental duplication has perhaps been a general trend in the evolution of the human genome. Once a gene has been duplicated and its level of expression has thereby been increased, secondary mutations can begin to change the duplicated gene without necessarily interfering with the original gene's function—since there is a redundant copy of the gene. It is postulated that much of the difference between humans and chimpanzees arose as a result of gene duplication.

Gene duplication is much more common in humans than in chimpanzees.[22] A significant fraction of the entire genome is duplicated in humans, and at least 60% of the overall difference between humans and chimpanzees is due to increased gene copy number in humans. Whether any of these gene duplications are related to speech or to the other cognitive skills that differ between humans and chimpanzees is, at present, unknown. But at least 33% of all duplications in the human genome are not reflected in the chimpanzee genome.

SOCIALITY AND INTELLIGENCE

The dominant idea in primate biology today is that competition and aggression are the keys to understanding social behavior in primates. Clearly, competition and aggression are inevitable within a group, but this focus fails to address the question of why social groups formed in the first place. If sociality only increases intragroup competition, why don't social groups simply break apart? The answer is apparently that sociality helps humans to avoid predation, and this is a more pressing evolutionary concern than is competition for limited resources. In simple terms, primates find refuge in sociality. They have not found refuge in size, the way elephants and giraffes have, and they have not found refuge in a cryptic habit, the way rats and mice have. Instead, they have found refuge in number, somewhat the way herding animals have, but with the further development of true social behavior. Primates seem to have decided that it is better to be alive and perhaps hungry than to have satisfied the hunger of a lion.

One can postulate that there must be a benefit to sociality, otherwise primates would not be as successful as they are in a competitive world. Primates band together to share food, to cooperate in infant care, to protect infants and sick members of the group, to search for food together, to coordinate the defense of range and resource, and to gain the benefit of enhanced vigilance that comes from many eyes searching for a predator.[23] Yet there has been little effort to determine how much time primates allocate to cooperation and to competition within a social context. This is a key issue because if more time is spent in competition than in cooperation, it would be hard to rationalize how animals could benefit from sociality. Recently, a review of 81 published studies was done to determine how 60 different primate species allocate individual time to affiliative and aggressive behaviors. Affiliative behavior was defined as food sharing, grooming, playing, huddling together, and acting in an alliance, whereas aggressive behavior was defined as fighting, visual or vocal threats, submissive behavior, and retreating from contact. Across 60 different primate species, about 10% of time was spent in affiliative social interactions, whereas less than 1% of time was spent in aggression. Among the great apes, 95% of social interactions were affiliative, and chimpanzees showed food-related aggression at a rate of about one incident per 143 hours of feeding. Thus, the vast majority of all social interactions were affiliative. Cooperative and peaceful interactions form a strong basis for social cohesion, helping troop members to weather the aggressive storms that must inevitably erupt.

Baboons have developed a sophisticated social intelligence that enables them to recognize and respond to the status of other members of the troop, thereby minimizing dangerous levels of aggression.[24] Typically, baboons live in large mixed-sex groups that are highly gregarious; females remain in the same natal troop for their entire life, whereas males disperse to different troops as they mature. Because of the stability of females within the troop, females establish matrilineal dominance hierarchies. This enables individual females to interact with other females of the troop in a stable way. Female-female relationships are built upon proximity, frequent grooming, and occasional supportive behavior in aggressive interactions. All female members of one matriline outrank—or are outranked by—all female members of another line. Aggressive interactions between females are usually limited only to threat vocalizations and submissive screams, where threats are virtually always directed toward subordinates and screams usually end a confrontation. However, in order for this system to work, all members of a troop must be able to identify who is dominant and who is submissive, otherwise there could be serious fights. Scientists recorded threats and screams from a baboon troop, then role-reversed vocalizations to make it seem that a subordinate female had evoked a submissive scream from a dominant female. Such altered calls were played back over loudspeakers to a free-ranging troop of baboons so that the response of individual baboons could be observed. The hypothesis was that quarrels between families would be much more disruptive than quarrels within families. In playback experiments, baboons responded more strongly to role reversals between families than to role reversals within families, showing that females were able to recognize family affiliations at the same time that they were responding to dominance hierarchies. Thus, baboons have a level of social sophistication and civility that is very nearly comparable to what is seen in Congress. Social interactions at this level of sophistication would probably favor the evolution of language, since language can be used to minimize social friction and modulate aggression.

OF WHAT VALUE IS SOCIALITY?

There is recent evidence that female baboons who are socially adept have a better chance of successfully raising infants.[25] Over the course of 16 years of observation of baboons in the Amboseli Basin at the foot of Mount Kilimanjaro, 108 adult female baboons were observed for a total of 633 female-years. This mountain of data was analyzed to determine if infant survival was affected by the sociality of the mother. Infant survival for each mother was measured as the proportion of her infants that survived to 1 year of age. Maternal sociality was measured as a composite of the time spent near other adults and the time spent grooming or being groomed. Mothers who were socially well integrated had a significantly greater likelihood of successfully raising infants—independent of the mother's dominance rank, group membership, or even the availability of food. Socially integrated females perhaps benefit because other females provide a more benign environment for raising infants, or perhaps because males can be induced to shield the mother and her infant from harassment, predation, or infanticide. This evidence that social connection enhances survival is very exciting, since it bears directly on the evolution of social behavior. If sociality is heritable, as surely it must be, the fact that highly social females have more offspring will mean that social beings should increase in the population over time.

The clearest evidence of primate social behavior is food sharing, which is shown by virtually no other animals. For food sharing to become established, two conditions must be met: individuals must continually interact with each other over a long period of time and individuals must be able to recognize one another and remember past patterns of interaction. This type of behavior can be studied in humans in an experimental situation by using functional magnetic resonance imaging (fMRI) to examine healthy subjects playing a gambling game.[26] The game, known as the iterated prisoner's dilemma game, requires two players to cheat or collaborate while gambling together, with the long-term pattern of interaction determining the payout for each player. Since the game is repeated over time, this makes it possible for a player who cheats to be "punished" by the other player. For players who collaborate well over time—thereby showing reciprocal altruism—the payout is somewhat less than maximal, but much more predictable. This game models food sharing, in that reciprocal altruism is rewarded, although at a lower rate than is cheating. For example, if a chimpanzee finds a new food source, he can share that food and have less to eat himself, assuming that his troop mate will share food in the future, or he can monopolize the food source, knowing that this makes food sharing by his troop mate less likely in the future.

The prisoner's dilemma game is regarded as an experimental model of reciprocal altruism.[27] In the fMRI version of the game, a subject in the MRI scanner plays a game with a subject outside the scanner, so that patterns of brain activation can be studied over time. This allows scientists to identify those parts of the brain that are concerned with cooperation in a social context. Since subjects play for money, there is some impetus to cheat. But the way that the payout is calculated also provides an impetus to collaborate. After each round, the winnings of each player are revealed, and the players play another round, to a minimum of at least 20 rounds. As players are gambling, images are acquired so as to determine which brain regions are involved in collaboration and which are used in cheating. This study found that the brain regions involved in collaboration are part of the natural "reward system" of the brain—the same brain regions that are subverted in addiction. Activation of this neural system may reward altruistic behavior and act as a powerful motivator of sociality. In other words, the intrinsic rewards of social interaction could help a person to resist those extrinsic rewards that could be obtained by cheating. If the brain's own reward system is sufficiently stimulated by social altruism, this would motivate a person to resist the temptation to accept but not reciprocate favors. That this system is involved in drug addiction is a fascinating connection that has not yet been fully explored.

THE SOCIAL INTELLIGENCE HYPOTHESIS

Sociality is why we have survived as a species. Truth be told, we are inadequate competitors away from the group. Humans without help are unable to kill all but the smallest or sickest prey and are subject to predation by all but the weakest predator. Given that the human lineage arose in Africa, home to some of the most savage animals on Earth, we could only have survived with help from other members of our species. The strongest evidence for this is that primates share a particularly intense form of social life that is, as far as we know, unlike that in any other animal. The answer to the question, "Why are we so social?" must be, "So that we can survive."

What unique capacities are required for the degree of sociality seen in primates and humans? We postulate that intelligence is either a direct outcome of natural selection for sociality or a preadaptation that made it possible for sociality to evolve. In any case, the level of sociality seen in primates and humans simply would not be possible without a relatively high degree of intelligence. It requires intelligence to recognize one another and to remember past patterns of interaction so that reciprocal altruism does not become exploitative. Intelligence is an adaptation that enables sociality, and highly developed sociality requires highly developed intelligence.

We have hypothesized that sociality is the main reason why the primate brain has evolved so differently than every other brain and why the human brain has become, in the end, so different from the primate brain. The most obvious human trait that facilitates sociality is language. The more sophisticated the social arrangement, the more critical it becomes that language be used to work out the details of social reciprocity. Language is clearly important for the transmission of information, but it also helps build social connections. If social connections increase the survival of baboon infants, it seems likely that sociality would also increase the survival of human infants.

But language is more than communication. It is also self-expression, and self-expression is responsible for much of what makes humans so very distinctive. Self-expression includes writing, painting, designing, building, making music, dancing, acting, and forming governments and societies. If sociality led to language and language to self-expression, it may be legitimate to think that sociality indirectly favors self-expression, perhaps as a way to make the self unique and, therefore, more worthy of support by the group. This can only be possible if there is a sense of self, a sense of uniqueness, and a consciousness of separateness, all of which are crucial to the theory of mind.

It is noteworthy that mental illness is often a breakdown of the social brain. Many of the key features of schizophrenia are expressed in language or are a direct result of some disturbance of language ability.[28] Patients with schizophrenia often find it difficult to understand speech; their speech may be very disorganized, they may have a reduced accuracy of comprehension, and they perform poorly compared to healthy subjects in most word tasks. Few schizophrenic patients are able to use language in a sophisticated way, and metaphor, irony, humor, sarcasm, and the emotional content of speech often don't register. Similarly, clinical depression usually involves a withdrawal from social contact and an experience that social contact is less rewarding and pleasurable. It may be that mental illness is a disorder of exactly those abilities that make us most human.

Sociality may be associated with the evolution of a broad range of human traits, perhaps even moral cognition—the type of thinking that helps us to recognize right from wrong. Research suggests that there is a neural basis for moral judgment.[29] Moral cognition arises from the integration of social knowledge with motivational and emotional states to achieve a sense of right and wrong. But morality is also the product of evolutionary pressures that have shaped social cognition and made reciprocal altruism a goal of social interaction. It is significant that moral development can be arrested by early damage to the prefrontal cortex. Patients with such damage can exhibit deficits in expression of embarrassment, pride, or regret. Several fMRI experiments show that specific brain regions are activated by visual images of a moral violation (e.g., a physical assault). One region that is commonly stimulated during such experiments is the limbic system—the brain structures concerned with reward and emotional response. Damage to these brain regions could account for psychopathic behavior, since such damage would likely reduce the emotional response to social transgression. It is interesting to note that many psychopaths have an intact theory of mind, since they can be quite devious, and deviousness requires clear insight into the motivations of others.

The social intelligence hypothesis holds that sociality is the main reason why the human brain has evolved so differently from every other brain. Yet one of the most obvious differences between humans and other primates is in the level of creativity and self-expression available to the human brain. Is creativity related to social behavior? One could postulate that creativity has evolved as a way for males to compete with one another for access to females. One has only to watch the way that teenage girls respond to the lead singer of a rock band to realize that such creativity, as trivial as it is, does have selective value. Whether the level of creativity shown by someone like Galileo or Einstein or Socrates would also lead to greater reproductive access is doubtful. But strong sexual selection for low-level creativity would be expected to produce rare individuals with high levels of creativity by chance alone.

THE EMERGENCE OF CREATIVITY

Creativity is perhaps the clearest example of emergent complexity—the idea that properties of the intact brain cannot be predicted, even from an intimate knowledge of the parts of the brain. Emergent complexity is essentially the idea that the system has strong synergy, that the whole is far more than the sum of its parts. No matter how deep our understanding of the brain at the level of the neuron, it may not be possible to describe how a collection of neurons can make art or build a civilization. Perhaps emergent complexity is inherent to every highly complex system, even if the system runs by rigid rules. Or perhaps creativity is simply a happy accident—a fortuitous combination of imagination, intuition, and a rigorous intellect—the product of uninhibited free association sorted and winnowed through stringent validity testing.

It is fascinating that creativity can occasionally emerge full-blown after brain injury.[30] Tommy McHugh was a former builder and heroin addict who had spent time in jail for violent offenses and who had no interest in art whatsoever. Then he suffered what was later diagnosed as a cerebral hemorrhage. The burst blood vessel in his brain was surgically repaired, but McHugh maintains that the stroke altered his personality and his life; he has called it the best thing that ever happened to him. Now, he is compulsively creative—painting, sculpting, and writing poetry in every available moment. He often draws faces within faces, which he describes as portraying the parallel lines of thought that run through his mind continually. Interestingly, McHugh has some cognitive impairment that accompanies his creativity; though he tests within the normal range of IQ, he has trouble when he switches trains of thought. When asked to list pieces of furniture, he talks about chairs and tables and so on, as can anyone else. However, if he is then asked to list animals, he tends to continue to talk about furniture. There is a perseveration that also reveals itself when McHugh talks, as sentences tumble out relentlessly until he is interrupted. Unfortunately, McHugh cannot be examined by MRI, as the surgical clip used to stop the bleeding might put him at grave risk in the strong magnetic field of the MRI.

However, there is speculation that McHugh has suffered damage to the left side of his frontal lobe. Typically, areas on the right side of the brain are thought to be used in artistic expression—skills such as visual processing, facial recognition, and spatial awareness—while areas on the left side of the brain moderate and inhibit the output of the right brain. If the inhibitory side of the brain is damaged by stroke, this could release the creative side of the brain from inhibition. The creativity that was always present would be freed from self-censorship, freed to explore. Such a fortunate outcome is highly unusual for a stroke patient. Yet McHugh's case suggests a vivid lesson for all of us: creativity is a matter of finding a balance between the vitality of mania and the vapidity of depression, between stimulation and inhibition, between imagination and self-censorship.

CHAPTER 17

TOWARD A THEORY OF EMERGENT COMPLEXITY

Knowledge of neuroscience is accruing so rapidly that it would be easy for the nonscientist—the citizen who pays the bills—to think that scientists know it all, or at least that we will know it all very soon. Yet it should be clear by now that our knowledge is fragmentary to the point of being idiosyncratic; we have learned in intimate detail some of the arcana of brain structure and function, but we have left unanswered some of the most important and most basic questions in the neurosciences.

As we stumble toward answers to the many challenging questions that relate to the brain, it would be well to remember some general trends in biology and in the neurosciences over the last several decades:

1. **Good research is hard to do.** As a scientist, you feel your way along in near-total darkness, knowing that your understanding is primitive, your intellect is limited, your research tools are flawed, and your grant funding is tenuous. It is far too easy to overinterpret your data or misread it entirely, and you can't check your answers in the back of the book because the book isn't written yet. There is a truism that if you know what you're doing, it isn't science.

Gone are the days when someone can have a good idea or a clear insight and publish it in the prestigious journal *Science* without spending a great deal of time and effort and money to gather new data. Good research will require collaboration across disciplines, since many of the most fascinating problems clearly do not fall under the aegis of any one discipline (see fig. 15). It is getting harder and harder to find funding for research, and the funding rate in all fields of science is in free fall. Easy research targets were never common and may now have nearly vanished. Were it not for the enormous number of scientists supported by vast sums of money from private enterprise, research progress would have slowed long ago, since the

Levels of Investigation

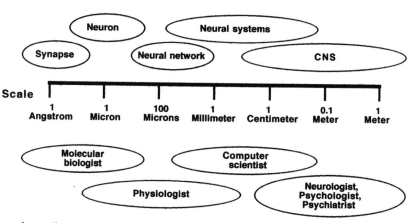

Figure 15. Levels of investigation in the nervous system. The central scale bar is in metric units of length. The upper part of the figure shows structures that typically have the dimensions shown in the scale bar, while the lower part of the figure shows the type of scientist that would focus at a particular scale. It is clear that there is a broad range of scales at which any particular structure can be investigated and a broad overlap in the type of scientist that may work at any given scale.

government has simply not allocated enough money for research. But private funding, whether from a pharmaceutical company or a private donor, can come with "strings" or a specific goal attached; consequently, private funding is not as likely to lead to true novelty as is government funding.

2. **The remaining research targets are complex, subtle, unusual, or all three.** So much knowledge has already been gained at the centers of established fields, where so many good scientists are working still, that progress is more likely to occur at the margins of fields. For example, it seems unlikely that anyone will make substantial new progress in studying a simple subject like how an action potential moves along a squid axon; this subject has been thoroughly explored. But studying how action potentials permute in an intact sense organ or in an animal brain is complex and very challenging, in part because granting agencies are reluctant to spend money on speculative or very difficult projects. Similarly, it is doubtful that scientists will make substantial progress in Alzheimer's disease by studying patients who become ill in their ninth decade of life; at that age, Alzheimer's is so common that many people show signs of it. Instead, progress in Alzheimer's is more likely to come by studying the rare subset of people who get the disease in their fifth decade of life.

3. **Brute force will no longer work.** Modern scientific methods generate too much data to analyze by brute force alone. Microarray analysis of gene expression in the brain can simultaneously measure levels of expression of 30,000 genes. A magnetic resonance imaging (MRI) study of the brain generates images with 50,000 pieces of data for every imaging sequence applied to the brain. Without computers to assist in data analysis, and without computer scientists to help design strategies for analysis, this would be an overwhelming amount of

information. The future of the neurosciences is thus inextricably linked to the computer sciences.

4. **Probabilistic methods of analysis will be required.** As we have seen, chance and random variation seem to play an alarmingly large role in brain function. If this is true, then the workings of the brain can be understood only by using computer simulation or modeling approaches. This will require scientific interactions among scientists with an increasingly diverse set of backgrounds.

How Can We Study the Brain?

The traditional approach to studying the brain is by atomistic experiments: a functioning brain is deconstructed to its subordinate parts and the parts are studied in isolation. This has led to massive efforts in single cell recording and in evaluating neural function in brain slices. No doubt, there are still useful experiments to be done using these simplistic approaches, but it is also time to start thinking about how the parts fit back together to make a functioning whole.

In short, it may be time to spend more effort studying the brain (and the mind) by holistic experiments. Such experiments could use medical imaging methods, in combination with other measures of brain function, to test how the intact brain perceives or responds to stimuli. For example, functional magnetic resonance imaging (fMRI) has enjoyed an explosive growth over the last few years. It combines the functional information that can be generated by a method such as positron emission tomography (PET) with the anatomical information that can be generated by traditional MRI studies (table 7). Currently, the drawback of fMRI is that it is too insensitive to use for studies of receptor density or drug binding in the brain. But PET and fMRI methods can be combined to gain the sensitivity of PET while retaining the structural detail of MRI and the temporal resolution of

Table 7. Characteristics of various brain imaging methods. MRA is magnetic resonance angiography, which is used to visualize blood vessels in the brain.

	Radioactive methods		Nonradioactive methods		
	PET	SPECT	MRI	fMRI	MRA
General uses	General neuronal activity	General neuronal activity	Brain anatomy	Local neuronal activity	Vascular anatomy
Typical use	Medical diagnosis	Research	Medical diagnosis	Research	Medical diagnosis
Sensitivity (detectable number of atoms)	Extremely high $< 10^{11}$ atoms	Very high $< 10^{12}$ atoms	Low $> 10^{17}$ atoms	Low $> 10^{17}$ atoms	Low $> 10^{17}$ atoms
Spatial resolution	4–6 mm	7–8 mm	< 1 mm	1–2 mm	< 1 mm
Isotopes studied	Oxygen, carbon, fluorine	Xenon, iodine, technetium	Virtually only proton	Proton	Proton
Cost per study	< $ 2,000	< $ 1,000	< $ 1,500	< $ 1,500	No added cost to MRI
Specific research uses	Cerebral blood flow Brain metabolic rate Brain receptor studies Drug binding Monitoring drug levels	Cerebral blood flow Brain metabolic rate Receptor studies possible	Causes of brain illness Correlates of brain illness	Loci of cognition	Causes of brain illness Correlates of brain illness

fMRI. If these methods were combined with sensitive behavioral measurements, especially in powerful MRI machines that can image small animals such as rats, it could be a new window into the mind.

It has always been possible to study brain function by altering it in an experimental manner. For example, both LSD and PCP have been used as models of psychosis in humans, and there has been some effort

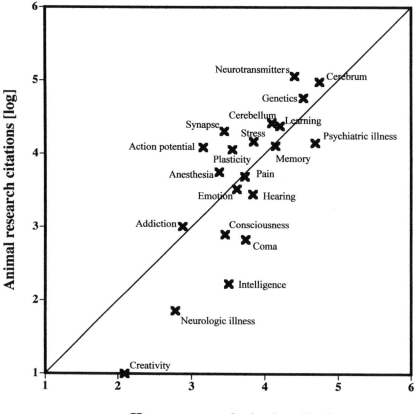

Figure 16. Plot of the number of published studies in animals and in humans. The logarithm of the number of PubMed citations that relate to human "neurologic illness" is plotted as a function of the number of citations that relate to animal "neurologic illness." A line of identity is drawn on the diagonal; topics that appear to the right of the diagonal, such as "neurologic illness," have been more thoroughly investigated in humans than in animals.

to gain insight into the abnormal mind by studying the impact of these drugs on the normal mind. Because of the emergence of both new drugs and new animal models of human disease, pharmacological approaches to the study of the brain will likely continue to be productive.

Do complex brains operate on the same principles as simple brains or is there some inherent difference that is a correlate of complexity? It is worth noting that complex brains have evolved repeatedly, so that many other vertebrates and even certain invertebrates have brains that could be thought of as complex.[1] Thus, it may be possible to compare these different types of brains to determine if there are any design principles common to all (or most) complex brains. This type of study has to ask the basic question, has evolution produced new principles that set complex brains apart from simple brains, or is complexity simply a matter of size and redundancy of function?

It is fair to ask what animal experimentation can tell us about the human mind. The answer to this question is not clear, since a great deal of research has been done on animals (see fig. 16), but sometimes this research provides relatively little insight into the human condition. For example, there may never be a legitimate animal model of schizophrenia, since schizophrenia is the most human of diseases. Schizophrenia impairs functions that most animal brains appear not to have, so it may be impossible to study schizophrenia in an animal. Although complex subjects often can be simplified for study in an animal model, it may be impossible to study psychotic behavior experimentally in a nonverbal animal.

Computational neurobiology—the approach of studying the brain and the mind by computer modeling—seems likely to have a major impact in the future. This field is still in its infancy, but there have already been intriguing results, as we shall see. There is reason to hope that modeling the function of neurons may point the way for future animal experiments.

What Is Complex Behavior?

Complex behavior is defined by its lack of predictability. If an organism—be it human or animal—responds to its environment in a way that defies easy prediction, this is complex behavior. If an identical input can lead to differing outputs, this is inherently unpredictable and complex; the more possible outcomes available, the more complex the behavior. But complex behavior also includes the situation in which an identical input produces an identical output, though by a different mechanism. For example, if an animal is frightened by a stimulus, the animal could "freeze" in place because it has chosen to freeze as part of an adaptive strategy or because it is simply overwhelmed and doesn't know what else to do.

The idea that there may be multiple pathways to the same endpoint can only be addressed if an experimental subject can describe the decision process so that the experimentalist will know the factors that went into a decision. This may mean that complex behaviors can be studied only in humans. But even then, having a verbal report by an experimental subject may not help, since people may be unable to say precisely why they did something. Alternatively, complex behaviors could be modeled on a computer, since the ability to report the decision process could be built into the program. For research purposes, minimally complex behavior should probably have at least three possible outcomes, since failure to respond at all must be considered an outcome.

The essence of complex behavior is nonpredictability. An intuitive example might be the decision of whether to loan someone your car. If a complete stranger asks, the answer will almost certainly be no. If a friend asks, the answer may still be no, but the decision process is more complex, since you will likely weigh your friend's driving record, the length of time that you would be without the car, the likelihood of a mishap on the part of your friend, and your own immediate and future needs for the car. For a distant friend, the answer may still predictably be no, and for your spouse, the answer will almost certainly be yes. But for fairly close friends, the answer may become an example of very complex behavior.

Complexity emerges from a brain that has grown sophisticated enough to support a mind. We have proposed that emergent complexity is a property not of ants but of the colony, and not of neurons but of the neuronal collective. The brain is a collection of neurons, each behaving in a specific way, but mind is a property of brain, which arises from the specific function of neurons. In short, *mind is what the brain does.* Mind is what interposes between brain and behavior; mind is the evidence of brain, but brain is the foundation of mind.

COMPLEX BEHAVIOR AND COMPUTATIONAL NEUROBIOLOGY

Complex behavior is the product of a brain that has grown sufficiently intricate that the output cannot be predicted from a thorough knowledge of the input. This might best be studied using the approach of a "probability engine," whereby probabilistic models of brain function would be used to model the mind on a computer.

In an astonishingly prescient paper about neural networks written in 1982, the computational neurobiologist J. J. Hopfield was perhaps the first person to describe emergent complexity. Hopfield proposed that

> computational properties of use to biological organisms or to the construction of computers can emerge as collective properties of systems having a large number of simple equivalent components (or neurons).[2]

This is a model of the mind's emergent complexity that can be tested by computer modeling. Both his original, simple model of neural function and some later models derived from it were criticized because they was unrealistic in some respects.[3] A neural network requires a "teacher," yet it is not apparent whether another part of the brain could adequately function as a teacher. The neural network functions best with a specific number of hidden neuronal units; too many hidden neurons and the network cannot generalize to new data, too few and the network simply cannot function. Yet it was never clear how the number of neurons devoted to a task can be adjusted, particularly in the absence of an expert "teacher." Neural networks seem to require the rapid transmission of information in both directions along the neuron, yet such bidirectional transmission likely cannot happen in the brain. A final problem is that a neural network quickly becomes cumbersome if several hidden layers of neurons are needed. It has not been established how many hidden layers are needed to replicate brain function, but the number must be large.

An interesting approach to modeling brain function has been to focus on the function of individual neurons and to use computer-based "neurons" to build networks.[4] Such work shows that cellular morphology is a critical feature that determines neuronal function: each neuron has a characteristic form, and this form has an influence on the way an action potential propagates down the neuron. If neurons are modeled in 3-dimensional space—with biologically plausible rules governing how an action potential moves down the neuron, how transmission across the synapse to the next neuron occurs, and how individual neurons are "grown" into an anatomically realistic network—such models could begin to reflect emergent properties of the brain. The morphology of individual neurons can be described with perhaps as few as 7 different parameters, each of which can be measured and incorporated into the model. This approach has been called computational neuroanatomy. Yet it should be noted that there are still very few experimental studies that link neuronal morphology to single cell function, so we cannot yet assess whether these models are accurate. Furthermore, the state of knowledge about human neuroanatomy is generally far worse than might be expected.[5] Because one cubic millimeter of the human visual cortex contains at least 40,000 neurons, our lack of detailed knowledge means that we have little hope of understanding how our brains work, except in the crudest possible way. Nevertheless, it may become possible to build a virtual rat brain within the next decade or so.

Computer models of individual neurons have begun to demonstrate properties characteristic of neurons in the brain.[6] Models now reveal how a neuron can sustain activity despite the presence of noise and distractors, how different types of membrane channels contribute to neural activity, and how the robustness of individual neurons factors into the robustness of a system. One model has predicted that as synapses between neurons are established and facilitated, so that a group of neurons gradually build a "cell assembly," there will be an abrupt change when the cell assembly is established, rather than a gradual emergence of the cell assembly. In a sense, this would be like an "Aha!" moment for a neuron. Another model has predicted that certain drugs may be able to stabilize a cell assembly too much, so that a subject might begin to show perseverative behavior—a slavish persistence in performing an action that does not yield the anticipated reward. Such behavior can be a symptom of damage to the frontal lobes of the brain, which may tell us something about the nature of frontal lobe injury in some patients.

Computer models have also been developed that focus on the cyclic or oscillatory behavior of large populations of neurons.[7] Studies suggest that correlated changes in neuronal activity in different parts of the brain may be important for governing attention and the flow of information. One computer model predicts that the output of a neuron is determined not only by the firing rate of input neurons but also by the degree of synchrony among input neurons. In other words, a neuron can be sensitive to temporal patterns among the neurons that activate it; synchronized inputs may be more effective at generating an action potential than nonsynchronized inputs. Oscillations arise naturally from intrinsic properties of neurons, but they may also be evoked by the recurrent or "closed-loop" nature of certain cortical circuits. A change in correlation between neurons can thus reveal changes in functional connectivity, and may facilitate the formation of new cell assemblies. Apparently, network oscillations can bias attention, link neurons into new assemblies, and facilitate synaptic plasticity. Subtle interactions between competing coalitions of neurons may be involved in most forms of learning. The importance of neuronal oscillations to learning, to perception, and even to consciousness has perhaps been underestimated.[8]

Recent efforts have attempted to identify the neural correlates of consciousness.[9] It is becoming clear that the brain is sensitive not only to simple correlations in the sensory input but also to correlations among correlations, making us exceedingly sensitive to change. Yet there is reason to suspect that "the best is the enemy of the good." It may be better to achieve a rapid but occasionally fallible response to the environment rather than to produce a slow but perfectly reliable response. In order to minimize response time, the brain has evolved to make use of imperfect or faulty information, to "jump to conclusions" in the interest of safety.

There is a key difference between computers and brains that is abundantly clear; computers are digital and they "think" using only zeros or ones, whereas brains are analog and they think using perhaps an infinite number of electrical states between zero and one. Because brains rely on continuously varying electrical states rather than on the clearly defined "on" or "off" of a computer, tiny variations in the electrical state of a neuron may enable that neuron to respond differently to identical inputs.[10] Thus, the analog nature of the brain may be a key that enables complexity to emerge despite the relatively inflexible wiring of neurons in the brain. A computer model that fails to incorporate analog features into the function of brain circuits is likely to be seriously in error. The analog nature of the brain may even explain the genesis of creativity, if creativity is defined as the ability to form new connections or to juxtapose novel ideas.

There is a basic principle that informs computer models of the brain.[11] Any model should start with a conceptually simple view of the system, adding details only when absolutely necessary. As a model is built and tested, it should become apparent where that model generates a poor description of reality, what testable predictions arise from the model, when more data are needed from experiments to constrain or specify the model, and when the model finally begins to generate a picture of reality that has detail enough to be satisfying. One clue to when a model is successful may be when emergent properties first become apparent; such properties could include the integration of signals across different time scales, the generation of distinct outputs that depend in sensitive ways upon the input, and the presence of feedback loops that behave in novel ways. Because the brain is not a machine that adheres to a well-defined design but, rather, is a machine that constantly rebuilds itself within a range of parameters, it will be an exceedingly difficult challenge to build an accurate computer model of the brain.

EMERGENT COMPLEXITY:
HOW A BRAIN DEVELOPS INTO A MIND

Perhaps the most important lesson from computational neurobiology is that it is probably the enormous number of similar neurons that allows properties of the mind to emerge from the brain. Even a bee's brain has roughly a million neurons, with stupendously intricate microcircuits that control extremely complex behaviors.[12] The human brain has at least 100 billion (10^{11}) neurons, so it is 100,000 times more complex than a bee brain. Each human neuron makes about 10,000 (10^4) synapses, meaning that there are perhaps as many as 10^{15} synapses or 1,000 trillion (1 quadrillion) interconnections between neurons in the human brain. Simple units performing simple tasks, bound together into a whole—this is the essence of a brain and the foundation of a mind. Despite the fact that each neuron performs rather simple tasks, the enormous number of similar units means that there can be specialization and a division of labor among neurons. Such a stable division of labor requires two properties of a neuron.

First, for a successful division of labor, neurons must be relatively robust and long-lived. If human neurons did not survive for roughly a human lifetime, then their synapses would be lost. If the information-rich interconnections between neurons were to be lost, then the information coded in those connections would be lost as well. For learning to last, for memory to endure, neurons must be able to survive (and to learn) for a relatively large fraction of a person's life.

Second, for a successful division of labor, neurons must be able to precisely modulate their interaction with other neurons. Were this not possible, complexity theory suggests that the aggregate of neurons would dissolve into an amalgam of individual cells, each competing with others for limited nutrients. Yet merely having a large number of neurons that successfully apportion labor among themselves is not a sufficient impetus for a mind to develop.

The evolution of the mind requires that there must be some sort of selective pressure favoring the elaboration of behavior. Were this not so, if there was no advantage to having a sophisticated brain, then there would be no reason for us to have a brain any more elaborate than that of a cockroach. The cockroach is one of the most successful organisms on Earth, and it has been recognizable as a cockroach for hundreds of millions of years. It could be that our position as top predator in the world requires that we be more clever than our prey; it could be that humans have warred against each other for so long that the ability to negotiate a peace has been selected for; it seems likely that social alliances are required to compensate for our fundamental ineptitude as a predator, our inability to exist in solitude. Whatever the nature of the selective pressure that has operated on the human brain, it is clear that without such pressure there would be no mind. Yet, even given a large number of similar neuronal units able to live and relate to each other for a long time, even given a strong selective pressure favoring the evolution of mind, this would not be enough to produce a mind.

For the mind to emerge from the brain, there must have been sufficient time for evolutionary experimentation. Time is the linchpin of evolution; without enormous periods of time for selection to act, no serious scientist could hold to the ideas of evolution. Yet, given sufficient time in combination with the other conditions that we have described, we believe that the emergence of the mind was inevitable.

In the evolution of the mind, complexity likely begat complexity. All other things being equal, an organism with a complex genome is more likely to develop a complex brain. This is a function of the way that evolution is thought to work in higher organisms. The entire genome of many different animal species has now been analyzed in detail, and one of the key findings to emerge from this work is that very different organisms share genes that are very similar in structure. Between 30% and 60% of genes may be shared between widely divergent organisms, which is a very powerful argument in favor of evolution.[13] It has been hypothesized that gene duplication is a major engine of evolution; this is the process whereby an existing gene is copied and co-opted for other purposes. If a newly copied gene gives rise to a functionally redundant protein, that protein is freed from whatever selective constraints were acting on it before, since the original protein is still there and still functional. Thus, the new protein is free to change, the duplicated gene is free to accumulate mutations and evolve, and the new protein is free to assume other cellular duties, all without harming the host.

This mechanism of gene duplication has been invoked to explain the gradual evolution of the immune system in vertebrates.[14] But it is possible that a similar mechanism of neuronal "replication" and co-option for other purposes could have given rise to the human brain as well. With neuronal duplication, duplicated neurons could serve a new function without shortchanging the original function, since the original neuron would remain to serve the original function. With neurons becoming more numerous, and with some of them freed from their original duties and co-opted for new duties, complexity would accrue. Just as proteins may acquire new functions to build an immune system, neurons may acquire novel functions to make an increasingly sophisticated mind.

WHAT PREDICTIONS ABOUT THE MIND ARISE FROM THE ANALOGY OF THE ANTHILL?

In the first chapter of this book, we postulated that the anthill is a viable analogy for the brain. Tiny neurons are like ants, performing circumscribed tasks in a predictable fashion, in a manner determined

by the role they were born to serve. Both neurons and ants look no farther than their nearest neighbor; they understand nothing of the larger
structure that results from their persistent action. Although there are
an enormous variety of tasks to perform, each neuron, like each ant,
does a limited number of tasks in a limited way. And yet complexity
emerges in an anthill as in a brain in ways that could not have been
predicted. Emergent properties are a result of subtle interaction, precise communication, and endless refinement, making the human mind
into what is arguably the most complex object in the universe. If this
view of the brain is correct, then we can expect that testable predictions about the brain should emerge from a careful consideration of the
analogy of the anthill:

1. **The process of self-organization can provide insights into
   complexity.** During the growth of an anthill or a brain, large
   numbers of units aggregate and self-organize to generate patterns and to store information.[15] Studying this stepwise process
   of self-organization is crucial. The gradual buildup from simple
   to complex could illuminate the overwhelming complexity of
   the finished product. It may even be true that the process of
   development will, to a limited extent, recapitulate and illuminate the process of evolution.

2. **An enormous number of neurons is required for functional specialization.** In social insects, large colony size is necessary for behavioral specialization and large colony size seems
   to be strongly associated with complex colony-level performance. This idea can be tested in ants as well as in brains by
   comparing across animal species. At issue is whether there is a
   relationship between the number of neurons and the complexity of neuronal function.

3. **Functional specialization is the key to emergent complexity.** Emergent complexity and task specialization go hand

in hand in social insects. This hypothesis cannot be tested in the human brain, where we already know that specialization and complexity coexist. Yet we can learn whether specialization is a prerequisite for complexity by studying small anthills and simple brains. This is the domain of the field biologist and the comparative physiologist, and it will be many years before these scientists become any less relevant than they have been through the history of the life sciences.

4. **In complex brains, neurons are smaller and more tightly packed.** In social insects, large colony size is associated with smaller insect body size. It would be interesting to compare across animal species to see whether there is an inverse trend between the number of neurons per brain and the physical size of neurons.

5. **Densely packed neurons have a lower per capita energy requirement.** Large colony size is associated with lower per capita productivity in social insects.[16] It is not yet clear whether energy constrains the function of individual neurons, and how energetic constraints might change as a function of brain size or neuronal packing density.

6. **Neuronal redundancy increases overall reliability.** In insect societies, an advantage of the colonial habit is that if one individual fails to complete a task, another may succeed; in a nonsocial insect, if a lone individual fails to perform a task, all of its effort will be lost. This may mean that neuronal redundancy, which guarantees that some neurons are available to perform a task if other neurons fail, is an important selective advantage of the complex brain.

7. **Neurons operate with probabilistic rules of thumb, despite a paucity of information.** We know that ants have an

imperfect knowledge of the anthill, but that each ant responds in a way that can be predicted from a clear understanding of certain probabilistic rules governing ant behavior.[17] But we do not yet know the extent to which single neurons in a network respond in a probabilistic sense. It is possible that all synaptic signals generate a probabilistic response, and that action potentials are less deterministic than our present understanding would suggest. This could be a fruitful subject for experimental and computational neuroscientists for years to come.

8. **The function of the neuron is not as simple as it would appear to be.** Communication between ants is an incredibly complex process. There are at least 10 different organs implicated in the secretion of trail-marking substances, and ants use many other forms of communication.[18] The apparent functional simplicity of neurons may be an illusion, the result of simplistic experimental systems and a rudimentary understanding of neuronal function. Is communication between neurons augmented by processes that we do not yet fully understand? Perhaps the cable properties of neurons affect the generation and conduction of an action potential, but this would never be recognized if one were to study a single synapse. It may be crucial for experimental neurobiologists to integrate across nearly as many synapses as the neuron does in order to understand neuronal function.

9. **Neuronal flexibility fosters stability but complicates an understanding of the brain.** If a neuron operates by probabilistic rule of thumb and has as much functional flexibility as we anticipate, then there can be many ways to accomplish the same task. Merely producing the same output in an experiment does not prove that the neuronal "decision process" was the same.

10. **Neuronal signals can potentially affect the entire brain.** Some signals in an anthill do not convey information any more detailed than off or on. For example, alarm pheromones are released by an ant as soon as an intruder is detected in the nest, and these pheromones can elicit an alarm response at a great distance from the site of release. Research has shown that complex social behaviors in an ant colony can be stimulated by the release of synthetic hormones into the air flowing through a colony.[19] The existence of neural signals that act at a distance to affect the entire brain is thus predicted. This may mean that the ability of scientists to control the social behavior of ants could have meaning for physicians trying to modulate how neurons interact in a complex brain illness such as schizophrenia.

11. **The chemical identity of the signal molecule provides behavioral specificity.** The complex social behavior of ants is mediated by chemoreceptors, which respond to chemicals released by other ants. The specific identity of the released chemical evokes a specific behavior, so that ant behavior can be precisely modulated. The behavior of ants is controlled by perhaps as few as 20 different chemicals.[20] In the same sense, we predict that there are specific behaviors associated with abundance or paucity of specific neurotransmitters in the human brain. Given that there may be 50 different neurotransmitters, there will be very complex interactions possible between and among neurochemicals in the brain.

12. **There is an established syntax of signal presentation.** Ants use a small number of different chemical signals to regulate behavior, but there is a rigid grammar as to the way such signals are used.[21] If there is any disruption of the rules that govern signal presentation, this can disturb the communication process. In the same way, it may be impossible for certain

neurons to respond to certain signals if signals are presented in a way that violates the "grammar" of the brain.

13. **Very small molecules may be more important in signaling than is presently known.** Fire ants use an exceptionally simple chemical to induce other fire ants to aggregate; carbon dioxide (CO_2) is released and other ants actually walk up the gas gradient to form groups. This is the simplest chemical signal known to be effective in any animal communication system.[22] It is already known that carbon monoxide (CO) is used as a signal molecule in the human brain, but scientists agree that CO is a very short-lived molecule, since it is so readily broken down. But perhaps short-lived or short-range signal molecules are more important than is presently known.

14. **Natural oscillations in neuronal activity may serve an organizing purpose.** Individual ants are quiescent most of the time, but when they do become active, it is usually in bursts of activity that occur 2 to 4 times per hour. These pulses, which tend to be coupled among nestmates, impose an order that may facilitate activity.[23] In the same sense, computer modeling suggests that neuronal oscillations can increase awareness and may be necessary for learning, but there is currently little experimental evidence in support of this idea.

15. **It may eventually become possible to predict the behavior of large coalitions of neurons.** In anthills, it will soon be possible to make a computer simulation such that the response of an entire anthill to a stressor can be predicted.[24] Similarly, it may become possible to predict the response of the intact brain to a drug or to a chemical delivered in the bloodstream.

16. **Tempo is one of the most basic properties of the human brain.** In ants, the inherent rate of activity varies enormously

from one species to another. Workers of some species walk slowly, with seeming deliberation and economy of movement, whereas other species walk in what seems a pointless hurry.[25] Different people have different characteristic energy levels, and it seems likely that their neurons have different inherent metabolic rates. A slow metabolic rate could be associated with clinical depression or a rapid metabolic rate could be characteristic of mania.

17. **The same environment can elicit different behaviors from different neurons.** Related ant species can show very different behaviors when exposed to the same environment. For example, a mound-building ant lives in the same forest as a similar ant species that does not build mounds. The mound-building species is able to retain nest heat, so these ants can be more active in cold weather.[26] We know also that the same environment elicits different behaviors from different people, yet it is not known if the same environment can elicit different behaviors from different neurons. This level of individuation has not been studied in depth, but it could potentially provide unique insights into the genesis of individual behaviors.

18. **Some neuronal behaviors are stereotypical and nonadaptive.** When ants are daubed with certain chemicals, this signals to other ants in the same nest that the individual is dead. The still alive but unprotesting ant may be picked up and deposited on a refuse pile as if it were a corpse. The ant may be able to clean itself off so that it will no longer be recognized as "dead" but, if it is not cleaned, if will continue to evoke stereotypical and nonadaptive behavior from nestmates.[27] Sometimes, a brain also shows responses that are stereotypical and nonadaptive; the more stereotypical, the less adaptive the behavior is likely to be. Perseverative behavior— slavish persistence in performing an action—is common in

patients with frontal lobe injury, but it is not known whether there can be perseverative behavior at the level of the neuron.

19. **Neurons may show rigidly stereotyped behaviors when they are about to die.** An ant that is injured or senescent will often remove itself from a nest when it is about to die. This has clear survival value for the nest but no individual benefit is known for the ant.[28] Yet there may be an analog for this behavior in the human brain: neurons that are about to die may use the cellular machinery associated with apoptosis or programmed cell death. This process is not well understood in the intact human brain, particularly during early development.

20. **Random noise may have a beneficial function.** In foraging ants, trail making is used as a strategy to recruit additional ants to help in hunting large prey items. Yet if ants were recruited to slavishly follow trails, a still-mobile prey item might never be found. Because ants wander away from trails as they follow them, their ability to reach their goal may be increased by using a new pathway.[29] It is possible that noise is responsible for generating variability in neuronal function, and this variability may ultimately be responsible for enhanced neuronal efficiency and even perhaps for generating creativity.

21. **The developmental trajectory of neural stem cells may be malleable.** In ant hills, the caste or body form of larval ants can be controlled artificially by subtle manipulation of the hormones to which larvae are exposed during maturation.[30] This is reminiscent of the ability of molecular neurobiologists to control the fate of stem cells in cell culture. This suggests that the developmental trajectory of neural stem cells in the brain might also be controlled so that stem cells could be induced to form any of several cell types that might have been damaged by stroke or other illness.

22. **Complexity is not always meaningful.** Among certain ant species, there is an elaborate ritual of interaction using the antennae, though changes in the ritual seem to have little or no effect on the outcome of an interaction. It has been postulated that this variation in antennal interaction, though itself meaningless, serves to keep the interaction fresh and to lessen the chance that one ant will become habituated to another and will therefore start to disregard the signals.[31] It is possible that such "empty rituals" could provide fodder for evolution in that new meaning could become added to standard interactions as time goes by. This may also be true in the brain. If there is a signaling mode that is underutilized, that mode could eventually be co-opted for other purposes and could come to serve other functions.

The anthill is a very deep analogy; many more predictions are therefore likely to emerge. Yet there is a crucial difference between ants and neurons, and this difference must be factored into any mathematical description of anthills and brains: ants and neurons differ in their potential degree of autonomy. An ant retains some degree of autonomy throughout its life span, whereas neurons are linked into a whole whose integrity probably cannot be safely compromised. Soldier ants and worker ants are all progeny of the same queen, so they share genetic identity, just as the neurons in a brain share genetic identity. Were a soldier ant to kill a worker, there would be no loss of genetic information from the colony, just as there would be no loss of genetic information from a brain if a neuron were to die. But something more is lost by the death of a neuron than by the death of an ant. The specificity of neuronal connections is such that a neuron that dies cannot be replaced by another neuron with precisely the same connections. In contrast, an ant in an anthill may be a perfectly replaceable part, whose function could be filled by literally thousands of other ants.

IMPLICATIONS OF EMERGENT COMPLEXITY

A major implication of emergent complexity comes from the perception that the same output can arise in response to many different inputs. As we have noted, if an animal is frightened, it could freeze in place because it has chosen to do so as part of an adaptive strategy or because it is just overwhelmed and doesn't know what else to do. To parse apart such complexity at the neuronal level, it will be necessary to study the simplest systems that show complex behaviors. Otherwise, we will lack sufficient insight to determine the relationship between input and output.

Thus, if the goal is to understand emergent complexity, it may be premature to study behavior in an intact organism. Instead, it seems worthwhile to study simple experimental systems that show emergent complexity. Science is the process of testing ideas against reality by making measurements precise enough to put the lie to false theory. One way to do this would be to place greater emphasis on using whole organ preparations from simple animals rather than simple nerve preparations from complex animals. The goal of this work would be to see how neurons in a native assemblage process information. Whole organ studies would be exciting if they could enable scientists to study the phenomenon of emergent complexity, which is very poorly reflected in experimental models so far. For example, in the locust's visual system, two specific neurons are sensitive to a looming visual target, and these neurons help the locust to avoid collisions.[32] But the behavior of a locust cannot be explained solely on the basis of information processed by these two cells. The sensitivity of a locust to a looming visual target is predictable only if such factors as neuronal shape, neuronal inhibition, and local characteristics of the neuronal pathways are also considered. Such unpredictability may be a simple analog for emergent complexity.

A second implication of emergent complexity is that we should regard computer models with some skepticism. Such models may convey more certainty than is warranted, even if they do a very good job of predicting outcomes. Models become particularly dangerous when they do not suggest a definitive experiment. A nonpredictive but appealing model can be very difficult to disprove, and such models tend to assume a life of their own. Eventually, such models may be regarded as true, even in the absence of a rigorous test. We must bear in mind that models are useful to organize knowledge but they can also limit vision. The psychiatry literature is full of complex wiring diagrams meant to explain how particular drugs alter brain chemistry so as to alleviate the symptoms of disease. These diagrams may accurately summarize experiments that have been done, but they won't necessarily predict the results of new experiments. Models can be exceedingly complex without being predictive, and they may be right only by sheerest accident.

Imagine that you take the back off of your television set and randomly add a few milliohms of resistance to some connection in the mess of exposed wiring. If you then tried to turn on your television, there is a chance that it wouldn't work at all: this would be analogous to a fatal change in brain chemistry. But there is also a chance that the television would simply have a somewhat degraded picture or even a perfectly normal picture at first that gradually degrades over time. No television repairman would be arrogant enough or foolish enough to diagnose the problem without getting into the guts of the television with an ohm-meter. Yet we diagnose mental illness all the time without probing neuronal function. This situation is forced upon us because it is unethical or impossible to probe neuronal function in most patients. In many cases, psychiatrists wouldn't even know what to look for, in terms of change. Nevertheless, we should be aware that it is probably premature to make explicit neuronal models of complex behaviors.

A third implication of the theory of emergent complexity is that the mind is probably inadequately engineered and prone to malfunction. This is because evolution to produce the human brain has been so rapid. Moreover, complexity is thought to arise from redundancy and a large genome, both of which would allow for the accumulation of "errors" or mutations. Such errors could be problematic, since there has been relatively little time for natural selection to remove them from the genome. Unless these errors are fatal, they would gradually accumulate and attain expression in certain children. It could be that the high rate of mental illness in human beings is a consequence of the rapid evolutionary process that led to our wondrous brains.

A fourth implication of the theory of emergent complexity is that the evolution of the human mind may well be an ongoing process. Given that the impetus for the evolution of the mind is unclear, it is possible that the selection pressure in favor of the mind's continuing evolution remains strong. Alternatively, it could be that our brain has reached a "tipping point" of such extensive redundancy and genomic size that continued evolution of the mind might come at a relatively low cost. Or it could even be that the continued evolution of the human mind is essential for the continued existence of our species, since we have achieved the ability to annihilate ourselves with ease.[33] Thus, continued brain evolution is expected, perhaps at a rapid rate.

A final implication of the theory of emergent complexity is that we are right to enshrine creativity and consciousness and an aesthetic and moral sense as the most uniquely human traits. Though creativity can potentially be explained as a function of loose neural connections and taut reality testing, it is something that is singularly human. Newton's calculus, Darwin's evolution, Rodin's *Burghers of Calais*, Einstein's relativity, Tolstoy's *Anna Karenina*, Watson and Crick's model of the DNA molecule, Picasso's *Guernica*—this is the stuff of humanity! Though it seems unlikely that we will ever fully explain creativity, we can certainly recognize and savor it as something that is unique to our wondrous species.

AFTERWORD

I f we entered a deserted Rome, knowing nothing whatsoever of its history, what would we make of it? Would we see St. Peter's Basilica and conclude that it is so beautiful that only God could have built it? As naïve as this seems, this is exactly what believers in intelligent design do when they argue that evolution cannot have built the human brain.

"INTELLIGENT DESIGN" V. EVOLUTION

Proponents of intelligent design argue that the odds are negligible that "random" evolution could have made the human mind. They argue that the probability of producing the brain by such a process is comparable to the probability of throwing a billion dice and having all of them come up "sixes." But, as we have seen, evolution is not a random process. In reality, the odds that the human brain evolved may be more analogous to throwing a billion dice, saving those that come up "sixes," and throwing the remaining dice again and again and again, each time saving "sixes." All that is required to produce a billion "sixes" by such a process is time, and time has been abundant: life on Earth has existed for nearly 5 billion years.

It is not possible to be a biologist without accepting the basic tenets of evolution, just as it is not possible to be an astrophysicist without accepting gravity. It is not that evolution is a belief system; evolution is an incredibly elegant, simple, powerful, direct, testable, and compelling set of ideas. Evolution is a "theory" in the same sense that gravity is a "theory"; though there are aspects of gravity that are not fully understood, we won't fly off into space while we argue the details.

A "theory" is knowledge that is systematically organized to explain a broad range of facts. A theory is not a new or controversial idea that is unsupported by evidence; it is an edifice that has been built and tested and modified so as to incorporate the best data and the best thinking of a wide spectrum of scientists. Gravitation is a theory. Relativity is a theory. Evolution is a theory. In contrast, there is no "theory" of intelligent design, there is just a stubborn refusal to consider the possibility of evolution. Intelligent design fails to advance any new ideas as an alternative to evolution—it is nothing more than a "just-so" story, devoid of intellectual content and lacking in any rationale save faith.

We are far closer to having a "proof" of evolution than we are to having a proof of intelligent design. In truth, we can definitively prove neither idea, nor will we ever be able to do so to the full satisfaction of the closed-minded. But there is so much evidence consistent with evolution that it is overwhelming: there are millions of fossils that show extinct forms of life, some of which could provide plausible ancestors for species that now exist; there are thousands of genes shared in common between similar species, many of which have been sequenced and integrated into an evolutionary framework; there are hundreds of proteins whose structure and function has been explained in terms of evolution; and there are millions of species whose similarities and differences can be understood in an evolutionary context. Intelligent design is not needed to explain any of these things and, in fact, can't explain any of these things.

Evolution is argued about in the details, but it is not controversial in any larger sense. To reject evolution now requires such an enormous leap of faith that most scientists would regard it as irrational. The media are, of course, obligated to report news that relates to "intelligent design," and such reports may be accompanied by interviews with certain individuals who are critical of evolution. But when reporters seek comment from "scientists" who don't believe in evolution, they inadvertently lower the bar for what it means to be a biologist. By seeking comment from a small cadre of vocal critics of evolution, the press unwittingly lends credence to a viewpoint that is not a cogent alternative to evolution. This is not to say that the theory of evolution is perfect: far from it. There is still much work to be done by scientists to explain how the human brain evolved and how the human mind emerged from it. But we should not be distracted by those members of the "Flat Earth Society" who linger at the fringes of the life sciences.

The Remaining Questions

Already sufficient to explain the brain in broad strokes, evolution has provided insight into a crucial question: What benefits arose from the evolutionary time and biological energy spent in developing the human brain, when very simple organisms with very simple brains have been so successful? Some selective advantage must result from the elaboration of human behavior. If there were no selective advantage to having a sophisticated brain, then there would have been no impetus for us to develop a brain any more complex than a cockroach. The nature of this selective pressure can never be proven conclusively, yet there is reason to think that sociality—the tendency of humans to associate with others to form tribes and societies—has been and still is the main driving force in the evolution of the human brain. Given a strong selective pressure for sociality and enough time to evolve novel solutions to those problems that are created by coexistence, evolution may have made the emergence of the human mind inevitable.

But other pressing questions remain. What is consciousness and what purpose does it serve? Can a machine or an animal show consciousness? What is creativity and how can we foster it? How much of who we are is learned and how much is locked in our genes? Is there a biological basis for empathy or religious experience? Are the esoteric functions that some attribute to mind, such as extrasensory perception or premonition, real? Can the brain ever truly know itself?

Such questions may seem heretical to some, especially perhaps to scientists who are stuck in a reductionist worldview, but the time has come to break down borders between fields, to attack resistant problems with innovative techniques. If great science is revolutionary, it follows that good science should at least be subversive.

NOTES

CHAPTER 1. THE ANTHILL OF THE BRAIN

1. B. Holldobler and E. O. Wilson, *The Ants* (Cambridge, MA: Belknap, 1990).

2. S. B. Laughlin and T. J. Sejnowski, "Communication in Neuronal Networks, *Science* 301 (2003): 1870–74.

CHAPTER 2. CONDUCTION AND CONNECTION

1. R. Eckert and D. Randall, *Animal Physiology: Mechanisms and Adaptations* (San Francisco: W. H. Freeman, 1983).

2. J. R. Cooper, F. E. Bloom, and R. H. Roth, *The Biochemical Basis of Neuropharmacology* (New York: Oxford University Press, 1996).

3. A. A. Faisal, J. A. White, and S. B. Laughlin, "Ion-Channel Noise Places Limits on the Miniaturization of the Brain's Wiring," *Current Biology*, 15 (2005): 1143–49.

4. M. A. Hofman, "Energy Metabolism, Brain Size and Longevity in Mammals," *Quarterly Review of Biology* 58 (1983): 495–512.

5. D. B. Chklovskii, B. M. Mel, and K. Svoboda, "Cortical Rewiring and Information Storage," *Nature* 431 (2004): 782–88.

6. P. W. Frankland and B. Bontempi, "The Organization of Recent and Remote Memories," *Nature Reviews Neuroscience* 6 (2005): 119–30.

7. S. B. Laughlin and T. J. Sejnowski, "Communication in Neuronal Networks," *Science* 301 (2003): 1870–74.

8. M. Gell-Mann, "Consciousness, Reduction, and Emergence: Some Remarks," *Annals of the New York Academy of Sciences* 929 (2001): 41–49.

9. H. G. Krapp and F. Gabbiani, "Spatial Distribution of Inputs and Local Receptive Field Properties of a Wide-Field, Looming Sensitive Neuron," *Journal of Neurophysiology* 93 (2005): 2240–53.

10. S. B. Laughlin, R. R. de Ruyter van Steveninck, and J. C. Anderson, "The Metabolic Cost of Neural Information," *Nature Neuroscience* 1 (1998): 36–41.

11. Laughlin and Sejnowski, "Communication in Neuronal Networks."

Chapter 3. The Sensation of Change

1. R. Dawkins, *The Selfish Gene*, 2nd ed. (New York: Oxford University Press, 1990).

2. C. Koch and G. Laurent, "Complexity and the Nervous System," *Science* 284 (1999): 96–98.

3. R. Eckert and D. Randall, *Animal Physiology: Mechanisms and Adaptations* (San Francisco: Freeman, 1983).

4. G. Beeli, M. Esslen, and L. Jancke, "Synaesthesia: When Coloured Sounds Taste Sweet, *Nature* 434 (2005): 38.

5. L. Buck and R. Axel, "A Novel Multigene Family May Encode Odorant Receptors: A Molecular Basis for Odorant Recognition," *Cell* 65 (1991): 175–87.

6. S. Firestein, "A Nobel Nose: The 2004 Nobel Prize in Physiology and Medicine," *Neuron* 45 (2005): 333–38.

7. D. P. Edward and L. M. Kaufman, "Anatomy, Development, and Physiology of the Visual System," *Pediatric Clinics of North America* 50 (2003): 1–23.

8. T. Ohyama et al., "Fraction of the Dark Current Carried by Ca^{2+} through cGMP-Gated Ion Channels of Intact Rod and Cone Photoreceptors," *Journal of General Physiology* 116 (2000): 735–53.

9. G. B. Arden et al., "Spare the Rod and Spoil the Eye," *British Journal of Ophthalmology* 89 (2005): 764–69.

10. D. H. Hubel, "Single Unit Activity in Lateral Geniculate Body and Optic Tract of Unrestrained Cats," *Journal of Physiology* 150 (1960): 91–104.

11. K. T. Brown and T. N. Wiesel, "Intraretinal Recording with Micropipette Electrodes in the Intact Cat Eye," *Journal of Physiology* 149 (1959): 537–62.

12. E. W. Jin and S. K. Shevell, "Color Memory and Color Constancy," *Journal of the Optical Society of America. Part A, Optics, Image Science and Vision* 13 (1996): 1981–91.

13. S. Watanabe, K. Miki, and R. Kakigi, "Mechanisms of Face Perception in Humans: A Magneto- and Electro-encephalographic Study," *Neuropathology* 25 (2005): 8–20.

14. K. Grammer, B. Fink, and N. Neave, "Human Pheromones and Sexual Attraction," *European Journal of Obstetrics Gynecology* 118 (2005): 135–42.

15. J. Friebely and S. Rako, "Pheromonal Influences on Sociosexual Behavior in Post-Menopausal Women," *Journal of Sex Research* 41 (2004): 372–80.

16. T. D. Smith et al., "Searching for the Vomeronasal Organ of Adult Humans: Preliminary Findings on Location, Structure, and Size," *Microscopy Research and Technique* 41 (1998): 483–91.

17. A. S. Levine, C. M. Kotz, and B. A. Gosnell, "Sugars: Hedonic Aspects, Neuroregulation, and Energy Balance," *American Journal of Clinical Nutrition* 78 (2003): 834S–842S.

18. F. W. Turek, C. Dugovic, and P. C. Zee, "Current Understanding of the Circadian Clock and the Clinical Implications for Neurological Disorders," *Archives of Neurology* 58 (2001): 1781–87.

19. C. S. D. Almond et al., "Hyponatremia among Runners in the Boston Marathon," *New England Journal of Medicine* 352 (2005): 1550–56.

CHAPTER 4. THE "DINOSAUR" BRAIN

1. T. A. Woolsey, J. Hanaway, and M. H. Gado, *The Brain Atlas: A Visual Guide to the Human Central Nervous System* (Hoboken, NJ: Wiley, 2003).

2. R. Apps and M. Garwicz, "Anatomical and Physiological Founda-

tions of Cerebellar Information Processing," *Nature Reviews Neuroscience* 6 (2005): 297–311.

3. W. T. Thach, H. P. Goodkin, and J. G. Keating, "The Cerebellum and the Adaptive Control of Movement," *Annual Review of Neuroscience* 15 (1992): 403–42.

4. J. Nolte, *The Human Brain: An Introduction to Its Functional Anatomy*, 3rd ed. (Baltimore: Mosby Yearbook, 1993).

5. D. M. Broussard and C. D. Kassardjian, "Learning in a Simple Motor System," *Learning and Memory* 11 (2004): 127–36.

6. E. S. Boyden, A. Katoh, and J. L. Raymond, "Cerebellum-Dependent Learning: The Role of Multiple Plasticity Mechanisms," *Annual Review of Neuroscience* 27 (2004): 581–609.

7. J. Doyon et al., "Experience-Dependent Changes in Cerebellar Contributions to Motor Sequence Learning," *Proceedings of the National Academy of Sciences of the United States of America* 99 (2002): 1017–22.

8. G. F. Alheid, W. K. Milsom, and D. R. McCrimmon, "Pontine Influences on Breathing: An Overview," *Respiratory Physiology & Neurobiology* 143 (2004): 105–114.

9. M. A. Nogues, A. J. Roncoroni, and E. Benarroch, "Breathing Control in Neurological Diseases," *Clinical Autonomic Research* 12 (2002): 440–49.

10. W. M. St. John and J. F. R. Paton, "Role of Pontine Mechanisms in the Neurogenesis of Eupnea," *Respiratory Physiology & Neurobiology* 143 (2004): 321–32.

11. H. C. Kinney et al., "Decreased Muscarinic Receptor Binding in the Arcuate Nucleus in Sudden Infant Death Syndrome," *Science* 269 (1995): 1446–50.

12. B. E. Jones, "Arousal Systems," *Frontiers in Bioscience* 8 (2003): s438–s451.

13. B. Carey, "Inside the Injured Brain, Many Kinds of Awareness," *New York Times*, April 5, 2005, http://www.nytimes.com/2005/04/05/health/05coma.html.

14. J. T. Giacino et al., "The Minimally Conscious State: Definition and Diagnostic Criteria," *Neurology* 58 (2002): 349–53.

15. S. Laureys et al., "Cerebral Processing in the Minimally Conscious State," *Neurology* 63 (2004): 916–18.

16. J. H. Adams, D. I. Graham, and B. Jennett, "The Neuropathology

of the Vegetative State after an Acute Brain Insult," *Brain* 123 (2000): 1327–38.

17. M. H. Lammi et al., "The Minimally Conscious State and Recovery Potential: A Follow-Up Study 2 to 5 Years after Traumatic Brain Injury," *Archives of Physical Medicine and Rehabilitation* 86 (2005): 746–54.

18. A. Farin et al., "Sex-Related Differences in Patients with Severe Head Injury: Greater Susceptibility to Brain Swelling in Female Patients 50 Years of Age and Younger," *Journal of Neurosurgery* 98 (2003): 32–36.

19. E. F. M. Wijdicks, "The Diagnosis of Brain Death," *New England Journal of Medicine* 344 (2001): 1215–21.

20. A. M. Capron, "Brain Death—Well Settled Yet Still Unresolved," *New England Journal of Medicine* 344 (2001): 1244–46.

21. I. de Beaufort, "Patients in a Persistent Vegetative State—A Dutch Perspective," *New England Journal of Medicine* 352 (2005): 2373–75.

22. G. J. Annas, "Culture of Life" Politics at the Bedside—The Case of Terri Schiavo," *New England Journal of Medicine* 352 (2005): 1710–15.

23. Adams et al., "The Neuropathology of the Vegetative State."

24. Carey, "Inside the Injured Brain."

CHAPTER 5. THE "HUMAN" BRAIN

1. S. W. Rogers, "Reconstructing the Behaviors of Extinct Species: An Excursion into Comparative Paleoneurology," *American Journal of Medical Genetics* 134A (2005): 349–56.

2. J. Nolte, *The Human Brain: An Introduction to Its Functional Anatomy*, 3rd ed. (Baltimore: Mosby Yearbook, 1993).

3. H. C. Kinney et al., "Neuropathological Findings in the Brain of Karen Ann Quinlan: The Role of the Thalamus in the Persistent Vegetative State," *New England Journal of Medicine* 330 (1994): 1469–75.

4. J. H. Adams, D. I. Graham, and B. Jennett, "The Neuropathology of the Vegetative State after an Acute Brain Insult," *Brain* 123 (2000): 1327–38.

5. B. Jennett, "Neuropathology in Vegetative and Severely Disabled Patients after Head Injury," *Neurology* 56 (2001): 486–90.

6. F. Crick and C. Koch, "A Framework for Consciousness," *Nature Neuroscience* 6 (2003): 119–26.

7. A. Zeman, "Consciousness," *Brain* 124 (2001): 1263–89.

8. H. Meeren et al., "Evolving Concepts on the Pathophysiology of Absence Seizures: The Cortical Focus Theory," *Archives of Neurology* 62 (2005): 371–76.

9. G. M. Edelman, *Wider Than the Sky: The Phenomenal Gift of Consciousness* (New Haven, CT: Yale University Press, 2004).

10. T. J. Sejnowski, "The Computational Self," *Annals of the New York Academy of Sciences* 1001 (2003): 262–71.

11. T. A. Woolsey, J. Hanaway, and M. H. Gado, *The Brain Atlas: A Visual Guide to the Human Central Nervous System* (Hoboken, NJ: Wiley, 2003).

12. S. B. Laughlin and T. J. Sejnowski, "Communication in Neuronal Networks," *Science* 301 (2003): 1870–74.

13. Ibid.

14. Ibid.

15. Ibid.

16. Ibid.

17. Ibid.

18. J. L. Ringo, "Neuronal Interconnection as a Function of Brain Size," *Brain, Behavior and Evolution* 38 (1991): 1–6.

19. C. Koch and G. Laurent, "Complexity and the Nervous System," *Science* 284 (1999): 96–98.

20. Ibid.

21. Laughlin and Sejnowski, "Communication in Neuronal Networks."

22. Koch and Laurent, "Complexity and the Nervous System."

23. R. R. Edelman, J. R. Hesselink, and M. B. Zlatkin, *Clinical Magnetic Resonance Imaging*, vol. 1 (Montreal: Saunders, 1996).

24. S. H. Greenblatt, "Phrenology in the Science and Culture of the 19th Century," *Neurosurgery* 37 (1995) 790–804.

25. N. C. Andreasen et al., "Intelligence and Brain Structure in Normal Individuals," *American Journal of Psychiatry* 150 (1993): 130–34.

26. M. Wilke et al., "Bright Spots: Correlations of Gray Matter Volume with IQ in a Normal Pediatric Population," *NeuroImage* 20 (2003): 202–15.

27. P. M. Thompson et al., "Genetic Influences on Brain Structure," *Nature Neuroscience* 4 (2001): 1253–58.

28. D. Posthuma et al., "Genetic Correlations between Brain Volumes and the WAIS-III Dimensions of Verbal Comprehension, Working Memory, Perceptual Organization, and Processing Speed," *Twin Research* 6 (2003): 131–39.

29. E. B. Isaacs et al., "Brain Morphometry and IQ Measurements in Preterm Children," *Brain* 127 (2004): 2595–2607.

30. Laughlin and Sejnowski, "Communication in Neuronal Networks."

31. G. Tononi and G. M. Edelman, "Consciousness and Complexity," *Science* 282 (1998): 1846–51.

32. C. A. Pedersen, "How Love Evolved from Sex and Gave Birth to Intelligence and Human Nature," *Journal of Bioeconomics* 6 (2004): 39–63.

CHAPTER 6. THE PSYCHOLOGY OF LEARNING

1. R. L. Gregory, *The Oxford Companion to the Mind*, 2nd ed. (New York: Oxford University Press, 2004).

2. M. D. Mauk and T. Ohyama, "Extinction as New Learning versus Unlearning: Considerations from a Computer Simulation of the Cerebellum," *Learning and Memory* 11 (2004): 566–71.

3. Gregory, *The Oxford Companion to the Mind*.

4. R. G. Steen, *DNA & Destiny: Nature and Nurture in Human Behavior* (New York: Plenum, 1996).

5. S. Maren, "Neurobiology of Pavlovian Fear Conditioning," *Annual Review of Neuroscience* 24 (2001): 897–931.

6. R. Adolphs et al., "Impaired Recognition of Emotion in Facial Expressions Following Bilateral Damage to the Human Amygdala," *Nature* 372 (1994): 669–72.

7. D. C. Knight et al., "Amygdala and Hippocampal Activity during Acquisition and Extinction of Human Fear Conditioning," *Cognitive, Affective & Behavioral Neuroscience* 4 (2004): 317–25.

8. S. Maren and G. J. Quirk, "Neuronal Signalling of Fear Memory," *Nature Reviews Neuroscience* 5 (2004): 844–52.

9. S. Mineka and A. Ohman, "Phobias and Preparedness: The Selective, Automatic, and Encapsulated Nature of Fear," *Biological Psychiatry* 52 (2002): 927–37.

10. J. L. Brown, *The Evolution of Behavior* (New York: Norton, 1975).

11. E. I. Knudsen, "Sensitive Periods in the Development of the Brain and Behavior," *Journal of Cognitive Neuroscience* 16 (2004): 1412–25.

12. P. K. Kuhl et al., "Cross-Language Analysis of Phonetic Units in Language Addressed to Infants," *Science* 277 (1997): 684–86.

13. P. K. Kuhl, "Early Language Acquisition: Cracking the Speech Code," *Nature Reviews Neuroscience* 5 (2004): 831–43.

14. C. D. Gilbert, M. Sigman, and R. E. Crist, "The Neural Basis of Perceptual Learning," *Neuron* 31 (2001): 681–97.

15. Ibid.

16. Ibid.

17. U. Wagner et al., "Sleep Inspires Insight," *Nature* 427 (2004): 352–55.

18. Gilbert, Sigman, and Crist, "The Neural Basis of Perceptual Learning."

19. D. O. Hebb, *The Organization of Behavior* (New York: Wiley, 1949), p. 62.

20. W. B. Scoville and B. Milner, "Loss of Recent Memory after Bilateral Hippocampal Lesions," *Journal of Neurology, Neurosurgery, and Psychology* 20 (1957): 11–21.

21. M. R. Delgado et al., "An fMRI Study of Reward-Related Probability Learning," *NeuroImage* 24 (2005): 862–73.

22. O. A. van den Heuvel et al., "Frontal-Striatal Dysfunction during Planning in Obsessive-Compulsive Disorder," *Archives of General Psychiatry* 62 (2005): 301–10.

23. Maren, "Neurobiology of Pavlovian Fear Conditioning."

24. L. T. Robertson, "Memory and the Brain," *Journal of Dental Education* 66 (2002): 30–42.

25. Ibid.

26. Ibid.

27. Gregory, *The Oxford Companion to the Mind*.

28. G. H. Southall, *Blind Tom, the Black Pianist-Composer; Continually Enslaved* (Lanham, MD: Scarecrow, 1999).

CHAPTER 7. THE BIOLOGICAL BASIS OF MEMORY

1. O. Jensen and J. E. Lisman, "Hippocampal Sequence-Encoding Driven by a Cortical Multi-item Working Memory Buffer," *Trends in Neurosciences* 28 (2005): 67–72.

2. R. Lorente de No, "Cerebral Cortex: Architecture, Intracortical Connections, Motor Projections," in *Physiology of the Nervous System*, ed. J. F. Fulton (London: Oxford University Press, 1949), pp. 288–330.

3. D. O. Hebb, *The Organization of Behavior* (New York: Wiley, 1949), p. 62.

4. Ibid., p. xix.

5. C. F. Stevens, "Presynaptic Function," *Current Opinion in Neurobiology* 14 (2004): 341–45.

6. Jensen and Lisman, "Hippocampal Sequence-Encoding."

7. D. Durstewitz, J. K. Seamans, and T. J. Sejnowski, "Neurocomputational Models of Working Memory," *Nature Neuroscience* 3 (2000): 1184–91.

8. S. Maren, "Neurobiology of Pavlovian Fear Conditioning," *Annual Review of Neuroscience* 24 (2001): 897–931.

9. F. Vargha-Khadem et al., "Differential Effects of Early Hippocampal Pathology on Episodic and Semantic Memory," *Science* 277 (1997): 376–80.

10. C. M. Temple and P. Richardson, "Developmental Amnesia: A New Pattern of Dissociation with Intact Episodic Memory," *Neuropsychologia* 42 (2004): 764–81.

11. H. Eichenbaum, "Hippocampus: Mapping or Memory?" *Current Biology* 10 (2000): R785–R787.

12. M. Shapiro, "Plasticity, Hippocampal Place Cells, and Cognitive Maps," *Archives of Neurology* 58 (2001): 874–81.

13. I. Fried et al., "Electric Current Stimulates Laughter," *Nature* 391 (1998): 650.

14. A. Pasupathy and E. K. Miller, "Different Time Courses of Learning-Related Activity in the Prefrontal Cortex and Striatum," *Nature* 433 (2005): 873–76.

15. G. N. Elston, "Cortex, Cognition and the Cell: New Insights into the Pyramidal Neuron and Prefrontal Function," *Cerebral Cortex* 13 (2003): 1124–38.

16. P. E. Schulz, "Long-Term Potentiation Involves Increases in the Probability of Neurotransmitter Release," *Proceedings of the National Academy of Sciences of the United States of America* 94 (1997): 5888–93.

17. M. A. Lynch, "Long-Term Potentiation and Memory," *Physiological Reviews* 84 (2004): 87–136.

18. Shapiro, "Plasticity, Hippocampal Place Cells, and Cognitive Maps."

19. Y. P. Tang et al., "Genetic Enhancement of Learning and Memory in Mice," *Nature* 401 (1999): 63–69.

20. C. L. M. Bon and J. Garthwaite, "On the Role of Nitric Oxide in

Hippocampal Long-Term Potentiation," *Journal of Neuroscience* 23 (2003): 1941–48.

21. M. V. L. Bennett and R. S. Zukin, "Electrical Coupling and Neuronal Synchronization in the Mammalian Brain," *Neuron* 41 (2004): 495–511.

22. W. S. McCulloch and W. Pitts, "A Logical Calculus of the Ideas Immanent in Nervous Activity," *Bulletin of Mathematical Biophysics* 5 (1943): 115–33.

23. K. J. Jeffery and I. C. Reid, "Modifiable Neuronal Connections: An Overview for Psychiatrists," *American Journal of Psychiatry* 154 (1997): 156–64.

24. D. J. Willshaw, O. P. Buneman, and H. C. Longuet-Higgins, "Nonholographic Associative Memory," *Nature* 22 (1969): 960–62.

25. F. Crick, "The Recent Excitement about Neural Networks," *Nature* 337 (1989): 129–32.

26. Ibid.

27. J. J. Hopfield, "Neural Networks and Physical Systems with Emergent Collective Computational Abilities," *Proceedings of the National Academy of Sciences of the United States of America* 79 (1982): 2554–58.

28. J. J. Hopfield, "Neurons with Graded Response Have Collective Computational Properties like Those of Two-State Neurons," *Proceedings of the National Academy of Sciences of the United States of America* 81 (1984): 3088–92.

29. J. J. Hopfield and C. D. Brody, "Learning Rules and Network Repair in Spike-Timing-Based Computation Networks," *Proceedings of the National Academy of Sciences of the United States of America* 101 (2004): 337–42.

30. Stevens, "Presynaptic Function."

31. C. Koch and I. Segev, "The Role of Single Neurons in Information Processing," *Nature Neuroscience* 3 (2000): 1171–77.

32. E. Loftus and K. Ketcham, *The Myth of Repressed Memory: False Memories and Allegations of Sexual Abuse* (New York: St. Martin's, 1994).

33. K. Sevcik, "Dirty Secrets," *Rolling Stone*, May 5, 2005.

34. Koch and Segev, "The Role of Single Neurons."

CHAPTER 8. BRAIN PLASTICITY AND NEURAL STEM CELLS

1. J. Altman, "Are New Neurons Formed in the Brains of Adult Mammals?" *Science* 135 (1962): 1127–28.

2. F. Nottebohm, "A Brain for All Seasons: Cyclical Anatomical Changes in Song Control Nuclei of the Canary Brain," *Science* 214 (1981): 1368–70.

3. A. Alvarez-Buylla, J. R. Kirn, and F. Nottebohm, "Birth of Projection Neurons in Adult Avian Brain May Be Related to Perceptual or Motor Learning," *Science* 249 (1990): 1444–46.

4. F. Nottebohm, "The Road We Travelled: Discovery, Choreography, and Significance of Brain Replaceable Neurons," *Annals of the New York Academy of Sciences* 1016 (2004): 628–58.

5. B. A. Reynolds and S. Weiss, "Generation of Neurons and Astrocytes from Isolated Cells of the Adult Mammalian Central Nervous System," *Science* 255 (1992): 1707–10.

6. C. Lois and A. Alvarez-Buylla, "Proliferating Subventricular Zone Stem Cells in the Adult Mammalian Forebrain Can Differentiate into Neurons and Glia," *Proceedings of the National Academy of Sciences of the United States of America* 90 (1993): 2074–77.

7. J. O. Suhonen et al., "Differentiation of Adult Hippocampus-Derived Progenitors into Olfactory Neurons In Vivo," *Nature* 383 (1996): 624–27.

8. F. H. Gage, J. Ray, and L. J. Fisher, "Isolation, Characterization, and Use of Stem Cells from the CNS," *Annual Review of Neuroscience* 18 (1995): 159–92.

9. C. Lois and A. Alvarez-Buylla, "Long-Distance Neuronal Migration in the Adult Mammalian Brain," *Science* 264 (1994): 1145–48.

10. C. Lois, J. M. Garcia-Verdugo, and A. Alvarez-Buylla, "Chain Migration of Neuronal Precursors," *Science* 271 (1996): 978–81.

11. T. R. Brazelton et al., "From Marrow to Brain: Expression of Neuronal Phenotypes in Adult Mice," *Science* 290 (2000): 1775–79.

12. E. Mezey et al., "Turning Blood into Brain: Cells Bearing Neuronal Antigens Generated In Vivo from Bone Marrow," *Science* 290 (2000): 1779–82; R. F. Castro et al., "Failure of Bone Marrow Cells to Transdifferentiate into Neural Cells In Vivo," *Science* 297 (2002): 1229.

13. Brazelton et al., "From Marrow to Brain."

14. E. Gould et al., "Regulation of Hippocampal Neurogenesis in Adulthood," *Biological Psychiatry* 48 (2000): 715–20.

15. H. van Praag et al., "Functional Neurogenesis in the Adult Hippocampus," *Nature* 415 (2002): 1030–34; S. Jessberger and G. Kempermann, "Adult-Born Hippocampal Neurons Mature into Activity-Dependent Responsiveness," *European Journal of Neuroscience* 18 (2003): 2707–12.

16. M. Alvarez-Dolado et al., "Fusion of Bone-Marrow-Derived Cells with Purkinje Neurons, Cardiomyocytes, and Hepatocytes," *Nature* 425 (2003): 968–73.

17. S. L. Florence, H. B. Taub, and J. H. Kaas, "Large-Scale Sprouting of Cortical Connections after Peripheral Injury in Adult Macaque Monkeys," *Science* 282 (1998): 1117–21.

18. E. G. Jones and T. P. Pons, "Thalamic and Brainstem Contributions to Large-Scale Plasticity of Primate Somatosensory Cortex," *Science* 282 (1998): 1121–24.

19. D. R. Kornack and P. Rakic, "Cell Proliferation without Neurogenesis in Adult Primate Neocortex," *Science* 294 (2001): 2127–30.

20. P. S. Eriksson et al., "Neurogenesis in the Adult Human Hippocampus," *Nature Medicine* 4 (1998): 1313–17.

21. Ibid.

22. M. R. Del Bigio, "Proliferative Status of Cells in Adult Human Dentate Gyrus," *Microscopy Research and Technique* 45 (1999): 353–58.

23. N. S. Roy et al., "Identification, Isolation, and Promoter-Defined Separation of Mitotic Oligodendrocyte Progenitor Cells from the Adult Human Subcortical White Matter," *Journal of Neuroscience* 19 (1999): 9986–95.

24. N. S. Roy et al., "In Vitro Neurogenesis by Progenitor Cells Isolated from the Adult Human Hippocampus," *Nature Medicine* 6 (2000): 271–77.

25. I. Blumcke et al., "Increase of Nestin-Immunoreactive Neural Precursor Cells in the Dentate Gyrus of Pediatric Patients with Early-Onset Temporal Lobe Epilepsy," *Hippocampus* 11 (2001): 311–21.

26. K. L. Spalding et al., "Retrospective Birth Dating of Cells in Humans," *Cell* 122 (2005): 133–43.

27. N. Sanal et al., "Unique Astrocyte Ribbon in Adult Human Brain Contains Neural Stem Cells but Lacks Chain Migration," *Nature* 427 (2004): 740–44.

28. Eriksson et al., "Neurogenesis in the Adult Human Hippocampus."

29. P. Rakic, "Neuroscience: Immigration Denied," *Nature* 427 (2004): 685–86.

30. E. Gould et al., "Learning Enhances Adult Neurogenesis in the Hippocampal Formation," *Nature Neuroscience* 2 (1999): 260–65.

31. H. van Praag, G. Kempermann, and F. H. Gage, "Running Increases Cell Proliferation and Neurogenesis in the Adult Mouse Dentate Gyrus," *Nature Neuroscience* 2 (1999): 266–70.

32. W. T. Greenough, N. J. Cohen, and J. M. Juraska, "New Neurons in Old Brains: Learning to Survive?" *Nature Neuroscience* 2 (1999): 203–205.

33. G. Kempermann, D. Gast, and F. H. Gage, "Neuroplasticity in Old Age: Sustained Fivefold Induction of Hippocampal Neurogenesis by Long-Term Environmental Enrichment," *Annals of Neurology* 52 (2002): 135–43.

34. Ibid.

35. T. J. Shors et al., "Neurogenesis in the Adult Is Involved in the Formation of Trace Memories," *Nature* 410 (2001): 372–76.

36. T. J. Shors et al., "Neurogenesis May Relate to Some but Not All Types of Hippocampal-Dependent Learning," *Hippocampus* 12 (2002): 578–84.

37. C. Rochefort et al., "Enriched Odor Exposure Increases the Number of Newborn Neurons in the Adult Olfactory Bulb and Improves Odor Memory," *Journal of Neuroscience* 22 (2002): 2679–89.

38. E. Drapeau et al., "Spatial Memory Performance of Aged Rats in the Water Maze Predict Levels of Hippocampal Neurogenesis," *Proceedings of the National Academy of Sciences of the United States of America* 100 (2003): 14385–90.

39. J. Raber et al., "Radiation-Induced Cognitive Impairments Are Associated with Changes in Indicators of Hippocampal Neurogenesis," *Radiation Research* 162 (2004): 39–47.

40. R. Rola et al., "Radiation-Induced Impairment of Hippocampal Neurogenesis Is Associated with Cognitive Deficits in Young Mice," *Experimental Neurology* 188 (2004): 316–30.

41. J. S. Snyder et al., "A Role for Adult Neurogenesis in Spatial Long-Term Memory," *Neuroscience* 130 (2005): 843–52.

42. E. Y. Snyder, G. Q. Daley, and M. Goodell, "Taking Stock and Planning for the Next Decade: Realistic Prospects for Stem Cell Therapies for the Nervous System," *Journal of Neuroscience Research* 76 (2004): 157–68.

43. Ibid.

44. Ibid.

45. Ibid.

46. Ibid.

47. A. Arvidsson et al., "Neuronal Replacement from Endogenous Precursors in the Adult Brain after Stroke," *Nature Medicine* 8 (2002): 963–70.

48. Ibid.

49. Sanal et al., "Unique Astrocyte Ribbon."

50. J. E. Malberg et al., "Chronic Antidepressant Treatment Increases Neurogenesis in Adult Rat Hippocampus," *Journal of Neuroscience* 20 (2000): 9104–10.

51. L. Santarelli et al., "Requirement of Hippocampal Neurogenesis for the Behavioral Effects of Antidepressants," *Science* 301 (2003): 805–809.

52. D. J. Selkoe, "Alzheimer's Disease Is a Synaptic Failure," *Science* 298 (2002): 789–91.

53. D. J. Selkoe, "Defining Molecular Targets to Prevent Alzheimer Disease," *Archives of Neurology* 62 (2005): 192–95.

54. M. H. Tuszynski et al., "A Phase 1 Clinical Trial of Nerve Growth Factor Gene Therapy for Alzheimer Disease," *Nature Medicine* 11 (2005): 551–55.

55. Ibid.

56. Snyder, Daley, and Goodell, "Taking Stock."

57. Ibid.

58. N. Rosenthal, "Prometheus's Vulture and the Stem-Cell Promise," *New England Journal of Medicine* 349 (2003): 267–74.

59. G. J. Annas, "Cloning and the U.S. Congress," *New England Journal of Medicine* 346 (2002): 1599–1602.

60. Ibid.

61. J. Alter, "The 'Pro-Cure' Movement," *Newsweek*, June 6, 2005.

62. I. L. Weissman, "Stem Cells—Scientific, Medical, and Political Issues," *New England Journal of Medicine* 346 (2002): 1576–79.

63. K. Evers, "European Perspectives on Therapeutic Cloning," *New England Journal of Medicine* 346 (2002): 1579–82.

64. G. Brumfiel, "Scientists Slam Bush Record," *Nature* 427 (2004): 663.

65. E. Blackburn, " Bioethics and the Political Distortion of Biomedical Science," *New England Journal of Medicine* 350 (2004): 1379–80.

CHAPTER 9. CONSCIOUSNESS

1. N. Block, "How Can We Find the Neural Correlate of Consciousness?" *Trends in Neurosciences* 19 (1996): 456–59.

2. R. L. Gregory, *The Oxford Companion to the Mind*, 2nd ed. (New York: Oxford University Press, 2004).

3. G. Miller, "What Is the Biological Basis of Consciousness?" *Science* 309 (2005): 79.

4. F. Crick and C. Koch, "A Framework for Consciousness," *Nature Neuroscience* 6 (2003): 119–26.

5. A. Chaudhuri and P. O. Behan, "Fatigue in Neurological Disorders," *Lancet* 363 (2004): 978–88.

6. A. Riehle et al., "Spike Synchronization and Rate Modulation Differentially Involved in Motor Cortical Function," *Science* 278 (1997): 1950–53.

7. P. N. Steinmetz et al., "Attention Modulates Synchronized Neuronal Firing in Primate Somatosensory Cortex," *Nature* 404 (2000): 187–90.

8. E. Salinas and T. J. Sejnowski, "Correlated Neuronal Activity and the Flow of Neural Information," *Nature Reviews Neuroscience* 2 (2001): 539–50.

9. S. A. Rose, J. F. Feldman, and J. J. Jankowski, "The Building Blocks of Cognition," *Journal of Pediatrics* 143 (2003): S54–S61.

10. Gregory, *The Oxford Companion to the Mind*.

11. F. Crick et al., "Consciousness and Neurosurgery," *Neurosurgery* 55 (2004): 273–81.

12. Ibid.

13. R. Srinivasan et al., "Increased Synchronization of Neuromagnetic Responses during Conscious Perception," *Journal of Neuroscience* 19 (1999): 5435–48.

14. Crick et al., "Consciousness and Neurosurgery."

15. Ibid.

16. A. Cowey, "Fact, Artefact, and Myth about Blindsight," *Quarterly Journal of Experimental Psychology* 57A (2004): 577–609.

17. F. Crick and C. Koch, "Are We Aware of Neural Activity in Primary Visual Cortex?" *Nature* 375 (1995): 121–23.

18. M. M. Mesulam, "From Sensation to Cognition," *Brain* 121 (1998): 1013–52.

19. Block, "How Can We Find the Neural Correlate of Consciousness?"

20. G. Tononi and G. M. Edelman, "Consciousness and Complexity," *Science* 282 (1998): 1846–51.

21. Crick and Koch, "Are We Aware of Neural Activity in Primary Visual Cortex?"

22. R. N. A. Henson, "Neuroimaging Studies of Priming," *Progress in Neurobiology* 70 (2003): 53–81.

23. C. N. Smith et al., "Acquisition of Differential Delay Eyeblink Classical Conditioning Is Independent of Awareness," *Behavioral Neuroscience* 119 (2005): 78–86.

24. P. J. Bayley, J. C. Frascino, and L. R. Squire, "Robust Habit Learning in the Absence of Awareness and Independent of the Medial Temporal Lobe," *Nature* 436 (2005): 550–53.

25. G. Egan et al., "Neural Correlates of the Emergence of Consciousness of Thirst," *Proceedings of the National Academy of Sciences of the United States of America* 100 (2003): 15241–46.

26. F. Crick, "The Recent Excitement about Neural Networks," *Nature* 337 (1989): 129–32.

27. L. F. Abbott and S. B. Nelson, "Synaptic Plasticity: Taming the Beast," *Nature Neuroscience* 3 (2000): 1178–83.

28. D. Durstewitz, J. K. Seamans, and T. J. Sejnowski, "Neurocomputational Models of Working Memory," *Nature Neuroscience* 3 (2000): 1184–91.

29. Salinas and Sejnowski, "Correlated Neuronal Activity."

30. G. Buzsaki and A. Draguhn, "Neuronal Oscillations in Cortical Networks," *Science* 304 (2004): 1926–29.

31. T. J. Sejnowski, "The Computational Self," *Annals of the New York Academy of Sciences* 1001 (2003): 262–71.

32. Tononi and Edelman, "Consciousness and Complexity."

33. A. Lutz et al., "Long-Term Meditators Self-Induce High-Amplitude Gamma Synchrony during Mental Practice," *Proceedings of the National Academy of Sciences of the United States of America* 101 (2004): 16369–73.

34. M. S. Gazzaniga, J. E. Bogen, and R. W. Sperry, "Some Functional Effects of Sectioning the Cerebral Commissues in Man," *Proceedings of the National Academy of Sciences of the United States of America* 48 (1962): 1765–69.

35. Ibid.

36. M. S. Gazzaniga, "Cerebral Specialization and Interhemispheric

Communication: Does the Corpus Callosum Enable the Human Condition?" *Brain* 123 (2000): 1293–1326.

37. Ibid.

38. D. R. Griffin and G. B. Speck, "New Evidence of Animal Consciousness," *Animal Cognition* 7 (2004): 5–18.

39. B. Hare et al., "The Domestication of Social Cognition in Dogs," *Science* 298 (2002): 1634–36.

40. C. D. L. Wynne, "Does Your Dog Understand You?" *Scientist* 18 (2004): 18–19.

41. J. Kaminski, J. Call, and J. Fischer, "Word Learning in a Domestic Dog: Evidence for 'Fast Mapping,'" *Science* 304 (2004): 1682–83.

42. Ibid.

43. Gregory, *The Oxford Companion to the Mind.*

44. Crick and Koch, "A Framework for Consciousness."

CHAPTER 10. ALTERED STATES

1. M. Boly et al., "Auditory Processing in Severely Brain Injured Patients: Differences between the Minimally Conscious State and the Persistent Vegetative State," *Archives of Neurology* 61 (2004): 233–38.

2. N. S. Miller and M. S. Gold, "Sedative-Hypnotics: Pharmacology and Use," *Journal of Family Practice* 29 (1989): 665–70.

3. S. M. Green and B. Krauss, "Clinical Practice Guideline for Emergency Departments: Ketamine Dissociative Sedation in Children," *Annals of Emergency Medicine* 44 (2004): 460–71.

4. A. Tung, "New Anesthesia Techniques," *Thoracic Surgery Clinics* 15 (2005): 27–38.

5. R. A. Veselis, R. A. Reinsel, and V. A. Feshchenko, "Drug-Induced Amnesia Is a Separate Phenomenon from Sedation: Electrophysiologic Evidence," *Anesthesiology* 95 (2001): 896–907.

6. B. Walder, M. R. Tramer, and M. Seeck, "Seizure-Like Phenomena and Propofol: A Systematic Review," *Neurology* 58 (2002): 1327–32.

7. R. L. Gregory, *The Oxford Companion to the Mind*, 2nd ed. (New York: Oxford University Press, 2004).

8. R. G. Steen, *DNA & Destiny: Nature and Nurture in Human Behavior* (New York: Plenum, 1996).

9. M. Manford and F. Andermann, "Complex Visual Hallucinations: Clinical and Neurobiological Insights," *Brain* 121 (1998): 1819–40.

10. Ibid.

11. A. S. Brown, "Looking at Deja Vu for the First Time," *Scientist* 19, no. 2 (2005): 20–21.

12. Ibid.

13. Ibid.

14. B. S. Chang and D. H. Lowenstein, "Epilepsy," *New England Journal of Medicine* 349 (2003): 1257–66.

15. Ibid.

16. Ibid.

17. J. M. Parent et al., "Dentate Granule Cell Neurogenesis Is Increased by Seizures and Contributes to Aberrant Network Reorganization in the Adult Rat Hippocampus," *Journal of Neuroscience* 17 (1997): 3727–38.

18. J. M. Parent, "Injury-Induced Neurogenesis in the Adult Mammalian Brain," *Neuroscientist* 9 (2003): 261–72.

19. B. Hattiangady, M. S. Rao, and A. K. Shetty, "Chronic Temporal Lobe Epilepsy Is Associated with Severely Declined Dentate Neurogenesis in the Adult Hippocampus," *Neurobiology of Disease* 17 (2004): 473–90.

20. C. M. Sinton and R. W. McCarley, "Neurophysiological Mechanisms of Sleep and Wakefulness: A Question of Balance," *Seminars in Neurology* 24 (2004): 211–23.

21. T. J. Balkin et al., "The Process of Awakening: A PET Study of Regional Brain Activity Patterns Mediating the Re-establishment of Alertness and Consciousness," *Brain* 125 (2002): 2308–19.

22. P. Maquet et al., "Functional Neuroanatomy of Human Rapid-Eye Movement Sleep and Dreaming," *Nature* 383 (1996): 163–66.

23. P. Maquet, "Experience-Dependent Changes in Cerebral Activation during Human REM Sleep," *Nature Neuroscience* 3 (2000): 831–36.

24. A. R. Braun et al., "Dissociated Pattern of Activity in Visual Cortices and Their Projections during Human Rapid Eye Movement Sleep," *Science* 279 (1998): 91–94.

25. R. Fosse, R. Stickgold, and J. A. Hobson, "Thinking and Hallucinating: Reciprocal Changes in Sleep," *Psychophysiology* 41 (2004): 298–305.

26. S. Chouinard et al., "Sleep in Untreated Patients with Schizophrenia: A Meta-Analysis," *Schizophrenia Bulletin* 30 (2004): 957–67.

27. R. Stickgold et al., "Sleep, Learning, and Dreams: Off-Line Memory Reprocessing," *Science* 294 (2001): 1052–57.

28. J. M. Siegel, "The REM Sleep-Memory Consolidation Hypothesis," *Science* 294 (2001): 1058–63.

29. T. J. Sejnowski and A. Destexhe, "Why Do We Sleep?" *Brain Research* 886 (2000): 208–23.

30. Ibid.

31. Ibid.

32. P. Maquet, "The Role of Sleep in Learning and Memory," *Science* 294 (2001): 1048–52.

33. M. A. Walker et al., "Practice with Sleep Makes Perfect: Sleep-Dependent Motor Skill Learning," *Neuron* 35 (2002): 205–11.

34. S. A. Mednick, K. Nakayama, and R. Stickgold, "Sleep-Dependent Learning: A Nap Is as Good as a Night," *Nature Neuroscience* 6 (2003): 697–98.

35. M. A. Walker et al., "Sleep-Dependent Motor Memory Plasticity in the Human Brain," *Neuroscience* 133 (2005): 911–17.

36. M. A. Walker and R. Stickgold, "Sleep-Dependent Learning and Memory Consolidation," *Neuron* 44 (2004): 121–33.

37. Stickgold et al., "Sleep, Learning, and Dreams."

38. Siegel, "The REM Sleep-Memory Consolidation Hypothesis."

39. R. Stickgold and M. P. Walker, "Memory Consolidation and Reconsolidation: What Is the Role of Sleep?" *Trends in Neurosciences* 28 (2005): 408–15.

40. D. S. Manoach et al., "A Failure of Sleep-Dependent Procedural Learning in Chronic, Medicated Schizophrenia," *Biological Psychiatry* 56 (2004): 951–56.

41. Sejnowski and Destexhe, "Why Do We Sleep?"

42. C. Cirelli, C. M. Gutierrez, and G. Tononi, "Extensive and Divergent Effects of Sleep and Wakefulness on Brain Gene Expression," *Neuron* 41 (2004): 35–43.

43. Ibid.

44. Ibid.

45. C. Cirelli, "A Molecular Window on Sleep: Changes in Gene Expression between Sleep and Wakefulness," *Neuroscientist* 11 (2005): 63–74.

46. P. Mazzarello, "What Dreams May Come?" *Nature* 408 (2000): 523.

47. M. Csikszentmihalyi, *Creativity: Flow and the Psychology of Discovery and Invention* (New York: HarperCollins, 1996).

48. U. Wagner et al., "Sleep Inspires Insight," *Nature* 427 (2004): 352–55.

49. R. Stickgold and M. A. Walker, "To Sleep, Perchance to Gain Creative Insight?" *Trends in Cognitive Sciences* 8 (2004): 191–92.

CHAPTER 11. EMOTION AND THE SOCIAL BRAIN

1. B. S. McEwen, "Protection and Damage from Acute and Chronic Stress: Allostasis and Allostatic Overload and Relevance to the Pathophysiology of Psychiatric Disorders," *Annals of the New York Academy of Sciences* 1932 (2004): 1–7.

2. D. S. Charney, "Psychobiological Mechanisms of Resilience and Vulnerability: Implications for Successful Adaptation to Extreme Stress," *American Journal of Psychiatry* 161 (2004): 195–216.

3. T. E. Seeman et al., "Allostatic Load as a Marker of Cumulative Biological Risk: MacArthur Studies of Successful Aging," *Proceedings of the National Academy of Sciences of the United States of America* 98 (2001): 4770–75.

4. I. S. Wittstein et al., "Neurohumoral Features of Myocardial Stunning Due to Sudden Emotional Stress," *New England Journal of Medicine* 352 (2005): 539–48.

5. C. W. Hoge et al., "Combat Duty in Iraq and Afghanistan, Mental Health Problems, and Barriers to Care," *New England Journal of Medicine* 351 (2004): 13–22.

6. M. Fazel, J. Wheeler, and J. Danesh, "Prevalence of Serious Mental Disorder in 7000 Refugees Resettled in Western Countries: A Systematic Review," *Lancet* 365 (2005): 1309–14.

7. M. Basoglu et al., "Psychiatric and Cognitive Effects of War in Former Yugoslavia: Association of Lack of Redress for Trauma and Posttraumatic Stress Reactions," *Journal of the American Medical Association* 294 (2005): 580–90.

8. S. Wessely, "Victimhood and Resilience," *New England Journal of Medicine* 353 (2005): 548–50.

9. M. A. Schuster et al., "A National Survey of Stress Reactions after the September 11, 2001, Terrorist Attacks," *New England Journal of Medicine* 345 (2001): 1507–12.

10. J. D. Bremner et al., "MRI and PET Study of Deficits in Hip-

pocampal Structure and Function in Women with Childhood Sexual Abuse and Posttraumatic Stress Disorder," *American Journal of Psychiatry* 160 (2003): 924–32.

11. R. J. Lindauer et al., "Smaller Hippocampal Volume in Dutch Police Officers with Posttraumatic Stress Disorder," *Biological Psychiatry* 56 (2004): 356–63.

12. E. Gould et al., "Proliferation of Granule Cell Precursors in the Dentate Gyrus of Adult Monkeys Is Diminished by Stress," *Proceedings of the National Academy of Sciences of the United States of America* 95 (1998): 3168–71.

13. E. Gould and P. Tanapat, "Stress and Hippocampal Neurogenesis," *Biological Psychiatry* 46 (1999): 1472–79.

14. C. Mirescu, J. D. Peters, and E. Gould, "Early Life Experience Alters Response of Adult Neurogenesis to Stress," *Nature Neuroscience* 7 (2004): 841–46.

15. J. E. Malberg and R. S. Duman, "Cell Proliferation in Adult Hippocampus Is Decreased by Inescapable Stress: Reversal by Fluoxetine Treatment," *Neuropsychopharmacology* 28 (2003): 1562–71.

16. G. Kempermann and G. Kronenberg, "Depressed New Neurons? Adult Hippocampal Neurogenesis and a Cellular Plasticity Hypothesis of Major Depression," *Biological Psychiatry* 54 (2003): 499–503.

17. R. S. Duman, "Depression: A Case of Neuronal Life and Death?" *Biological Psychiatry* 56 (2004): 140–45.

18. T. R. Insel and R. D. Fernald, "How the Brain Processes Social Information: Searching for the Social Brain," *Annual Review of Neuroscience* 27 (2004): 697–722.

19. T. R. Insel, " Is Social Attachment an Addictive Disorder?" *Physiology & Behavior* 79 (2003): 351–57.

20. B. Buwalda et al., "Long-Term Effects of Social Stress on Brain and Behavior: A Focus on Hippocampal Functioning," *Neuroscience and Biobehavioral Reviews* 29 (2005): 83–97.

21. Insel, " Is Social Attachment an Addictive Disorder?"

22. M. Battaglia et al., "Influence of the Serotonin Transporter Promotor Gene and Shyness on Children's Cerebral Responses to Facial Expressions," *Archives of General Psychiatry* 62 (2005): 85–94.

23. S. M. Korte et al., "The Darwinian Concept of Stress: Benefits of Allostasis and Costs of Allostatic Load and the Trade-Offs in Health and Disease," *Neuroscience and Biobehavioral Reviews* 29 (2005): 3–38.

24. G. Seematter et al., "Relationship between Stress, Inflammation and Metabolism," *Current Opinion in Clinical Nutrition and Metabolic Care* 7 (2004): 169–73.

25. McEwen, "Protection and Damage from Acute and Chronic Stress."

26. A. Aron et al., "Reward, Motivation, and Emotion Systems Associated with Early-Stage Intense Romantic Love," *Journal of Neurophysiology* 94 (2005): 327–37.

27. Ibid.

28. Ibid.

29. P. Shaw et al., "The Impact of Early and Late Damage to the Human Amygdala on 'Theory of Mind' Reasoning," *Brain* 127 (2004): 1535–48.

30. Ibid.

31. A. D. Rowe et al., "'Theory of Mind' Impairments and Their Relationship to Executive Functioning Following Frontal Lobe Excisions," *Brain* 124 (2001): 600–16.

32. C. M. Bird et al., "The Impact of Extensive Medial Frontal Lobe Damage on 'Theory of Mind' and Cognition," *Brain* 127 (2004): 914–28.

CHAPTER 12. MOTIVATION AND THE SOCIAL MIND

1. R. A. Depue and J. V. Morrone-Strupinsky, "A Neurobehavioral Model of Affiliative Bonding: Implications for Conceptualizing a Human Trait of Affiliation," *Behavioral and Brain Sciences* 28 (2005): 313–95.

2. F. Benedetti and M. Amanzio, "The Neurobiology of Placebo Analgesia: From Endogenous Opioids to Cholecystokinin," *Progress in Neurobiology* 51 (1997): 109–25.

3. L. E. Sullivan, D. A. Fiellin, and P. G. O'Connor, "The Prevalence and Impact of Alcohol Problems in Major Depression: A Systematic Review," *American Journal of Medicine* 118 (2005): 330–41.

4. E. M. Jones, D. Knutson, and D. Haines, "Common Problems in Patients Recovering from Chemical Dependency," *American Family Physician* 68 (2003): 1971–78.

5. J. A. Byars et al., "Naltrexone Augments the Effects of Nicotine Replacement Therapy in Female Smokers," *Journal of Addictive Diseases* 24 (2005): 49–60.

6. M. Sofuoglu et al., "Effects of Naltrexone and Isradipine, Alone or in Combination, on Cocaine Responses in Humans," *Pharmacology, Biochemistry, and Behavior* 75 (2003): 801–808.

7. P. G. O'Connor and D. A. Fiellin, "Pharmacologic Treatment of Heroin-Dependent Patients," *Annals of Internal Medicine* 133 (2000): 40–54.

8. D. J. Drobes et al., "Effects of Naltrexone and Nalmefene on Subjective Response to Alcohol among Non-treatment-seeking Alcoholics and Social Drinkers," *Alcoholism, Clinical and Experimental Research* 28 (2004): 1362–70.

9. J. D. Loeser and R. Melzack, "Pain: An Overview," *Lancet* 353 (1999): 1607–1609.

10. R. H. Dworkin, "Pain Insensitivity in Schizophrenia: A Neglected Phenomenon and Some Implications," *Schizophrenia Bulletin* 20 (1994): 235–48.

11. L. Diatchenko et al., "Genetic Basis for Individual Variations in Pain Perception and the Development of a Chronic Pain Condition," *Human Molecular Genetics* 14 (2005): 135–43.

12. Loeser and Melzack, "Pain."

13. A. Hrobjartsson and P. C. Gotzsche, "Is the Placebo Powerless? An Analysis of Clinical Trials Comparing Placebo with No Treatment," *New England Journal of Medicine* 344 (2001): 1594–1602.

14. Ibid.

15. A. Hrobjartsson and P. C. Gotzsche, "Is the Placebo Powerless? Update of a Systematic Review with 52 New Randomized Trials Comparing Placebo with No Treatment," *Journal of Internal Medicine* 256 (2004): 91–100.

16. L. Colloca and F. Benedetti, "Placebos and Painkillers: Is Mind as Real as Matter?" *Nature Reviews Neuroscience* 6 (2005): 545–52.

17. M. Amanzio et al., "Response Variability to Analgesics: A Role for Non-specific Activation of Endogenous Opioids," *Pain* 90 (2001): 205–15.

18. Ibid.

19. F. Benedetti, M. Amanzio, and G. Maggi, "Potentiation of Placebo Analgesia by Proglumide," *Lancet* 346 (1995): 1231.

20. B. T. Walsh et al., "Placebo Response in Studies of Major Depression: Variable, Substantial, and Growing," *Journal of the American Medical Association* 287 (2002): 1840–47.

21. J. D. Levine et al., "Role of Pain in Placebo Analgesia," *Proceedings*

of the National Academy of Sciences of the United States of America 76 (1979): 3528–31.

22. J. D. Levine, N. C. Gordon, and H. L. Fields, "The Mechanism of Placebo Analgesia," *Lancet* 2, no. 8091 (1978): 654–57.

23. NIH Consensus Development Panel on Acupuncture, *Journal of the American Medical Association* 280 (1998): 1518–24.

24. Ibid.

25. J. Shen et al., "Electroacupuncture for Control of Myeloablative Chemotherapy-Induced Emesis: A Randomized Controlled Trial," *Journal of the American Medical Association* 284 (2000): 2755–61.

26. D. Irnich et al., "Randomised Trial of Acupuncture Compared with Conventional Massage and 'Sham' Laser Acupuncture for Treatment of Chronic Neck Pain," *British Medical Journal* 322 (2001): 1–6.

27. N. Kotani et al., "Preoperative Intradermal Acupuncture Reduces Postoperative Pain, Nausea and Vomiting, Analgesic Requirement, and Sympathoadrenal Responses," *Anesthesiology* 95 (2001): 349–56.

28. K. Streitberger et al., "Acupuncture Compared to Placebo-Acupuncture for Postoperative Nausea and Vomiting Prophylaxis: A Randomised Placebo-Controlled Patient and Observer Blind Trial," *Anaesthesia* 59 (2004): 142–49.

29. Ibid.

30. N. P. Assefi et al., "A Randomized Clinical Trial of Acupuncture Compared with Sham Acupuncture in Fibromyalgia," *Annals of Internal Medicine*143 (2005): 10–19.

31. K. Linde et al., "Acupuncture for Patients with Migraine: A Randomized Controlled Trial," *Journal of the American Medical Association* 293 (2005): 2118–25.

32. R. B. Bausell et al., "Is Acupuncture Analgesia an Expectancy Effect?" *Evaluation & the Health Professions* 28 (2005): 9–26.

33. A. J. Vickers et al., "Acupuncture for Chronic Headache in Primary Care: Large, Pragmatic, Randomised Trial," *British Medical Journal* 328 (2004): 744.

34. J. Tierney, "Debunking the Drug War," *New York Times*, August 9, 2005, http://www.nytimes.com/2005/08/09/opinion/09tierney.

35. Ibid.

36. R. Room, T. Babor, and J. Rehm, "Alcohol and Public Health," *Lancet* 365 (2005): 519–30.

37. Ibid.

38. E. V. Nunes and F. R. Levin, "Treatment of Depression in Patients with Alcohol or Other Drug Dependence: A Meta-Analysis," *Journal of the American Medical Association* 291 (2004): 1887–96.

39. K. Nixon and F. T. Crews, "Temporally Specific Burst in Cell Proliferation Increases Hippocampal Neurogenesis in Protracted Abstinence from Alcohol," *Journal of Neuroscience* 24 (2004): 9714–22.

40. T. E. Robinson and B. Kolb, "Structural Plasticity Associated with Exposure to Drugs of Abuse," *Neuropharmacology* 47 (2004): 33–46.

41. B. A. Johnson et al., "Oral Topiramate for Treatment of Alcohol Dependence: A Randomised Controlled Trial," *Lancet* 361 (2003): 1677–85.

42. D. I. Lubman, M. Yucel, and C. Pantelis, "Addiction, a Condition of Compulsive Behavior? Neuroimaging and Neuropsychological Evidence of Inhibitory Dysregulation," *Addiction* 99 (2004): 1491–1502.

43. Ibid.

44. L. Cardenas et al., "Brain Reward System Activity in Major Depression and Comorbid Nicotine Dependence," *Journal of Pharacology and Experimental Therapeutics* 302 (2002): 1265–71.

45. Depue and Morrone-Strupinsky, "A Neurobehavioral Model of Affiliative Bonding."

CHAPTER 13. GENES, ENVIRONMENT, AND HUMAN BEHAVIOR

1. R. J. Herrnstein and C. Murray, *The Bell Curve: Intelligence and Class Structure in American Life* (New York: Free Press, 1994).

2. R. G. Steen, *DNA & Destiny: Nature and Nurture in Human Behavior* (New York: Plenum, 1996).

3. J. N. Giedd, "Structural Magnetic Resonance Imaging of the Adolescent Brain," *Annals of the New York Academy of Sciences* 1021 (2004): 77–85.

4. R. G. Steen et al., "Brain Volume in Pediatric Patients with Sickle Cell Disease: Evidence of Volumetric Growth Delay?" *American Journal of Neuroradiology* 26 (2005): 455–62.

5. R. G. Steen et al., "Age-Related Changes in the Pediatric Brain: Quantitative Magnetic Resonance (qMRI) Provides Evidence of Maturational Changes during Adolescence," *American Journal of Neuroradiology* 18 (1997): 819–28.

6. P. L. Yakovlev and A. R. Lecours, "The Myelogenetic Cycles of Regional Maturation in the Brain," in *Regional Development of the Brain in Early Life*, ed. A. Minkowski (Oxford: Blackwell, 1967), pp. 3–70.

7. Y. Ge et al., "Age-Related Total Gray Matter and White Matter Changes in Normal Adult Brain. Part I: Volumetric MR Imaging Analysis," *American Journal of Neuroradiology* 23 (2002): 1327–33.

8. N. Raz et al., "Differential Aging of the Human Striatum: Longitudinal Evidence," *American Journal of Neuroradiology* 24 (2003): 1849–56.

9. N. Gogtay et al., "Dynamic Mapping of Human Cortical Development during Childhood through Early Adulthood," *Proceedings of the National Academy of Sciences of the United States of America* 101 (2004): 8174–79.

10. C. Beaulieu, "Imaging Brain Connectivity in Children with Diverse Reading Ability," *NeuroImage* 25 (2005): 1266–71.

11. Z. Nagy, H. Westerberg, and T. Klingberg, "Maturation of White Matter Is Associated with the Development of Cognitive Functions during Childhood," *Journal of Cognitive Neuroscience* 16 (2004): 1227–33.

12. C. R. Gale et al., "Critical Periods of Brain Growth and Cognitive Function in Children," *Brain* 127 (2004): 321–29.

13. Herrnstein and Murray, *The Bell Curve*.

14. A. J. Reynolds et al., "Long-Term Effects of an Early Childhood Intervention on Educational Achievement and Juvenile Arrest: A 15-Year Follow-Up of Low-Income Children in Public Schools," *Journal of the American Medical Association* 285 (2001): 2339–46.

15. P. D. Evans et al., "Microcephalin, a Gene Regulating Brain Size, Continues to Evolve Adaptively in Humans," *Science* 309 (2005): 1717–20.

16. Ibid.

17. N. Mekel-Bobrov et al., "Ongoing Adaptive Evolution of *ASPM*, a Brain Size Determinant in *Homo Sapiens*," *Science* 309 (2005): 1720–22.

18. G. H. Mochida and C. A. Walsh, "Genetic Basis of Developmental Malformations of the Cerebral Cortex," *Archives of Neurology* 61 (2004): 637–40.

19. M. Caceres et al., "Elevated Gene Expression Levels Distinguish Human from Non-human Primate Brains," *Proceedings of the National Academy of Sciences of the United States of America* 100 (2003): 13030–35.

20. T. Hayakawa, "A Human-Specific Gene in Microglia," *Science* 309 (2005): 1693.

21. L. Krubitzer and D. M. Kahn, "Nature versus Nurture Revisited: An Old Idea with a New Twist," *Progress in Neurobiology* 70 (2003): 33–52.

22. S. F. Oster, "Ganglion Cell Axon Pathfinding in the Retina and Optic Nerve," *Seminars in Cell & Developmental Biology* 15 (2004): 125–36.

23. E. S. Ruthazer, "You're Perfect, Now Change—Redefining the Role of Developmental Plasticity," *Neuron* 45 (2005): 825–28.

24. Steen, *DNA & Destiny*, pp. 23–29.

25. Ibid., pp. 31.

26. T. J. Bouchard et al., "Sources of Human Psychological Differences: The Minnesota Study of Twins Reared Apart," *Science* 250 (1990): 223–28.

27. R. W. Pickens et al., "Heterogeneity in the Inheritance of Alcoholism: A Study of Male and Female Twins," *Archives of General Psychiatry* 48 (1991): 19–28.

28. Steen, *DNA & Destiny*, pp. 66–67.

29. B. M. Hicks et al., "Family Transmission and Heritability of Externalizing Disorders: A Twin-Family Study," *Archives of General Psychiatry* 61 (2004): 922–28.

30. G. R. Uhl and R. W. Grow, "The Burden of Complex Genetics in Brain Disorders," *Archives of General Psychiatry* 61 (2004): 223–29.

31. Ibid.

32. Steen, *DNA & Destiny*, pp. 49–62.

33. C. Ikonimidou et al., "Neurotransmitters and Apoptosis in the Developing Brain," *Biochemical Pharmacology* 62 (2001): 401–405.

34. V. L. Kvigne et al., "Characteristics of Children Who Have Full or Incomplete Fetal Alcohol Syndrome," *Journal of Pediatrics* 145 (2004): 635–40.

35. Steen, *DNA & Destiny*, pp. 1–282.

36. Ibid.

Chapter 14. Neurology and Illnesses of the Brain

1. J. B. Martin, "The Integration of Neurology, Psychiatry, and Neuroscience in the 21st Century," *American Journal of Psychiatry* 159 (2002): 695–704.

2. S. C. Yudofsky and R. E. Hales, "Neuropsychiatry and the Future of Psychiatry and Neurology," *American Journal of Psychiatry* 159 (2002): 1261–64.

3. J. F. R. Kerr, A. H. Wyllie, and A. R. Currie, "Apoptosis: A Basic Biological Phenomenon with Wide-Ranging Implications in Tissue Kinetics," *British Journal of Cancer* 26 (1972): 239–57.

4. J. P. Taylor, J. Hardy, and K. H. Fischbeck, "Toxic Proteins in Neurodegenerative Disease," *Science* 296 (2002): 1991–95.

5. C. Warlow et al., "Stroke," *Lancet* 362 (2003): 1211–24.

6. R. G. Steen et al., "Brain Injury in Children with Hemoglobin SS Sickle Cell Disease: Prevalence and Etiology," *Annals of Neurology* 54 (2003): 564–72.

7. P. M. Rothwell et al., "Change in Stroke Incidence, Mortality, Case-Fatality, Severity, and Risk Factors in Oxfordshire, UK from 1981 to 2004 (Oxford Vascular Study)," *Lancet* 363 (2004): 1925–33.

8. A. Arvidsson, "Neuronal Replacement from Endogenous Precursors in the Adult Brain after Stroke," *Nature Medicine* 8 (2002): 963–70.

9. Z. Nadareishvili and J. Hallenbeck, "Neuronal Regeneration after Stroke," *New England Journal of Medicine* 348 (2003): 2355–56.

10. J. W. Langston et al., "Chronic Parkinsonism in Humans Due to a Product of Meperidine-Analog Synthesis," *Science* 219 (1983): 979–80.

11. C. M. Tanner et al., "Parkinson Disease in Twins: An Etiologic Study," *Journal of the American Medical Association* 281 (1999): 341–46.

12. C. R. Gale et al., "Mortality from Parkinsons Disease and Other Causes in Men Who Were Prisoners of War in the Far East," *Lancet* 354 (1999): 2116–18.

13. M. P. Mattson, "Gene-Diet Interactions in Brain Aging and Neurodegenerative Disorders," *Annals of Internal Medicine* 139 (2003): 441–44.

14. A. Samii, J. G. Nutt, and B. R. Ransom, "Parkinson's Disease," *Lancet* 363 (2004): 1783–93.

15. Ibid.

16. C. R. Freed et al., "Transplantation of Embryonic Dopamine Neurons for Severe Parkinson's Disease," *New England Journal of Medicine* 344 (2001): 710–19.

17. J. G. Nutt et al., "Randomized, Double-Blind Trial of Glial Cell Line–Derived Neurotrophic Factor (GDNF) in PD," *Neurology* 60 (2003): 69–73.

18. N. K. Patel et al., "Intraputamenal Infusion of Glial Cell Line–Derived Neurotrophic Factor in PD: A Two-Year Outcome Study," *Annals of Neurology* 57 (2005): 298–302.

19. J. A. Girault and P. Greengard, "The Neurobiology of Dopamine Signaling," *Archives of Neurology* 61 (2004): 641–44.

20. D. A. Snowdon, " Healthy Aging and Dementia: Findings from the Nun Study," *Annals of Internal Medicine* 139 (2003): 450–54.

21. K. Ritchie and S. Lovestone, "The Dementias," *Lancet* 360 (2002): 1759–66.

22. M. M. Esiri et al., "Pathological Correlates of Late-Onset Dementia in a Multicentre, Community-Based Population in England and Wales," *Lancet* 357 (2001): 169–75.

23. S. E. Vermeer et al., "Silent Brain Infarcts and the Risk of Dementia and Cognitive Decline," *New England Journal of Medicine* 348 (2003): 1215–22.

24. C. M. Clark and J. H. T. Karlawish, "Alzheimer Disease: Current Concepts and Emerging Diagnostic and Therapeutic Strategies," *Annals of Internal Medicine* 138 (2003): 400–10.

25. Esiri et al., "Pathological Correlates of Late-Onset Dementia."

26. Snowdon, "Healthy Aging and Dementia."

27. D. A. Snowdon et al., "Linguistic Ability in Early Life and Cognitive Function and Alzheimer's Disease in Late Life: Findings from the Nun Study," *Journal of the American Medical Association* 275 (1997): 528–32.

28. R. S. Wilson et al., "Participation in Cognitively Stimulating Activities and Risk of Incident Alzheimer Disease," *Journal of the American Medical Association* 287 (2002): 742–48.

29. P. M. Thompson et al., "Mapping Cortical Change in Alzheimer's Disease, Brain Development, and Schizophrenia," *NeuroImage* 23 (2004): S2–S18.

30. Ritchie and Lovestone, "The Dementias."

31. W. E. Klunk et al., "Imaging Brain Amyloid in Alzheimer's Disease with Pittsburgh Compound-B," *Annals of Neurology* 55 (2004): 306–19.

32. I. Casserly and E. Topol, "Convergence of Atherosclerosis and Alzheimer's Disease: Inflammation, Cholesterol, and Misfolded Proteins," *Lancet* 363 (2004): 1139–46.

33. M. Etminan, S. Gill, and A. Samii, "Effect of Non-steroidal Anti-inflammatory Drugs on Risk of Alzheimer's Disease: Systematic Review and Meta-analysis of Observational Studies," *British Medical Journal* 327 (2003): 128–32.

34. H. Kaduszkiewicz et al., "Cholinesterase Inhibitors for Patients with Alzheimer's Disease: Systematic Review of Randomised Clinical Trials," *British Medical Journal* 331 (2005): 321–27.

35. R. C. Petersen et al., "Vitamin E and Donepezil for the Treatment of Mild Cognitive Impairment," *New England Journal of Medicine* 352 (2005): 2379–88.

36. D. A. Bennett et al., "Neurofibrillary Tangles Mediate the Association of Amyloid Load with Clinical Alzheimer Disease and Level of Cognitive Function," *Archives of Neurology* 61 (2004): 378–84.

37. K. SantaCruz et al., "Tau Suppression in a Neurodegenerative Mouse Model Improves Memory Function," *Science* 309 (2005): 476–81.

38. M. F. Weiner et al., "Early Behavioral Symptoms and Course of Alzheimer's Disease," *Acta Psychiatrica et Neurologica Scandinavica* 111 (2005): 367–71.

39. H. Tost et al., "Huntington's Disease: Phenomenolgical Diversity of a Neuropsychiatric Condition That Challenges Traditional Concepts in Neurology and Psychiatry," *American Journal of Psychiatry* 161 (2004) 28–34.

40. A. Ebringer, T. Rashid, and C. L. Wilson, "Bovine Spongiform Encephalopathy, Multiple Sclerosis, and Creutzfeld-Jacob Disease Are Probably Autoimmune Diseases Evoked by *Acinebacter* Bacteria," *Annals of the New York Academy of Sciences* 1050 (2005): 417–28.

41. N. Sanai, A. Alvarez-Buylla, and M. S. Berger, "Neural Stem Cells and the Origin of Gliomas," *New England Journal of Medicine* 353 (2005): 811–22.

42. S. Koponen et al., "Axis I and II Psychiatric Disorders after Traumatic Brain Injury: A 30-Year Follow-Up Study," *American Journal of Psychiatry* 159 (2002): 1315–21.

43. S. J. Collins, V. A. Lawson, and C. L. Masters, "Transmissable Spongiform Encephalopathies," *Lancet* 363 (2004): 51–61.

44. W. M. Cowan and E. R. Kandel, "Prospects for Neurology and Psychiatry," *Journal of the American Medical Association* 285 (2001): 594–600.

45. M. Menken, "Demystifying Neurology," *British Medical Journal* 324 (2002): 1469–70.

46. Ibid.

CHAPTER 15. PSYCHIATRY AND ILLNESSES OF THE MIND

1. R. L. Gregory, *The Oxford Companion to the Mind*, 2nd ed. (New York: Oxford University Press, 2004).

2. B. A. Palmer, V. S. Pankratz, and J. M. Bostwick, "The Lifetime Risk of Suicide in Schizophrenia: A Reexamination," *Archives of General Psychiatry* 62 (2005): 247–53.

3. H. Heila et al., "Mortality among Patients with Schizophrenia and

Reduced Psychiatric Hospital Care," *Psychological Medicine* 35 (2005): 725–32.

4. G. R. Thompson and J. Partridge, "Coronary Calcification Score: The Coronary-Risk Impact Factor," *Lancet* 363 (2004): 557–59.

5. T. S. Szasz, "The Myth of Mental Illness," *American Psychologist* 15 (1960): 113–18.

6. Ibid.

7. B. Carey, "Snake Phobias, Moodiness and a Battle in Psychiatry," *New York Times*, June 13, 2005.

8. Ibid.

9. A. Frances et al., *Diagnostic and Statistical Manual of Mental Disorders (DSM-IV)*, 4th ed. (Washington, DC: American Psychiatric Association, 1994).

10. R. C. Kessler et al., "Prevalence and Treatment of Mental Disorders, 1990 to 2003," *New England Journal of Medicine* 352 (2005): 2515–23.

11. Ibid.

12. Frances et al., *Diagnostic and Statistical Manual,* p. 327.

13. Ibid., pp. 320–26.

14. J. J. Mann, "The Medical Management of Depression," *New England Journal of Medicine* 353 (2005): 1819–31.

15. J. March et al., "Fluoxetine, Cognitive-Behavioral Therapy, and Their Combination for Adolescents with Depression," *Journal of the American Medical Association* 292 (2004): 807–20.

16. B. T. Walsh et al., "Placebo Response in Studies of Major Depression: Variable, Substantial, and Growing," *Journal of the American Medical Association* 287 (2002): 1840–47.

17. I. Elkin et al., "National Institute of Mental Health Treatment of Depression Collaborative Research Program: General Effectiveness of Treatments," *Archives of General Psychiatry* 46 (1989): 971–82.

18. A. I. Scott and C. P. Freeman, "Edinburgh Primary Care Depression Study: Treatment Outcome, Patient Satisfaction, and Cost after 16 Weeks," *British Medical Journal* 304 (1992): 883–87; L. M. Mynors-Wallis et al., "Randomised Controlled Trial Comparing Problem Solving Treatment with Amitriptyline and Placebo for Major Depression in Primary Care," *British Medical Journal* 310 (1995): 441–45.

19. H. C. Schulberg et al., "Treating Major Depression in Primary Care Practice. Eight-Month Clinical Outcomes," *Archives of General Psychiatry* 53 (1996): 913–19; C. F. Reynolds et al., "Nortriptyline and Interpersonal Psy-

chotherapy as Maintenance Therapies for Recurrent Major Depression: A Randomized Controlled Trial in Patients Older Than 59 Years," *Journal of the American Medical Association* 281 (1999): 39–45.

20. R. B. Jarrett et al., "Treatment of Atypical Depression with Cognitive Therapy or Phenelzine: A Double-Blind, Placebo-Controlled Trial," *Archives of General Psychiatry* 56 (1999): 431–37.

21. J. W. Williams et al., "Treatment of Dysthymia and Minor Depression in Primary Care: A Randomized Controlled Trial in Older Adults," *Journal of the American Medical Association* 284 (2000): 1519–26; R. J. DeRubeis et al., "Cognitive Therapy vs Medications in the Treatment of Moderate to Severe Depression," *Archives of General Psychiatry* 62 (2005): 409–16; D. A. Revicki et al., "Cost-Effectiveness of Evidence-Based Pharmacotherapy or Cognitive Behavioral Therapy Compared with Community Referral for Major Depression in Predominantly Low-Income Minority Women," *Archives of General Psychiatry* 62 (2005): 868–75.

22. March et al., "Fluoxetine, Cognitive-Behavioral Therapy, and Their Combination."

23. N. Casacalenda, J. C. Perry, and K. Looper, "Remission in Major Depressive Disorder: A Comparison of Pharmacotherapy, Psychotherapy, and Control Conditions," *American Journal of Psychiatry* 159 (2002): 1354–60.

24. Revicki et al., "Cost-Effectiveness of Evidence-Based Pharmacotherapy."

25. Williams et al., "Treatment of Dysthymia and Minor Depression"; C. Chilvers et al., "Antidepressant Drugs and Generic Counselling for Treatment of Major Depression in Primary Care: Randomised Trial with Patient Preference Arms," *British Medical Journal* 322 (2001): 1–5.

26. W. Styron, *Darkness Visible: A Memoir of Madness* (New York: Random House, 1990).

27. DeRubeis et al., "Cognitive Therapy vs Medications."

28. M. Olfson et al., "National Trends in the Outpatient Treatment of Depression," *Journal of the American Medical Association* 287 (2002): 203–209.

29. March et al., "Fluoxetine, Cognitive-Behavioral Therapy, and Their Combination."

30. M. B. Keller et al., "A Comparison of Nefazadone, the Cognitive Behavioral-Analysis System of Psychotherapy, and Their Combination for the Treatment of Chronic Depression," *New England Journal of Medicine* 342 (2000): 1462–70.

31. S. Pampallona et al., "Combined Pharmacotherapy and Psychological Treatment for Depression: A Systematic Review," *Archives of General Psychiatry* 61 (2004): 714–19.

32. Reynolds et al., "Nortriptyline and Interpersonal Psychotherapy."

33. G. A. Fava et al., "Six-Year Outcome of Cognitive Behavior Therapy for Prevention of Recurrent Depression," *American Journal of Psychiatry* 161 (2004): 1872–76.

34. D. A. Brent, "Antidepressants and Pediatric Depression—The Risk of Doing Nothing," *New England Journal of Medicine* 351 (2004): 1598–1601.

35. J. A. Grunbaum et al., "Youth Risk Behavior Surveillance—United States, 2003," http://www.cdc.gov/healthyyouth/yrbs.

36. Brent, "Antidepressants and Pediatric Depression."

37. M. L. Murray, I. C. Wong, and M. Thompson, "Do Selective Serotonin Reuptake Inhibitors Cause Suicide?" *British Medical Journal* 330 (2005): 1151.

38. R. D. Gibbons et al., "The Relationship between Antidepressant Medication Use and Rate of Suicide," *Archives of General Psychiatry* 62 (2005): 165–72.

39. C. Martinez et al., "Antidepressant Treatment and the Risk of Fatal and Non-fatal Self Harm in First Episode Depression: Nested Case-Control Study," *British Medical Journal* 330 (2005): 389–95.

40. Brent, "Antidepressants and Pediatric Depression"; T. K. Richmond and D. S. Rosen, "The Treatment of Adolescent Depression in the Era of the Black Box Warning," *Current Opinion in Pediatrics* 17 (2005): 466–72.

41. Ibid.

42. H. S. Mayberg et al., "Deep Brain Stimulation for Treatment-Resistant Depression," *Neuron* 45 (2005): 651–60.

43. Ibid.

44. Frances et al., *Diagnostic and Statistical Manual*, p. 285.

45. Ibid., pp. 274–83.

46. E. Q. Wu et al., "The Economic Burden of Schizophrenia in the United States in 2002," *Journal of Clinical Psychiatry* 66 (2005): 1122–29.

47. P. F. Sullivan, K. S. Kendler, and M. C. Neale, "Schizophrenia as a Complex Trait: Evidence from a Meta-analysis of Twin Studies," *Archives of General Psychiatry* 60 (2003): 1187–92.

48. P. F. Sullivan, "The Genetics of Schizophrenia," *PLoS Medicine* 2 (2005): e212.

49. R. G. Steen et al., "Brain Volume in First-Episode Schizophrenia: Systematic Review and Meta-analysis of Magnetic Resonance Imaging Studies," *British Journal of Psychiatry* 188 (2006): 510–18; R. G. Steen, R. M. Hamer, and J. A. Lieberman, "Measurement of Brain Metabolites by ^{1}H Magnetic Resonance Spectroscopy in Patients with Schizophrenia: A Systematic Review and Meta-Analysis," *Neuropsychopharmacology* 30 (2005): 1949–62.

50. N. C. Andreasen, "Linking Mind and Brain in the Study of Mental Illnesses: A Project for a Scientific Psychopathology," *Science* 275 (1997): 1586–93.

51. J. A. Lieberman et al., "Effectiveness of Antipsychotic Drugs in Patients with Chronic Schizophrenia," *New England Journal of Medicine* 353 (2005): 1209–23.

52. Ibid.

53. G. Tononi and G. M. Edelman, "Schizophrenia and the Mechanisms of Conscious Integration," *Brain Research Reviews* 31 (2000): 391–400.

54. L. F. Jarskog et al., "Apoptotic Mechanisms in the Pathophysiology of Schizophrenia," *Progress in Neuro-Psychopharmacology & Biological Psychiatry* 29 (2005): 846–58.

55. Ibid.

56. Ibid.

57. M. Brune, "Schizophrenia—An Evolutionary Enigma?" *Neuroscience and Biobehavioral Reviews* 28 (2004): 41–53.

58. Ibid.

59. F. Post, "Creativity and Psychopathology: A Study of 291 World-Famous Men," *British Journal of Psychiatry* 165 (1994): 22–34.

60. J. L. Karlsson, "Mental Abilities of Male Relatives of Psychotic Patients," *Acta Psychiatrica Scandinavica* 104 (2001): 466–68.

61. N. J. Andreasen and P. S. Powers, "Creativity and Psychosis: An Examination of Conceptual Style," *Archives of General Psychiatry* 32 (1975): 70–73.

62. G. E. Vaillant, "Mental Health," *American Journal of Psychiatry* 160 (2003): 1373–84.

63. Ibid.

64. Ibid.

65. Ibid.

66. Ibid.

67. Ibid.

CHAPTER 16. INTELLIGENCE AND SOCIALITY

1. P. Pica et al., "Exact and Approximate Arithmetic in an Amazonian Indigene Group," *Science* 306 (2004): 499–503.

2. M. Hayashi, Y. Mizuno, and T. Matsuzawa, "How Does Stone Tool Use Emerge? Introduction of Stones and Nuts to Naive Chimpanzees in Captivity," *Primates* 46 (2005): 91–102.

3. M. Hauser, "Our Chimpanzee Mind," *Nature* 437 (2005): 60–63.

4. Ibid.

5. N. Kawai and T. Matsuzawa, "Numerical Memory Span in a Chimpanzee," *Nature* 403 (2000): 39–40.

6. Pica et al., "Exact and Approximate Arithmetic."

7. P. Gordon, "Numerical Cognition without Words: Evidence from Amazonia," *Science* 306 (2004): 496–99.

8. R. Gelman and C. R. Gallistel, "Language and the Origin of Numerical Concepts," *Science* 306 (2004): 441–43.

9. M. D. Hauser, N. Chomsky, and W. T. Fitch, "The Faculty of Language: What Is It, Who Has It, and How Did It Evolve?" *Science* 298 (2002): 1569–79.

10. N. Chomsky, *Reflections on Language* (New York: Pantheon, 1975).

11. R. S. Fouts and G. S. Waters, "Chimpanzee Sign Language and Darwinian Continuity: Evidence for a Neurological Continuity for Language," *Neurological Research* 23 (2001): 787–94.

12. N. Chomsky, "Linguistics and Cognitive Science: Problems and Mysteries," in *The Chomskyan Turn*, ed. A. Kasher (Cambridge, MA: Blackwell, 1991).

13. S. Pinker and R. Jackendoff, "The Faculty of Language: What's Special about It?" *Cognition* 95 (2005): 201–36.

14. J. K. Rilling and T. R. Insel, "The Primate Neocortex in Comparative Perspective Using Magnetic Resonance Imaging," *Journal of Human Evolution* 37 (1999): 191–223.

15. H. H. Stedman et al., "Myosin Gene Mutation Correlates with Anatomical Changes in the Human Lineage," *Nature* 428 (2004): 415–19.

16. E. A. Nimchinsky et al., "A Neuronal Morphologic Type Unique to Humans and Great Apes," *Proceedings of the National Academy of Sciences of the United States of America* 96 (1999): 5268–73.

17. E. Kohler et al., "Hearing Sounds, Understanding Actions: Action Representation in Mirror Neurons," *Science* 297 (2002): 846–48.

18. W. Enard et al., "Intra- and Interspecific Variation in Primate Gene Expression Patterns," *Science* 296 (2002): 340–43.

19. J. Gu and X. Gu, "Induced Gene Expression in Human Brain after the Split from Chimpanzee," *Trends in Genetics* 19 (2003): 63–65.

20. S. Dorus et al., "Accelerated Evolution of Nervous System Genes in the Origin of *Homo sapiens,*" *Cell* 119 (2004): 1027–40.

21. P. Khaitovich et al., "Regional Patterns of Gene Expression in Human and Chimpanzee Brains," *Genome Research* 14 (2004): 1462–73.

22. Z. Cheng et al., "A Genome-Wide Comparison of Recent Chimpanzee and Human Segmental Duplications," *Nature* 437 (2005): 88–92.

23. R. W. Sussman, P. A. Garber, and J. M. Cheverud, "Importance of Cooperation and Affiliation in the Evolution of Primate Sociality," *American Journal of Physical Anthropology* 128 (2005): 84–97.

24. T. J. Bergman et al., "Hierarchical Classification by Rank and Kinship in Baboons," *Science* 302 (2003): 1234–36.

25. J. B. Silk, S. C. Alberts, and J. Altmann, "Social Bonds of Female Baboons Enhance Infant Survival," *Science* 302 (2003): 1231–34.

26. J. K. Rilling et al., "A Neural Basis for Social Cooperation," *Neuron* 35 (2002): 395–405.

27. Ibid.

28. R. L. C. Mitchell and T. J. Crow, "Right Hemisphere Language Functions and Schizophrenia: The Forgotten Hemisphere?" *Brain* 128 (2005): 963–78.

29. J. Moll et al., "The Neural Basis of Human Moral Cognition," *Nature Reviews Neuroscience* 6 (2005): 799–809.

30. J. Giles, "Change of Mind," *Nature* 430 (2004): 14.

CHAPTER 17. TOWARD A THEORY OF EMERGENT COMPLEXITY

1. T. H. Bullock, "How Are Complex Brains Different? One View and an Agenda for Comparative Neurobiology," *Brain, Behavior and Evolution* 41 (1993): 88–96.

2. J. J. Hopfield, "Neural Networks and Physical Systems with Emer-

gent Collective Computational Abilities," *Proceedings of the National Academy of Sciences of the United States of America* 79 (1982): 2554–58.

3. F. Crick, "The Recent Excitement about Neural Networks," *Nature* 337 (1989): 129–32.

4. G. A. Ascoli, "Progress and Perspectives in Computational Neuroanatomy," *Anatomical Record* 257 (1999): 195–207.

5. F. Crick and E. Jones, "Backwardness of Human Neuroanatomy," *Nature* 361 (1993): 109–10.

6. D. Durstewitz, J. K. Seamans, and T. J. Sejnowski, "Neurocomputational Models of Working Memory," *Nature Neuroscience* 3 (2000): 1184–91.

7. E. Salinas and T. J. Sejnowski, "Correlated Neuronal Activity and the Flow of Neural Information," *Nature Reviews Neuroscience* 2 (2001): 539–50.

8. G. Buzsaki and A. Draguhn, "Neuronal Oscillations in Cortical Networks," *Science* 304 (2004): 1926–29.

9. F. Crick and C. Koch, "A Framework for Consciousness," *Nature Neuroscience* 6 (2003): 119–26.

10. A. Watson, "Why Can't a Computer Be More Like a Brain?" *Science* 277 (1997) 1934–36.

11. G. Weng, U. S. Bhalla, and R. Iyengar, "Complexity in Biological Signaling Systems," *Science* 284 (1999): 92–96.

12. C. Koch and G. Laurent, "Complexity and the Nervous System," *Science* 284 (1999): 96–98.

13. D. C. Presgraves, "Evolutionary Genomics: New Genes for New Jobs," *Current Biology* 15 (2005): R52–R53.

14. J. Klein and N. Nikolaidis, "The Descent of the Antibody-Based Immune System by Gradual Evolution," *Proceedings of the National Academy of Sciences of the United States of America* 102 (2005): 169–74.

15. J. K. Parrish and L. Edelstein-Keshet, "Complexity, Pattern, and Evolutionary Trade-Offs in Animal Aggregation," *Science* 284 (1999): 99–101.

16. I. Karsai and J. W. Wenzel, "Productivity, Individual-Level and Colony-Level Flexibility, and Organization of Work as Consequences of Colony Size," *Proceedings of the National Academy of Sciences of the United States of America* 95 (1998): 8665–69.

17. B. Holldobler and E. O. Wilson, *The Ants* (Cambridge, MA: Belknap, 1990), p. 253.

18. Ibid., p. 275.

19. Ibid., p. 171.

20. Ibid., p. 247.

21. Ibid., p. 252.

22. Ibid., p. 289.

23. Ibid., p. 295.

24. Ibid., p. 253.

25. Ibid., p. 365.

26. Ibid., p. 174.

27. Ibid., p. 297.

28. Ibid.

29. Ibid., p. 272.

30. Ibid., p. 174.

31. Ibid., p. 259.

32. H. G. Krapp and F. Gabbiani, "Spatial Distribution of Inputs and Local Receptive Field Properties of a Wide-Field, Looming Sensitive Neuron," *Journal of Neurophysiology* 93 (2005): 2240–53.

33. C. A. Pedersen, "How Love Evolved from Sex and Gave Birth to Intelligence and Human Nature," *Journal of Bioeconomics* 6 (2004): 39–63.

INDEX